Animal Biotechnology 2

Heiner Niemann • Christine Wrenzycki
Editors

Animal Biotechnology 2

Emerging Breeding Technologies

 Springer

Editors
Heiner Niemann
Institute of Farm Animal Genetics
Friedrich-Loeffler-Institut (FLI)
Mariensee
Germany

Christine Wrenzycki
Faculty of Veterinary Medicine
Justus-Liebig-University Giessen
Giessen
Germany

ISBN 978-3-030-06412-9   ISBN 978-3-319-92348-2   (eBook)
https://doi.org/10.1007/978-3-319-92348-2

This Springer imprint is published by the registered company Springer International Publishing AG part
of Springer Nature.
The registered company address is: Gewerbestrasse 11, 6330 Cham, Switzerland

Preface

The domestication of farm animals was a seminal advance that laid the foundation stone for agriculture as it is known today. Compelling evidence is now available that the domestication process started about 10–15,000 years ago at various locations in the world. Choice of breeding stock was initially made by visual selection for specific phenotypes and/or traits, science-based selection only emerging in the sixteenth to nineteenth century with the advance in statistical and genetic knowledge. Progress in selection and propagation of superior genotypes by conventional breeding practices was glacially slow and remained so until the introduction of assisted reproductive technology (ART), most notably Artificial insemination (AI), in the first half of the twentieth century. Artificial insemination remains the most widely used of these technologies and has and continues to play a central role in the dissemination of valuable male genetics around the globe. A means of increasing the rate of propagation of female genomes was only achieved relatively recently with the development of multiple ovulation (MO) and embryo transfer technology (ET) in the immediate post world-war II period. The full potential of MOET is still yet to be realized as it plays a key enabling role in the development of the next generation of technologies, including in vitro production of embryos, somatic cloning and precise genetic modification. Advances in DNA methodology in this century have been truly remarkable. The result is genomic maps now being available for all the major farm animals together with tools that allow precise genome editing at specific genomic loci at even the single base pairs. When combined with ART, this integration of molecular and reproductive technologies has resulted in the development of an impressive range of innovative breeding concepts aimed at improving genetic gain through precise editing of the genome and its rapid dissemination made possible through a dramatically shortened generation interval. In addition to the enormous potential of these advances in agriculture they also open up the prospect of generating new animal products, for example the provision of new models of disease for the health sciences or recombinant pharmaceutical proteins and even regenerative tissue or functional xenografts for medicine. Arguably, the only limit to the scope of animal biotechnology is the human imagination.

However, experience has revealed that translation of these developments into product is not straight-forward; their transformative potential raising many expected and unexpected ethical and legal questions that have already sparked a heated public debate, much of it ill-informed.

This book is designed to provide the reader with the information needed to fully appreciate what is being achieved through the exciting advances in research and application of animal biotechnology with specific focus on the key developments in reproductive and molecular biology that underpin these advances. The book also seeks to address the major issues of concern raised by the public in relation to the social impact of these new methodologies together with the many legal and ethical aspects emerging from this. Gaining a broader public understanding and acceptance of animal biotechnology is seen by the authors as critical to the full realization of the potential of the remarkable scientific advances to address the challenges to food security raised by the ever-accelerating growth in human demand within the production constraints imposed by the diminishing availability of arable land and climate change.

The editors trust that a better appreciation of these technologies and their potential, when applied responsibly, to combat the looming agricultural challenges faced by mankind, will enhance rational debate on these issues.

The editors are extremely grateful to Susanne Tonks who provided major assistance in preparation of this book.

Mariensee, Germany   Heiner Niemann
Giessen, Germany   Christine Wrenzycki

Contents

Cloning of Livestock by Somatic Cell Nuclear Transfer

1

Kenneth R. Bondioli

Abstract

Since the cloning of "Dolly" by somatic cell nuclear transfer in 1996, numerous articles have been published concerning the application of this technology to a large variety of mammalian species including all the major livestock species. While live births have been obtained for many species, the efficiency of cellular reprogramming essential for success has not been significantly improved. This chapter will attempt to address the inputs utilized for this procedure and the major manipulation steps with the objective of identifying the major factors which might affect this efficiency of reprogramming and some of the studies addressing these factors. Finally, the challenging task of setting optimum endpoints for experiments involving domestic species will be discussed.

The first nuclear transfer experiments were conducted in amphibians (Briggs and King 1952) and involved the transfer of a nucleus from a differentiated cell into an enucleated mature oocyte. The essence of nuclear transfer is to reprogram the genome of a differentiated cell, and development from the one-cell zygote through to term is recapitulated. The first somatic cell nuclear transfer (SCNT) in a mammal was achieved in sheep with the birth of "Dolly" (Wilmut et al. 1997). This achievement was remarkable because a fully differentiated somatic cell from an adult animal was used as the donor nucleus. Complete reprogramming of a differentiated cell by nuclear transfer had previously been opinioned to be "biologically impossible" (McGrath and Solter 1984). Except for this dramatic difference involving the donor cell nucleus, the techniques used in this experiment were very similar to those previously used to clone livestock utilizing embryonic cells (Willadsen 1986; Prather et al. 1987; Bondioli

K. R. Bondioli
School of Animal Sciences, Louisiana State University, Agricultural Center,
Baton Rouge, LA 70803, USA
e-mail: kbondioli@agcenter.lsu.edu

© Springer International Publishing AG, part of Springer Nature 2018
H. Niemann, C. Wrenzycki (eds.), *Animal Biotechnology 2*,
https://doi.org/10.1007/978-3-319-92348-2_1

1

et al. 1990). Since the birth of "Dolly," SCNT has resulted in live birth in a large number of mammalian species (Edwards and Schrick 2015). It appears that SCNT can be successful in terms of resulting in live birth in all mammalian species. While SCNT is very successful in this respect, the overall efficiency of the procedure remains low in all species and generally is less than 5% when calculated as proportion of cells fused to oocytes resulting in live birth surviving more than a few days. This efficiency is in comparison to the more than 40% birth rate obtained with embryos for in vitro fertilization using the same in vitro procedures of in vitro oocyte maturation and in vitro embryo culture. In addition, offspring, especially those from livestock species, have a high incidence of abnormalities including large offspring syndrome (LOS), severe placental abnormalities, respiratory problems, prolonged gestation, and dystocia (Young et al. 1998; Yang et al. 2007; Edwards and Schrick 2015). This overall low efficiency and incidence of abnormal development are likely due to incomplete and/ or incorrect nuclear epigenetic reprogramming.

In all livestock species, the procedures are basically the same, consisting of enucleation of a mature metaphase II oocyte, fusion of an intact somatic cell by electroporation followed by some sort of activation treatment, and in vitro culture prior to transfer into a recipient for development to term. Exact protocols for SCNT in pigs (Giraldo et al. 2012) and cattle (Ross and Cibelli 2010) are published and will not be repeated here. This chapter will discuss the key inputs and major steps, highlighting different approaches (between and within species), and how each of these inputs and steps may contribute to the overall low efficiency observed for this procedure.

1.1   Oocytes Used for SCNT

The oocytes used in nuclear transfer are a key biological component for the procedure. Largely unknown components in the cytoplasm of the oocyte are responsible for the genomic reprogramming allowing development to be directed by the genome of a differentiated somatic cell. The reprogramming events in SCNT have a lot in common with those occurring after fertilization. All evidence points to the fact that the same characteristics of an oocyte important for developmental competence after SCNT are the same as those required for developmental competence following fertilization. The oocytes used in the nuclear transfer procedure represent a major input that usually accounts for the greatest cost associated with application of the technology. The oocytes also contribute one of the largest sources of variability. Since the procedure has a low overall efficiency, a relatively large number of oocytes are required to produce live offspring or conduct meaningful experiments. This requirement for large numbers and a general lack of knowledge concerning what is a "good oocyte" makes acquiring this input problematic and variable. In contrast to the situation in rodents, essentially all SCNT in livestock use abattoir-derived in vitro matured oocytes. Completely in vivo matured oocytes are not a particularly viable option for livestock. Recovery of mature oocytes with expanded cumulus cells is very difficult in livestock species, and attempts to recover in vivo matured oocytes by ultrasound guided aspiration in cattle (Sarmiento 2014) resulted in low efficiency. A few studies in pigs (Bondioli et al. 2001) and goats (Reggio et al. 2001) have utilized oocytes recovered from midsize to large follicles following gonadotropin stimulation. Oocytes recovered in this manner are still immature (germinal vesicle stage) and require in vitro maturation but are from larger more mature follicles than typically recovered from abattoir-sourced ovaries. There are too few studies of this nature to determine if this partial in vivo maturation yields a higher efficiency for SCNT, and the vast majority of procedures both commercial and research have relied upon the abattoir-derived in vitro matured oocytes.

Both abattoir sourcing and in vitro maturation can contribute to the variability and overall low efficiency of SCNT. When ovaries obtained from an abattoir are used for follicular aspiration, little to nothing is known about the animals they came from. This can lead to extreme variability due to seasonal differences, nutritional status, animal age, and management practices (culling rate). An example of the latter might be that when milk or calf prices are high, producers cull fewer animals, and those that are culled may be older or reproductively unfit. The majority of pigs slaughtered in the USA and in Europe are prepubertal at the time of slaughter. While oocytes from prepubertal or peripubertal pigs have been used for in vitro fertilization and pronuclear injection, they are not very suitable for the type of manipulation required for SCNT. Oocytes for porcine SCNT are preferably recovered from older sows, which significantly complicates the task of abattoir sourcing. The small ruminant livestock industries are not well developed in the USA; thus slaughter facilities for these species are generally small and not uniformly distributed regionally with the USA. Slaughter of horses for food is no longer allowed in the USA, so abattoir-sourced equine oocytes are not an option in the USA. All of these factors combine to make abattoir-sourced oocytes of sufficient quantity for SCNT problematic and a source of significant variability. The ability to cryopreserve oocytes could alleviate many of the difficulties. While cryopreservation of human oocytes by vitrification and subsequent fertilization or intracytoplasmic sperm injection has become a common clinical procedure, far less research has been conducted with cryopreservation of oocytes from domestic species. Oocytes cryopreserved by vitrification have been used for SCNT in cattle (Hou et al. 2005; Yang et al. 2008) and sheep (Moawad et al. 2011). Use of vitrified bovine oocytes resulted in live birth (Hou et al. 2005) and late-term pregnancy (Yang et al. 2008), and the experiments in sheep produced blastocysts. In each case in vitro and in vivo development rates were lower with vitrified oocytes compared to non-cryopreserved oocytes.

The difference in developmental competence between in vitro matured oocytes and in vivo matured oocytes has been established, with in vivo matured oocytes reaching higher rates of embryo development following fertilization than their in vitro matured counterparts. (Labrecque and Sirard 2014). While embryo development following fertilization is not the same as embryo development following nuclear transfer, there are many similarities. Both situations require extensive genome reprogramming, and nonnuclear cytoplasmic organelles are crucial for both. The sub-optimum developmental competence of in vitro matured oocytes compared to in vivo matured oocytes can certainly contribute to the overall low efficiency of SCNT. Factors affecting the developmental competence of in vitro matured oocytes have been a major area of investigation because they also contribute to poor developmental potential of in vitro fertilized oocytes in domestic animal breeding. These factors and their optimization particularly for bovine oocytes are dealt with in other parts of this volume and will not be repeated here. It is likely that improvements made for in vitro maturation of immature oocytes will enhance developmental potential for SCNT embryos as well as in vitro fertilized embryos.

In addition to the genomic reprogramming function of the oocyte cytoplasm, the oocyte utilized in SCNT is also the source of numerous cytoplasmic organelles crucial for embryo development. Of particular significance in this regard are the mitochondria contributed by the oocyte in a nuclear transfer procedure. The special role of mitochondria in this discussion results from two factors: (1) the important role of mitochondria in controlling cellular metabolism and the proposed link between metabolism, pluripotency, and reprogramming following nuclear transfer (Folmes et al. 2011; Esteves et al. 2012) and (2) the fact that mitochondrial function requires coordinated activity between mitochondria factors and nuclear-encoded proteins. Abnormal mitochondria function has been proposed to directly impact reprogramming in SCNT (Hiendleder et al. 2005), and aberrant mitochondria-nucleus crosstalk is a contributing aspect of this disturbed function (Lloyd et al. 2006). In the case of SCNT, not only is there a possibility of aberrant cross-talk because of the difference between somatic cell mitochondria function and embryonic mitochondria function but also a very real possibility of aberrant cross-talk due to genetic distance between the oocyte donor and the nuclear donor. Some degree of genetic distance is essentially guaranteed in the case of outbred domestic animals and accentuated by the likelihood of there being dramatic breed differences between oocyte and nuclear donors. When bovine oocytes are abattoir sourced for SCNT, they frequently are collected from the ovaries of dairy (primarily Holstein in the USA) breeds. This is simply because production dairy cattle are most commonly culled because of milk production, rather than reproductive failure making oocyte recovery more effecient. Production beef cattle on the other hand are frequently culled because of sub-optimum reproductive performance and are generally older, making oocyte recovery less effecient. Many beef breeds that may be used for nuclear donor cells are hybrids of *Bos taurus* and *Bos indicus* breeds and thus would represent considerable genetic distance if these donor cells are fused into an oocyte from a Holstein cow. An example of extreme genetic distance between the donor nucleus and the recipient oocyte could exist for porcine SCNT as well. In some cases, it is preferable to create porcine biomedical models in one of the breeds of miniature pigs that have been established (Cho et al. 2007). When these cells from the miniature pigs are used for SCNT following genetic manipulation, they would likely be fused into oocytes recovered for domestic pigs. This would most likely represent considerable genetic difference between the donor nucleus and the recipient oocyte and may lead to aberrant cytoplasmic-nuclear cross-talk. The possible effect of genetic distance between nuclear donor cells and recipient oocytes on cytoplasmic-nuclear cross-talk is an understudied potential complication of SCNT in domestic species.

The nuclear transfer procedure most commonly used for livestock involves the fusion of an intact cell with the enucleated oocyte. This procedure creates the possibility of mitochondria heteroplasmy or a mosaic mitochondria population in nuclear transfer-derived embryos and offspring. The fate of the somatic mitochondria from the donor cell in the oocyte cytoplasm has been studied in livestock species (Meirelles et al. 2001). While some degree of mitochondrial heterplasmy has been detected in nuclear transfer embryos and offspring, in the majority of cases, the mitochondria population of the oocyte predominates, and the mitochondria from the donor cell do not replicate.

1.2   Donor Cells Used for SCNT

The first example of SCNT in a mammal, the birth of "Dolly," utilized a mammary epithelial cell as a donor cell. This choice of cell type was driven more by a commercial interest than a biological choice of what cell type would most likely be successful. There are more than 200 cell types distinguishable by morphology in mammals, and less than 5% of these have been tested as nuclear donors. Of those tested all support development to blastocysts, but some repeatedly failed to generate viable offspring (Kato et al. 2000; Wakayama and Yanagimachi 2001; Oback and Wells 2002). The large majority of SCNT experiments in livestock have been conducted with skin fibroblasts, and the majority of these experiments have used skin fibroblasts recovered from early pregnancy fetuses. Very little specific information is available concerning what makes the ideal donor cell. The decision of what cell type to use is generally made from three considerations: (1) the objective of the SCNT procedure, 2) the ease of collecting the tissue from the donor animal, and 3) the ability to culture various cell types in a particular laboratory environment.

To date the majority of SCNT procedures with domestic livestock have been experimental in nature aimed at investigating factors that may affect the efficiency of the procedure. In these cases, the objective does not greatly influence the choice of donor cell type and cells that have been selected on an assumption of which cells will be most successful. In these cases, the choice has generally been skin fibroblasts from early pregnancy fetuses. A second objective, primarily for porcine (Polejaeva et al. 2016) but in a few cases for bovine (Kuroiwa et al. 2002), has been to use SCNT to support genetic manipulation particularly for gene knockouts in order to create biomedical models. This is similar to the situation for purely experimental objectives, and fetal fibroblasts have generally been used. A very different situation exists if the objective is to use SCNT in animal breeding to duplicate a specific genome demonstrated to be of value. This situation virtually dictates the use of cells from an adult animal and perhaps from an aged or even deceased animal. There have been reports of using skin fibroblast cells from adult animals for SCNT (Li et al. 2013), but most of the information we have about the procedure has been derived from experiments utilizing fetal fibroblasts. Factors to consider which could be very different in cells from aged animals include incidence of genetic mutations, altered epigenetic profiles, and lack of maintenance of parental imprinting patterns. If the goal of using SCNT in livestock breeding programs is to be realized, these factors will need to be investigated in the context of which cell types should be used as donor cells.

Ease of collecting the tissue has certainly been a factor in the choice of skin fibroblasts particularly if tissue from adult animals is collected. Skin biopsies such as a simple "ear notch" are readily collected from livestock species and seldom require any sort of anesthesia. The initial difficulty of collecting fetal tissue is understood, but once a fetus is recovered, the collection of skin fibroblasts from that fetus is simple and straightforward. Related to the ease of collection is the relative ease and success of establishing cultures from skin tissue. Viable cell cultures can be established from skin by either enzymatic digestion or simple outgrowth for tissue pieces. There does not seem to be any difference in the viability of cultures established by these two methods in cattle (Giraldo et al. 2007a). Once established skin fibroblast cultures are easily maintained with the use of standard tissue culture media (such as DMEM or TCM 199), supplemented with 10–15% bovine serum and passage by trypsinization. These cells also survive cryopreservation by routine methods very well. What is not assured by these routine cell culture methods is long-term culture without development of chromosomal abnormalities. It is not entirely clear what the average incidence of chromosomal abnormalities is in cultured fibroblasts, but the incidence of chromosomal abnormalities in the form of aneuploidy can be high in cell populations after repeated passage (Giraldo et al. 2007a). This can be a very important factor if genetic manipulation in the donor cell population is conducted prior to SCNT because these procedures often require long-term culture. If a donor cell with a gross chromosome abnormality such as aneuploidy is used for SCNT, the resulting embryo would have little or no chance of development. It is not clear how much this contributes to the inefficiency of the procedure.

Related to in vitro culture conditions are the dynamics of the cell cycle. When skin fibroblasts are cultured with the conditions described above, they display a long cell cycle (2–3 days) with up to 70% of the cells at any time being in the G1 phase of the cell cycle (Giraldo et al. 2007b). This is advantageous for their use as donor cells for SCNT because G1/G0 is the preferred cell cycle stage. The work of Keith Campbell and associates (Campbell et al. 1996a) established the importance of cell cycle synchrony between the donor cell and the oocyte in nuclear transfer procedures with embryonic and somatic cells. Cellular reprograming is enhanced if the donor nucleus is exposed to the reprogramming factors of the oocyte cytoplasm immediately after transfer. This is accomplished by fusion of donor cells into oocytes when MPF is high which leads to nuclear envelope breakdown and premature chromosome condensation (Campbell et al. 1993). A decrease in MPF consistent with oocyte activation will lead to DNA replication in preparation for the first mitotic division. If the transferred nucleus has begun (S phase) or completed (G2 phase) DNA replication, this replication will be reinitiated (Johnson and Rao 1970) which leads to a chromosome content inconsistent with a normal mitotic division. When a donor cell is fused with an oocyte with high MPF levels, it is optimum for that cell to be in G1 or the quiescent G0 stage of its cell cycle to ensure a normal DNA replication producing a 4C nucleus consistent with a normal first mitotic division in the nuclear transfer embryo. The skin fibroblast cultured under normal conditions with approximately 70% of cells in G1 at any time produces this condition with minimal manipulation.

Of particular interest as donor cells for SCNT are stem cells. It has been proposed that stem cells would be more reprogrammable and thus lead to greater efficiency when used as donor cells. At least one study in mice (Rideout et al. 2000) resulted in higher cloning efficiency with embryonic stem cells. Attempts to isolate embryonic stem cells comparable to those used in the mice experiments (capable of generating germline chimeras) have not been successful in livestock. Recent descriptions of induced pluripotent stem cells which are essentially the same as embryonic stem cells in mice have likewise led to the hypothesis that these cells would more reprogrammable when used at donor cells. These cells have not been thoroughly described for livestock species, and this hypothesis has yet to be tested. Somatic stem cells for many tissues from various species including livestock have been described. While these cells display some cell-type plasticity in culture, "lineage plasticity" of these cells remains controversial, and the use of these cells as nuclear donors has not led to any increase in cloning efficiency (Oback 2008).

1.3 Treatment of Donor Cells Prior to Fusion

A number of treatments have been applied to donor cells prior to fusion for nuclear transfer. The first such treatment for SCNT in livestock was inducing cells to exit the growth cycle and arrest in the G0 quiescence state by culture in low serum conditions, "serum starvation" as described by Keith Campbell (Campbell et al. 1996b; Wilmut et al. 1997). The value of this treatment is unclear (Kasinathan et al. 2001), and one study (Kues et al. 2000) has shown that serum starvation can induce DNA fragmentation in bovine fibroblasts. As discussed above, fibroblasts in culture have an elongated G1 phase, and culturing these cells to confluence can create a population with a high incidence of G1 without serum starvation. A recent report has shown that induction of quiescence by serum starvation results in hypomethylation of DNA and lysines 4, 9, and 27 of histone H3 resulting in a more relaxed chromatin structure and enhanced reprogramming following nuclear transfer (Kallingappa et al. 2016).

The process of cellular reprogramming is an epigenetic event, and the incomplete and/or incorrect reprogramming as the root of low efficiency with SCNT involves incomplete and/or incorrect epigenetic remodeling. Epigenetic marks, being a posttranslational modification, are created and altered by enzymatic reactions. A variety of molecules and procedures have been developed which affect the activity of these enzymes and are referred to as epigenetic modulating agents. A variety of these agents has been used to treat somatic cells prior to fusion with oocytes in an attempt to correct epigenetic marks. One of the most prevalent examples of these is trichostatin A (TSA), a potent inhibitor of the histone deacetylase enzymes (HDACs). Inhibition of the deacetylase enzyme would be expected to increase the level of histone acetylation which is an epigenetic mark which increases gene expression. A second prevalent epigenetic modulator is 5-aza-2′-deoxycytidine (5azadC) which is a DNA methyltransferase inhibitor. Inhibition of DNA methyltransferase (DNMT) especially the "maintenance enzyme" DNMT1 will decrease DNA methylation, which is also an epigenetic mark usually consistent with increased gene expression. If cellular differentiation is characterized as sequential inhibition of gene expression, particularly those genes necessary for early embryo development, removal of those epigenetic marks to allow those genes to be expressed would be fundamental for the reprogramming during SCNT. These agents alone or in combination have been used to treat bovine cells (Enright et al. 2003, 2005; Giraldo et al. 2007b; Ding et al. 2008; Wang et al. 2011; Fig. 1.1).

Fibroblasts cultured for an extended period of time have increased levels of acetylated histones and decreased levels of methylated DNA (Enright et al. 2003; Wilson and Jones 1983). Giraldo et al. (2008) compared bovine fibroblasts after 5 and 35 population doublings (PD). Cells at 35 PD had reduced levels of transcripts for DNMT 1 and 3A with constant levels of transcript for DNMT 3B. Fibroblasts at 35 PD also had lower levels of methylated DNA than at 5 PD. A higher proportion of SCNT embryos from PD35 donor cells developed beyond the 8–16-cell stage. When day 7 SCNT embryos were transferred to recipients and recovered at day 13, a higher proportion of those reconstituted with PD35 donor cells showed subsequent development with larger conceptuses.

In addition to extended culture or chemical inhibitors of epigenetic-modifying enzymes, small interference RNA (siRNA) can be used to reduce transcript levels encoding these enzymes and thus reduce the levels of the modifying enzymes (Giraldo and Bondioli 2011). This approach utilizing siRNA directed against DNMT1 in bovine fibroblasts resulted in a reduction of the transcript encoding this enzyme and hypomethylation in the treated cells (Giraldo et al. 2009). Use of these cells as donor nuclei in nuclear transfer resulted in reduced methylation levels in early SCNT embryos (Fig. 1.2).

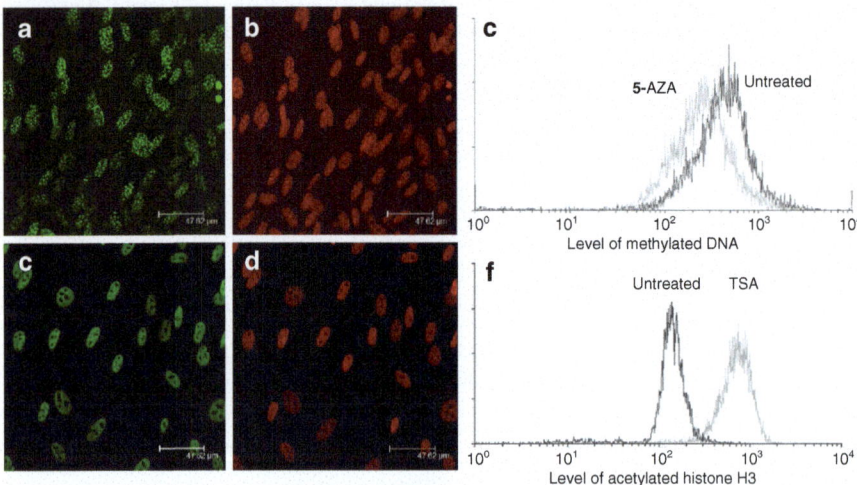

Fig. 1.1 Bovine cells incubated with anti-methylcytidine antibody, labeled with Alexa Fluor 488 (**a**) and counterstained with propidium iodide (**b**). Level of methylated DNA after incubation with 1 μM of 5-azacitidine for 48 h (**c**). Cells incubated with anti-acetyl-histone H3, labeled with Alexa Fluor 488 (**d**) and counter stained with propidium iodide (**e**). Level of acetylated histone after incubation with 1 μM of trichostatin (TSA) for 12 h (**f**). From Giraldo et al. (2007b)

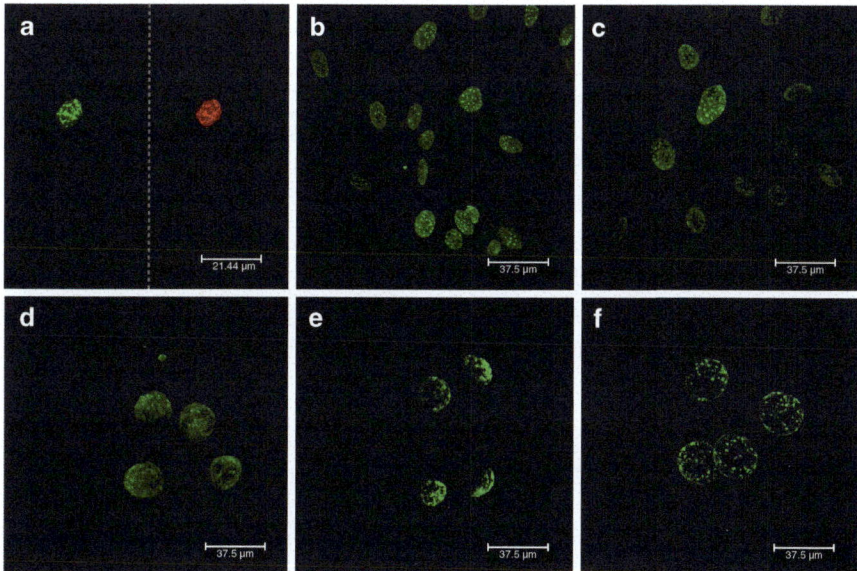

Fig. 1.2 (**a**) Bovine cells incubated with anti-methylcytidine antibody and (left) labeled with Alexa Fluor 488 and (right) counterstained with propidium iodide. Methylation patterns of fibroblast cells treated with (**b**) non-silencing and (**c**) DNMT1-specific siRNA 24 h post-transfection. Level of DNA methylation of 4-cell-stage embryos produced by (**d**) IVF and nuclear transfer using donor cells treated with (**e**) non-silencing or (**f**) DNMT1-specific siRNA. From Giraldo et al. (2009)

The use of chemical agents such as TSA for inhibition of HDACs in cultured cells was discussed above. These chemical agents are inhibitors of the entire family of 18 different HDACS which have been shown to regulate many cellular functions including cell proliferation, differentiation, and development (Yang and Seto 2008). Staszkiewicz et al. (2013) used siRNA targeting specific HDACs in bovine fibroblasts and studied the effect on expression of pluripotency genes. Their data suggests that reduction in the activity of SIRT3, one of five members of the sirtuin family of HDACs could play a role in upregulation of the Oct4-Sox2-Nanog transcriptional network. SIRT3 is preferentially localized to mitochondria and is associated with energy metabolism (Ahn et al. 2008). Energy metabolism and promotion of glycolysis have been linked to establishment of pluripotency and cellular reprogramming (Folmes et al. 2011; Esteves et al. 2012).

1.4   Oocyte Enucleation

For the majority of nuclear transfer procedures, oocytes are enucleated as mature metaphase II oocytes, with the manipulation setup depicted in Fig. 1.3 using the approach depicted in Fig. 1.4. The process involves location of the extruded first polar body, puncturing the zona pellucida with a beveled pipet and aspiration of the

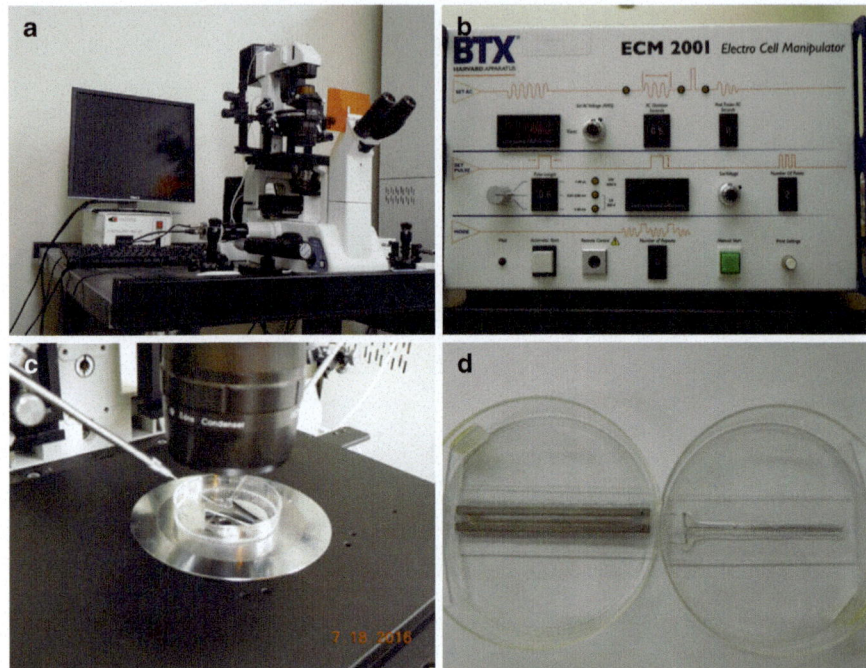

Fig. 1.3 Manipulation equipment used for somatic cell nuclear transfer. (**a**) Inverted microscope equipped with hydraulic-controlled micromanipulators. (**b**) Orientation of micromanipulator-controlled micropipettes used for oocyte enucleation and insertion of donor cells. (**c**) Square wave pulse generator used for fusion of enucleated oocytes and donor cells. (**d**) Two examples of fusion chambers used with pulse generator shown in **c**

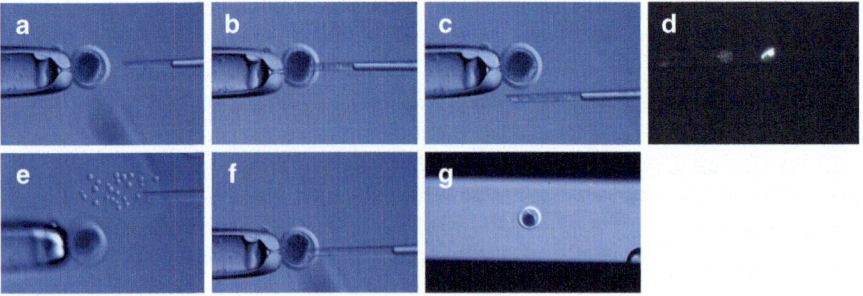

Fig. 1.4 Porcine somatic cell nuclear transfer. (**a**) Mature unfertilized porcine oocyte on holding pipette with the polar body in the "3 o'clock position" and beveled micropipette used for enucleation. (**b**) Enucleation of Hoechst 33342-stained oocyte by withdrawing the polar body and adjacent cytoplasm. Polar body and adjacent cytoplasm in enucleation pipette. (**c**) Illumination with visible light. (**d**) Illuminated with UV light. (**e**) Porcine fetal fibroblasts. (**f**) Injection of fibroblast donor cell between the enucleated oocyte and the zona pellucida. (**g**) Enucleated oocyte and donor cell in fusion chamber (see Fig. 1.1d) with oocyte and donor cell oriented for fusion

polar body and a small amount of adjacent oocyte cytoplasm. The process is complicated for oocytes of domestic species due to the inability to visualize the metaphase spindle. Enucleation in this manner relies upon the assumption that the metaphase spindle will be adjacent to the first polar body. This assumption is true for a period after extrusion of the polar body but becomes less likely as oocytes age post-maturation. Prior to enucleation oocytes are stained with a membrane-permeable fluorescent DNA stain such as bisbenzimide (Hoechst 33342) by incubation at a low concentration for 10–15 min. It is possible to visualize the metaphase chromosomes directly by excitation with UV, but this is usually avoided because of concern for damaging effects to the oocyte from the energy released in the form of heat with UV excitation (Li et al. 2004b). Alternatively, "blind" enucleation is attempted without UV excitation and then confirmed by visualization of the spindle in the pipet as depicted in Fig. 1.2, and UV excitation of the oocyte is avoided. If the spindle is not observed, a second attempt can be made, or the oocyte is rejected. If oocytes are enucleated immediately after extrusion of the first polar body, this approach will be highly successful. A limited number of studies have been conducted to determine the effect of staining and UV illumination on the viability of oocytes with varying results dependent upon species and length of UV irradiation. In a study conducted with porcine mature oocytes, the combination of exposure to Hoechst 33342 and UV irradiation decreased subsequent development following in vitro fertilization and was more pronounced with increased exposure to the UV illumination (Maside et al. 2011). The in vitro fertilization model is clearly different from the SCNT procedure because of the likely effect on the oocyte nuclear component, which is essential for development after fertilization but not essential for nuclear transfer. In a study of bovine nuclear transfer utilizing embryonic cells, Westhusin and colleagues (Westhusin et al. 1992) found that development to term was not affected by either exposure to the DNA stain or irradiation with UV.

An alternative method of enucleation is utilized in the zona-free and manipulator-free system of cloning referred to as "handmade cloning" (Vajta et al. 2005). In this system the oocyte zona pellucida is removed, and a small segment of the oocyte adjacent to the polar body is cut with a handheld razor blade. One of the possible consequences of this approach to oocyte enucleations is a decrease in oocyte cytoplasmic volume. This could affect the ability of the oocyte to reprogramming the donor nucleus and affect the nuclear/cytoplasmic volume ratio of blastomeres in the developing embryo. Attempts at chemical enucleation and PolScope microscopy enucleation have been described but have not been adapted to a repeatable method of livestock SCNT (Li et al. 2004a).

1.5 Oocyte and Donor Cell Fusion and Activation

For livestock species, fusion of the oocyte and donor cell is accomplished by electroporation of the two adjacent cells. A typical electrofusion instrument and fusion chambers are shown in Fig. 1.3. Electroporation instruments used for cell fusion are square wave generators that produce a square-shaped pulse as opposed to the exponential decay-type curves normally used for cell electroporation. In most cases, the donor cell is placed between the oocyte and the zona pellucida, and the zona pellucida is important to ensure contact. The electrofusion instrument shown in Fig. 1.3 has an alternating current (AC) cell alignment function, but this function is of questionable use for nuclear transfer because of the large difference in cell size. For SCNT, alignment of the oocyte and donor cell fusion plane perpendicular to the electrical field is essential but is accomplished manually. The AC alignment function may serve to enhance contact between the oocyte and donor cell once proper alignment is accomplished manually. In the case of the zona-free "handmade cloning" described earlier, contact is enhanced by the use of the lectin phytohemagglutinin (PHA). No advantage has been reported for the use of PHA for fusion in the presence of a zona pellucida. An alternative cell fusion system is available and is generally referred to as the "Chop Sticks" approach. With this system, fusion is conducted with the aid of micromanipulators, and the electrodes are placed next to the oocyte/cell combination that is held in proper alignment on a holding pipet. This system can be highly successful but requires that fusion be conducted one couplet at a time as opposed to fusion in a chamber where multiple oocyte/cell couplets can be fused simultaneously. Fusion is accomplished in a nonionic solution consisting primarily of a sugar such as mannitol, sucrose, or sorbitol. These solutions typically have a low concentration of calcium. If ionic strength of the fusion medium is too high, it will generate heat which can be detrimental to the embryos. The strength of the electrical field, which induces electroporation, is expressed in volts per centimeter where volts are the voltage applied and centimeter is the distance between the electrodes. Numerous fusion protocols have been reported for livestock species, which vary between laboratories and between species. In general, electrical fields of 1.2–1.5 kV/Cm are applied in single or multiple pulses of less than 100 microsecond duration.

It is possible at least for some species to induce oocyte activation with the electroporation pulse used for fusion if the calcium concentration in the fusion medium is high enough. In most cases fusion is induced with conditions (low calcium concentration) that do not induce activation. This allows for MPF levels in the oocyte to induce nuclear envelope breakdown and exposure of the donor chromatin to the oocyte cytoplasm. Oocyte activation is then induced several hours later. For porcine SCNT, this activation is induced with the same electroporation pulse used for fusion but with a higher concentration of calcium in the medium. For ruminant species, especially cattle, some sort of chemical activation is required. This normally involves a two-step process consisting of brief exposure to a calcium transporter such as ionomycin followed by an extended exposure to the protein kinase inhibitor 6-dimethylaminopurine (6-DMAP) (Susko-Parrish et al. 1994) or the protein synthesis inhibitor cycloheximide (Presicce and Yang 1994). The effect of these chemical activation procedures on subsequent embryo development is not known. One study utilized cRNA encoding the sperm activation factor phospholipase C zeta (Ross et al. 2009) for bovine SCNT. Further studies utilizing alternative activation procedures for bovine SCNT are warranted.

1.6 Treatment of Nuclear Transfer Embryos Post Fusion and Activation

There have been a number of investigations involving treating nuclear transfer-derived embryos after fusion and activation with epigenetic modulating agents. The majority of these treatments have involved inclusion of histone deacetylase inhibitors in culture medium after fusion and activation during the first cell cycle of the reconstructed nuclear transfer embryo. Histone acetylation appears to be a key epigenetic modification for reprogramming. Hyperacetylation of histones brought on by inhibition of the deacetylase enzymes corresponds to chromatin relaxation creating a transcriptionally permissive state (Rybouchkin et al. 2006; Zhao et al. 2010). While creation of the transcriptionally permissive state is not sufficient for reprogramming, it is likely that it facilitates and may be necessary for reprogramming to occur. A number of histone deacetylase inhibitors (HDACi) such as TSA (Cervera et al. 2009; Ding et al. 2008; Enright et al. 2003; Wee et al. 2007), valproic acid (VPA) (Costa-Borges et al. 2010; Huang et al. 2011), Scriptaid (Wang et al. 2011; Zhao et al. 2010), sodium butyrate (Das et al. 2010; Shi et al. 2003), suberoylanilide hydroxamic acid (Ono et al. 2010), m-carboxycinnamic acid bishydroxamide (Dai et al. 2010), and oxamflatin (Hou et al. 2014; Mao et al. 2015) have been applied to nuclear transfer-derived embryos post fusion and activation to modify the epigenetic pattern of the donor chromatin and enhance in vitro and/or in vivo development. The majority of these studies have investigated in vitro development to the blastocyst stage, but some (Zhao et al. 2010; Mao et al. 2015) have shown modest improvements in in vivo development of porcine SCNT embryos. The optimum duration of treatment has been reported to be between 14 and 16 h for VPA and Scriptaid (Huang et al. 2011; Whitworth et al. 2011), and exposure to TSA for longer than 14 h has been detrimental to cloning efficiency (Kishigami et al. 2006).

1.7 Culture and Transfer of Cloned Embryos

Post fusion in vitro culture and transfer of cloned embryos are related topics for domestic animal species because the type of transfer for each species determines the length of in vitro culture. For those species which nonsurgical embryo transfers are efficient (cattle and horses), SCNT embryos are cultured in vitro to the late morula or blastocyst stage and transferred to the uterus of recipients. Gestation and delivery of twins frequently result in life-threatening results for mother and/or offspring in both cattle and horses; thus the best procedure is to transfer only one embryo per recipient. In vitro culture to the blastocyst stage allows for some selection of SCNT embryos prior to transfer which is an important efficiency consideration. Despite the fact that cloned embryos maintain some metabolic features of somatic cells rather than early-stage embryos, embryos from SCNT are cultured to the blastocyst stage using media shown to be most efficient for culture of mammalian embryos produced by in vitro fertilization (Ross and Cibelli 2010; Arias et al. 2013).

In domestic species for which nonsurgical embryo transfer is not an option such as the pig, SCNT embryos are transferred to recipients shortly after fusion. In pigs it is common to transfer large numbers (over 100) of SCNT embryos to a single recipient to produce litters of approximately eight piglets (Giraldo et al. 2012).

1.8 Evaluation of Cloning Efficiency

Probably the most difficult aspect of research concerning SCNT is how to evaluate efficiency and determine if specific treatments or conditions enhance reprogramming and subsequent embryo development. Clearly, the result of a successful SCNT procedure is the birth of full-term offspring with normal survival and lacking any abnormalities. Studies with full-term birth and survival as endpoints are limited even with laboratory animals and extremely limited with livestock species. Most of the studies with domestic animal SCNT that include full-term development result in less than five surviving offspring spread over multiple treatments. While such a result does demonstrate that full-term development is possible, this is not a statistically meaningful result concerning the treatments and involves a significant expenditure of time and resources.

A large number of SCNT experiments have in vitro development to the blastocyst stage as an endpoint for evaluation of reprogramming efficiency. In vitro development to blastocysts following parthenogenic activation alone at rates similar to those for SCNT has been reported for cattle (Wang et al. 2008) and pigs (Cheng et al. 2007). Since these blastocysts have no chance of producing full-term offspring upon transfer, it is highly questionable if this endpoint accurately reflects developmental potential for SCNT embryos. It is conceivable that some cases of incomplete reprogramming might enhance development to blastocysts yet decrease the efficiency of full-term development. Many of the studies that have used in vitro development as an endpoint have included total blastocyst cell numbers and/or the inner cell mass to trophectoderm cell ratio. While this added data has some value, there is no data showing that a statistically significant increase in cell number at the blastocyst stage will lead to a statistically higher rate of development to term. In the bovine commercial embryo transfer industry, extensive data have shown essentially normal pregnancy rates and full-term development for half embryos created by embryo splitting (Gray et al. 1991). In addition to cell number data, many studies ending at the blastocyst stage have included gene expression data at the blastocyst stage. Most of these studies have measured transcript levels for the major pluripotency genes, Oct 4, Nanog, and Sox 2. This is potentially valuable information but requires careful interpretation. In addition to the reservation that transcript levels do not necessarily equate to protein levels, particularly during early embryo development, there is no clear picture of what are "normal" expression levels for these genes in embryos, and it is not clear if statistically significant increase reflects higher development potential (Radzisheuskaya et al. 2013).

One of the characteristics of SCNT procedures is a high rate of loss between the blastocyst stage and early fetal development at or around the time of implantation

(Edwards and Schrick 2015). The long gestation period for livestock species and the cost of transfer of embryos to be carried to term make any methods that extend development beyond the blastocyst stage valuable tools for determining SCNT efficiency. Important to this aim is the ability to evaluate a large enough sample to achieve statistical significance. This is crucial for bovine nuclear transfer because single embryos are transferred to each recipient if the embryos are to be carried to term. Giraldo and colleagues (Giraldo et al. 2008) transferred up to 25 SCNT embryos to single synchronous recipients at the blastocyst stage and recovered elongated and ovoid embryos (Fig. 1.5) at day 13 or 14 by nonsurgical flushing. Survival of the SCNT embryos was estimated from the number of advanced-stage embryos collected and the presence of an embryonic disc. These embryos can also be used for gene expression analysis, and the embryonic disc can be separated from the trophectoderm and analyzed separately. There is clearly extensive embryo loss beyond the elongation stage for SCNT embryos, but for cattle, this system can be an efficient method to extend the development period beyond the blastocyst stage without requiring a large number of live animals. In pigs where nonsurgical embryo transfers and recovery are not an option, large numbers of SCNT embryos can be

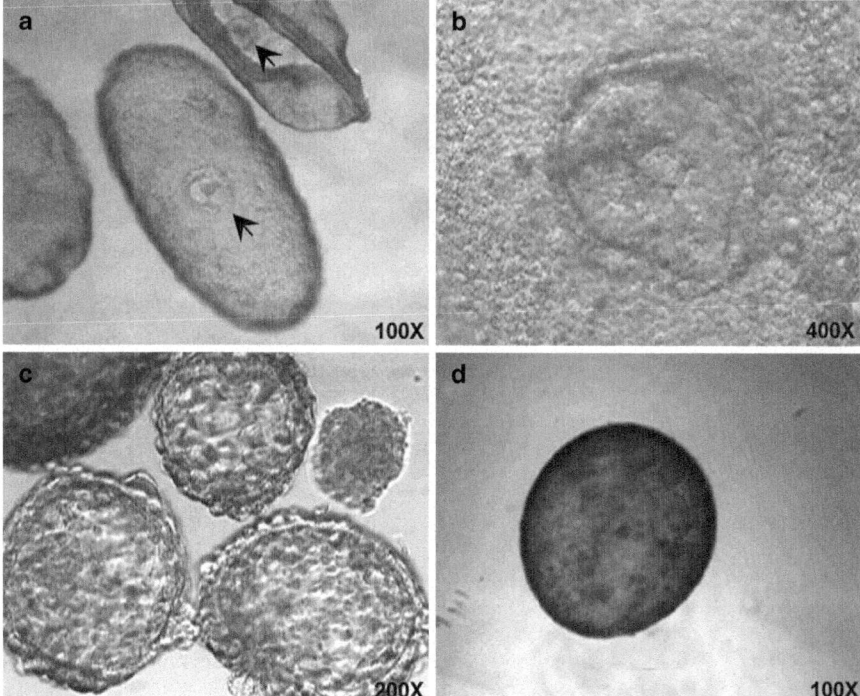

Fig. 1.5 Bovine embryos collected on day 13 postestrus. Elongating in vivo embryos (**a**) with an identifiable embryonic disc (**b**). Spherical and ovoid cloned embryos reconstructed using cells with high and low levels of DNMT1 mRNA, respectively (**c**, **d**). Arrows indicate the presence of an embryonic disc in in vivo embryos. From Giraldo et al. (2008)

transferred and conceptuses recovered by euthanasia at approximately day 25. At this stage of development, the number of surviving conceptuses and the presence or absence of a normal amount of fetal tissue within the conceptus can be evaluated.

1.9 Future Directions

Application of SCNT in domestic livestock is limited by the low efficiency and the high incidence of developmental abnormalities leading to low survival. It is very likely that these two factors are closely related, and as efficiency increases, the incidence of developmental abnormalities will decrease. It is clear that the low efficiency and developmental abnormalities result from incomplete and/or incorrect epigenetic reprogramming of the donor cell genome. Improvements in reprogramming of the donor cell genome can be achieved in the donor cell prior to fusion and in the oocyte after fusion. The ability to improve reprogramming after fusion is limited by the low number of cell divisions before the embryonic genome is activated. Major improvements in reprogramming will come from reprogramming in the donor cell prior to fusion. The success with induced pluripotency of somatic cells in some species suggests that this will be a fruitful approach for SCNT in livestock. Increased expression of key pluripotency genes, facilitated by epigenetic modifiers, is a promising approach. These epigenetic modifiers can also facilitate reprogramming in the oocyte after fusion of an induced pluripotent cell. It is important to consider that as somatic cells are induced into pluripotency, it can be expected that the cell cycle will be more like that of embryonic stem cells and cell cycle synchrony between the donor cell and the oocyte will become more critical and challenging.

Improvement in the developmental competence of in vitro matured oocytes also has the potential of significantly improve the efficiency of SCNT. Developmental competency of oocytes and the ability to reprogram an incoming genome probably have multiple pathways in common. Controlling the genetic distance between the donor cell and the recipient oocyte is an important consideration. Refinement in procedures for oocyte enucleation, fusion, activation, and culture can impact the efficiency of SCNT, but they are secondary to the central issue of reprogramming the somatic cell nucleus.

References

Ahn BH, Kim HS, Song S, Lee IH, Liu J, Vassilopoulos A, Deng CX, Finkel T (2008) A role for the mitochondrial deacetylase Sirt3 in regulating energy homeostasis. Proc Natl Acad Sci U S A 105:14447–14452

Arias ME, Ross PJ, Felmer RN (2013) Culture medium composition affects the gene expression pattern and in vitro development potential of bovine somatic cell nuclear transfer (SCNT) embryos. Biol Res 46(4):452–462

Bondioli KR, Westhusin ME, Looney CR (1990) Production of identical bovine offspring by nuclear transfer. Theriogenology 33:165–174

Bondioli K, Ramsoondar J, Williams B, Costa C, Fodor W (2001) Cloned pigs generated from cultured skin fibroblasts derived from a h-transferase transgenic boar. Mol Reprod Dev 60:189–195

Briggs R, King TJ (1952) Transplantation of living nuclei from blastula cells into enucleated frogs' eggs. Proc Natl Acad Sci U S A 38:455–463

Campbell KH, Ritchie WA, Wilmut I (1993) Nuclear-cytoplasmic interactions during the first cell cycle of nuclear transfer reconstructed bovine embryos: implications for deoxyribonucleic acid replication and development. Biol Reprod 49:933–942

Campbell K, Loi P, Otaegui P, Wilmut I (1996a) Cell cycle co-ordination in embryo cloning by nuclear transfer. Rev Reprod 1:40–46

Campbell KHS, McWhir J, Ritchie WA, Wilmut I (1996b) Sheep cloned by nuclear transfer from a cultured cell line. Nature 380:64–66

Cervera RP, Martí-Gutiérrez N, Escorihuela E, Moreno R, Stojkovic M (2009) Trichostatin a affects histone acetylation and gene expression in porcine somatic cell nucleus transfer embryos. Theriogenology 72:1097–1110

Cheng WM, Sun XL, An L, Zhu SE, Li XH, Li Y, Tian JH (2007) Effect of different parthenogenetic activation methods on the developmental competence of in vitro matured porcine oocytes. Anim Biotechnol 18(2):131–141

Cho PS, Lo DP, Wikiel KJ, Rowland HC, Coburn RC, McMorrow IM, Goodrich JG, Arn JS, Billiter RA, Houser SL, Shimizu A, Yang YG, Sachs DH, Huang CA (2007) Establishment of transplantable porcine tumor cell lines derived from MHC inbred miniature swine. Blood 110:3996–4004

Costa-Borges N, Santalo J, Ibanez E (2010) Comparison between the effects of valproic acid and trichostatin a on in vitro development, blastocyst quality, and full-term development of mouse somatic cell nuclear transfer embryos. Cell Reprogram 12:437–446

Dai X, Hao J, Hou XJ, Hai T, Fan Y, Yu Y, Jouneau A, Wang L, Zhou Q (2010) Somatic nucleus reprogramming is significantly improved by m-carboxycinnamic acid bishydroxamide, a histone deacetylase inhibitor. J Biol Chem 285:31002–31010

Das ZC, Gupta MK, Uhm SJ, Lee HT (2010) Increasing histone acetylation of cloned embryos, but not donor cells, by sodium butyrate improves their in vitro development in pigs. Cell Reprogram 12:95–104

Ding X, Wang Y, Zhang D, Guo Z, Zhang Y (2008) Increased pre-implantation development of cloned bovine embryos treated with 5-aza-2′-deoxycytidine and trichostatin a. Theriogenology 70:622–630

Edwards JL, Schrick FN (2015) Cloning by somatic cell nuclear transfer. In: Hopper RM (ed) Bovine reproduction. John Wiley & Sons, Inc, Hoboken, NJ, pp 771–783

Enright BP, Kubota C, Yang X, Tian XC (2003) Epigenetic characteristics and development of embryos cloned from donor cells treated by trichostatin a or 5-aza-2′-deoxycytidine. Biol Reprod 69:896–901

Enright BP, Sung LY, Chang CC, Yang X, Tian XC (2005) Methylation and acetylation characteristics of cloned bovine embryos from donor cells treated with 5-aza-2′-deoxycytidine. Biol Reprod 72:944–948

Esteves TC, Psathaki OE, Pfeiffer MJ, Balbach ST, Zeuschner D, Shitara H, Yonekawa H, Siatkowski M, Fuellen G, Boiani M (2012) Mitochondrial physiology and gene expression analyses reveal metabolic and translational dysregulation in oocyte-induced somatic nuclear reprogramming. PLoS One 7:e36850

Folmes CD, Nelson TJ, Martinez-Fernandez A, Arrell DK, Lindor JZ, Dzeja PP, Ikeda Y, Perez-Terzic C, Terzic A (2011) Somatic oxidative bioenergetics transitions into pluripotency-dependent glycolysis to facilitate nuclear reprogramming. Cell Metab 14:264–271

Giraldo AM, Bondioli KR (2011) Inhibition of DNA methy lation in somatic cells epigenetics protocols. In: Tollefsbol TO (ed) Methods in molecular biology™ no. 791. Humana Press, New York, pp 145–156

Giraldo AM, Lynn JW, Godke RA, Bondioli KR (2007a) Proliferative characteristics and chromosomal stability of bovine donor cells for nuclear transfer. Mol Reprod Dev 73:1230–1238

Giraldo AM, Lynn JW, Purpera MN, Godke RA, Bondioli KR (2007b) DNA methylation and histone acetylation patterns in cultured bovine fibroblasts for nuclear transfer. Mol Reprod Dev 74:1514–1524

Giraldo, A.M., D.A. Hylan, C.B. Ballard, M.N. Purpera, T.D. Vaught, J.W. Lynn, , R.A. Godke, K.R. Bondioli. 2008. Effect of epigenetic modifications of donor somatic cells on the subsequent chromatin remodeling of cloned bovine embryos. Biol Reprod 78:832–840

Giraldo AM, Lynn JW, Purpera MN, Vaught TD, Ayares DL, Godke RA, Bondioli KR (2009) Inhibition of DNA Methyltransferase 1 expression in bovine fibroblast cells used for nuclear transfer. Reprod Fertil Dev 21:785–795

Giraldo AM, Ball S, Bondioli KR (2012) Production of transgenic and knockout pigs by somatic cell nuclear transfer. Methods Mol Biol 885:105–123

Gray KR, Bondioli KR, Betts CL (1991) The commercial application of embryo splitting in beef cattle. Theriogenology 35:37–44

Hiendleder S, Zakhartchenko V, Wolf E (2005) Mitochondria and the success of somatic cell nuclear transfer cloning: from nuclear-mitochondrial interactions to mitochondrial complementation and mitochondrial DNA recombination. Reprod Fertil Dev 17:69–83

Hou Y-P, Dai Y-P, Zhu S-E, Zhu H-B, Wu T-Y, Gong G-C, Wang H-P, Wang L-L, Liu Y, Li R, Wan R, Li N (2005) Bovine oocytes vitrified by the open pulled straw method and used for somatic cell cloning supported development to term. Theriogenology 64:1381–1391

Hou L, Ma F, Yang J, Riaz H, Wang Y, Wu W, Xia X, Ma Z, Zhou Y, Zhang L, Ying W, Xu D, Zuo B, Ren Z, Xiong Y (2014) Effects of histone deacetylase inhibitor oxamflatin on in vitro porcine somatic cell nuclear transfer embryos. Cell Reprogram 16:253–265

Huang Y, Tang X, Xie W, Zhou Y, Li D, Yao C, Zhou Y, Zhu J, Lai L, Ouyang H, Pang D (2011) Histone deacetylase inhibitor significantly improved the cloning efficiency of porcine somatic cell nuclear transfer embryos. Cell Reprogram 13:513–520

Johnson RT, Rao PN (1970) Mammalian cell fusion: studies on the regulation of DNA synthesis and mitosis. Nature 225:159–164

Kallingappa PK, Turner PM, Eichenlaub MP, Green AL, Oback FC, Chibnall AM, Wells DN, Oback B (2016) Quiescence loosens epigenetic constraints in bovine somatic cells and improves their reprogramming into totipotency. Biol Reprod 95(16):11–10

Kasinathan P, Knott JG, Wang Z, Jerry DJ, Robl JM (2001) Production of calves from G1 fibroblasts. Nat Biotechnol 19:1176–1178

Kato Y, Tani T, Tsunoda Y (2000) Cloning of calves from various somatic cell types of male and female adult, newborn and fetal cows. J Reprod Fertil 120:231–237

Kishigami S, Mizutani E, Ohta H, Hikichi T, Thuan NV (2006) Significant improvement of mouse cloning technique by treatment with trichostatin a after somatic nuclear transfer. Biochem Biophys Res Commun 340(1):183–189

Kues WA, Anger M, Carnwath JW, Paul D, Motlik J, Niemann H (2000) Cell cycle synchronization of porcine fetal fibroblasts: effects of serum deprivation and reversible cell cycle inhibitors. Biol Reprod 62:412–419

Kuroiwa Y, Kasinathan P, Choi YJ, Naeem R, Tomizuka K, Sullivan EJ, Knott JG, Duteau A, Goldsby RA, Osborne BA, Ishida I, Robl JM (2002) Cloned transchromosomic calves producing human immunoglobulin. Nat Biotechnol 20:889–894

Labrecque R, Sirard MA (2014) The study of mammalian oocyte competence by transcriptome analysis: progress and challenges. Mol Hum Reprod 20:103–116

Li GP, Bunch TD, White KL, Aston KI, Meerdo LN, Pate BJ, Sessions BR (2004a) Development, chromosomal composition, and cell allocation of bovine cloned blastocyst derived from chemically assisted enucleation and cultured in conditioned media. Mol Reprod Dev 68:189–197

Li GP, White KL, Bunch TD (2004b) Review of enucleation methods and procedures used in animal cloning: state of the art. Cloning Stem Cells 6:5–13

Li Z, Shi J, Liu D, Zhou R, Zeng H, Zhou X, Mai R, Zeng S, Luo L, Yu W, Zhang S, Wu Z (2013) Effects of donor fibroblast cell type and transferred cloned embryo number on the efficiency of pig cloning. Cell Reprogram 15:35–42

Lloyd RE, Lee JH, Alberio R, Bowles EJ, Ramalho-Santos J, Campbell KH, John JCS (2006) Aberrant nucleo-cytoplasmic cross-talk results in donor cell mtDNA persistence in cloned embryos. Genetics 172:2515–2527

Mao J, Zhao M-T, Whitworth KM, Spate LD, Walters EM, O'Gorman C, Lee K, Samuel MS, Murphy CN, Wells K, Rivera RM, Prather RS (2015) Oxamflatin treatment enhances cloned porcine embryo development and nuclear reprogramming. Cell Reprogram 17:28–40

Maside C, Gil MA, Cuello C, Sanchez-Osorio J, Parrilla I, Lucas X, Caamano JN, Vazquez JM, Roca J, Martinez EA (2011) Effects of Hoechst 33342 staining and ultraviolet irradiation on the developmental competence of in vitro-matured porcine oocytes. Theriogenology 76:1667–1675

McGrath J, Solter D (1984) Inability of mouse blastomere nuclei transferred to enucleated zygotes to support development in vitro. Science 226:1317–1319

Meirelles FV, Bordignon V, Watanabe Y, Watanabe M, Dayan A, Lobo RB, Garcia JM, Smith LC (2001) Complete replacement of the mitochondrial genotype in a Bos indicus calf reconstructed by nuclear transfer to a Bos taurus oocyte. Genetics 158(1):351–356

Moawad AR, Choi I, Zhu J, Campbell KH (2011) Ovine oocytes vitrified at germinal vesicle stage as cytoplast recipients for somatic cell nuclear transfer (SCNT). Cell Reprogram 13:289–296

Oback B (2008) Cloning from stem cells: different lineages, different species, same story. Reprod Fertil Dev 21:83–94

Oback B, Wells D (2002) Donor cells for nuclear cloning: many are called, but few are chosen. Cloning Stem Cells 4:147–168

Ono T, Li C, Mizutani E, Terashita Y, Yamagata K, Wakayama T (2010) Inhibition of class IIb histone deacetylase significantly improves cloning efficiency in mice. Biol Reprod 83:929–937

Polejaeva IA, Rutigliano HM, Wells KD (2016) Livestock in biomedical research: history, current status and future prospective. Reprod Fertil Dev 28:112–124

Prather RS, Barnes FL, Sims MM, Robl JM, Eyestone WH, First NL (1987) Nuclear transplantation in the bovine embryo: assessment of donor nuclei and recipient oocyte. Biol Reprod 37(4):859–866

Presicce GA, Yang X (1994) Nuclear dynamics of parthenogenesis of bovine oocytes matured in vitro for 20 and 40 hours and activated with combined ethanol and cycloheximide treatment. Mol Reprod Dev 37:61–68

Radzisheuskaya A, Le Bin Chia G, dos Santos RL, Theunissen TW, Castro LFC, Nichols J, Silva JCR (2013) A defined Oct4 level governs cell state transitions of pluripotency entry and differentiation into all embryonic lineages. Nat Cell Biol 15:579–590

Reggio BC, James AN, Green HL, Gavin WG, Behboodi E, Echelard Y, Godke RA (2001) Cloned transgenic offspring resulting from somatic cell nuclear transfer in the goat: oocytes derived from both follicle-stimulating hormone-stimulated and nonstimulated abattoir-derived ovaries. Biol Reprod 65(5):1528–1533

Rideout WM 3rd, Wakayama T, Wutz A, Eggan K, Jackson-Grusby L, Dausman J, Yanagimachi R, Jaenisch R (2000) Generation of mice from wild-type and targeted ES cells by nuclear cloning. Nat Genet 24:109–110

Ross PJ, Cibelli JB (2010) Bovine somatic cell nuclear transfer. Methods Mol Biol 636:155–177

Ross PJ, Rodriguez RM, Iager AE, Beyhan Z, Wang K, Ragina NP, Yoon SY, Fissore RA, Cibelli JB (2009) Activation of bovine somatic cell nuclear transfer embryos by PLCZ cRNA injection. Reproduction 137(3):427–437

Rybouchkin A, Kato Y, Tsunoda Y (2006) Role of histone acetylation in reprogramming of somatic nuclei following nuclear transfer. Biol Reprod 74:1083–1089

Sarmiento JA (2014) The role of histone methyltransferases in determining developmental potential of bovine oocytes. Louisiana State University. PhD dissertation. http://etd.lsu.edu/docs/available/etd-11052014-192543/

Shi W, Hoeflich A, Flaswinkel H, Stojkovic M, Wolf E, Zakhartchenko V (2003) Induction of a senescent-like phenotype does not confer the ability of bovine immortal cells to support the development of nuclear transfer embryos. Biol Reprod 69:301–309

Staszkiewicz J, Power RA, Harkins LL, Barnes CW, Strickler KL, Rim JS, Bondioli KR, Eilersten KJ (2013) Silencing histone deacetylase-specific isoforms enhances expression of pluripotency genes in bovine fibroblasts. Cell Reprogram 15:397–404

Susko-Parrish JL, Liebfried-Rutledge ML, Northey DL, Schutzkus V, First NL (1994) Inhibition of protein kinases after an induced calcium transient causes transition of bovine oocytes to embryonic cycles without meiotic completion. Dev Biol 166:729–739

Vajta G, Kragh PM, Mtango NR, Callesen H (2005) Hand-made cloning approach: potentials and limitations. Reprod Fertil Dev 17(1–2):97–112

Wakayama T, Yanagimachi R (2001) Mouse cloning with nucleus donor cells of different age and type. Mol Reprod Dev 58:376–383

Wang ZG, Wang W, Yu SD, Xu ZR (2008) Effects of different activation protocols on preimplantation development, apoptosis and ploidy of bovine parthenogenetic embryos. Anim Reprod Sci 105(3–4):292–301

Wang Y, Su J, Wang L, Xu W, Quan F, Liu J, Zhang Y (2011) The effects of 5-aza-2′- deoxycytidine and trichostatin a on gene expression and DNA methylation status in cloned bovine blastocysts. Cell Reprogram 13:297–306

Wee G, Shim JJ, Koo DB, Chae JI, Lee KK, Han YM (2007) Epigenetic alteration of the donor cells does not recapitulate the reprogramming of DNA-methylation in cloned embryos. Reproduction 134:781–787

Westhusin ME, Levanduski MJ, Scarborough R, Looney CR, Bondioli KR (1992) Viable embryos and normal calves after nuclear transfer into Hoechst stained enucleated demi-oocytes of cows. J Reprod Fertil 95(2):475–480

Whitworth KM, Zhao J, Spate LD, Li R, Prather RS (2011) Scriptaid corrects gene expression of a few aberrantly reprogrammed transcripts in nuclear transfer pig blastocyst stage embryos. Cell Reprogram 13:191–204

Willadsen SM (1986) Nuclear transplantation in sheep embryos. Nature 320:63–65

Wilmut I, Schnieke AE, McWhir J, Kind AJ, Campbell KHS (1997) Viable offspring derived from fetal and adult mammalian cells. Nature 385:810–813

Wilson VL, Jones PA (1983) DNA methylation decreases in aging but not in immortal cells. Science 220:1055–1057

Yang XJ, Seto E (2008) The Rpd3/Hda1 family of lysine deacetylases: from bacteria and yeast to mice and men. Nat Rev Mol Cell Biol 9:206–218

Yang X, Smith SL, Tian XC, Lewin HA, Renard JP, Wakayama T (2007) Nuclear reprogramming of cloned embryos and its implications for therapeutic cloning. Nat Genet 39:295–302

Yang B-C, Im G-S, Kim D-H, Yang B-S, Oh H-J, Park H-S, Seong H-H, Kim S-W, Ka H-H, Lee C-K (2008) Development of vitrified–thawed bovine oocytes after in vitro fertilization and somatic cell nuclear transfer. Anim Reprod Sci 103:25–37

Young LE, Sinclair KD, Wilmut I (1998) Large offspring syndrome in cattle and sheep. Rev Reprod 3:155–163

Zhao J, Hao Y, Ross JW, Spate LD, Walters EM, Samuel MS, Rieke A, Murphy CN, Prather RS (2010) Histone deacetylase inhibitors improve in vitro and in vivo developmental competence of somatic cell nuclear transfer porcine embryos. Cell Reprogram 12:75–83

Commercial Applications of SCNT in Livestock

2

Mark Walton

Abstract

For most people, the report of the birth of Dolly the sheep in 1996 was their first inclination that cloning animals was not science fiction. Dolly's arrival was both a seminal moment in the science of reproduction and another step in the evolution of Advanced Reproduction Technologies (ART) in livestock. Somatic cell nuclear transfer (SCNT, aka "cloning"), the process that led to Dolly, is a form of ART that results in exact genetic copies of the donor animal. Livestock produced through cloning are no different than any other animal, and livestock products, e.g., meat and milk, from cloned animals, were found by regulators from Europe, Japan, and the United States to be identical to products from conventional animals. Companies in China, Australia, South America, and the United States provide cloning services to livestock producers and breeders. Cost, burdensome regulatory processes, and public antipathy, especially in Europe, have limited the impact of cloning.

2.1 Introduction

For most people the report of the birth of Dolly (Wilmut et al. 1997) was their first inclination that cloning animals was not science fiction. Although the general public was largely unaware of the full body of work that preceded Dolly's birth, her existence represented the culmination of nearly a century of research in animal reproduction and reproductive physiology. Dolly's arrival was both a seminal moment in the science of reproduction and another step in the evolution of advanced reproductive technologies (ARTs) in livestock.

M. Walton
MWalton Enterprises llc, Austin, TX, USA
e-mail: mark.walton@mwaltonenterprisesllc.com

© Springer International Publishing AG, part of Springer Nature 2018
H. Niemann, C. Wrenzycki (eds.), *Animal Biotechnology 2*,
https://doi.org/10.1007/978-3-319-92348-2_2

The work that eventually led to Dolly's birth in 1996 began in 1891 when Hans Driesch divided an early-stage embryo from sea urchin to produce twins (Sunderland 2012) and contributed to the work of Hans Spemann who proposed a "fantastical" experiment, to remove the nucleus from an unfertilized egg and replace it with the nucleus from a differentiated cell, that would answer the question of whether or not a differentiated nucleus could direct the development of an undifferentiated zygote (Spemann 1938).

The term "clone" was coined in 1903 by Herbert Webber, a plant physiologist, and JBS Haldane is credited with being the first to use the word in the context of making a genetic copy of an animal (Palca 2009; Thomas 2012).

Briggs and King developed the techniques required for successful nuclear transfer and demonstrated the technique using embryonic tissue in leopard frogs (Briggs and King 1952), and the first animal to be cloned from an adult cell was a frog (Gurdon 1962). The last 20 years of the twentieth century saw a flurry of activity including Willadsen's reports of producing sheep and calves from cloned embryos in 1986 and 1989, respectively, and advances in multigenerational embryo cloning (Barnes et al. 1990; Stice and Keefer 1993; Willadsen 1986, 1989). It was from this lineage of work that Dolly was born.

By 2007, 10 years after Dolly's birth, an additional 18 mammalian species had been cloned from adult cells using somatic cell nuclear transfer (SCNT) including pig, cow, goat, and horse (Baguisi et al. 1999; Galli et al. 2003; Polejaeva et al. 2000; Wells et al. 1999). The ability to produce genetic copies of adult animals was of interest in agriculture and the biomedical industry, and Dolly's birth stimulated commercial interest in cloning in both sectors.

2.2   SCNT in Livestock Production

2.2.1   Before Dolly

To more fully appreciate the excitement Dolly's arrival generated in the livestock industry, it is helpful to consider the role of ARTs in livestock breeding and production.

Successful livestock operations have the same basic goal as every other business: to produce a product consumers want at a cost that enables them to sell the product at a profitable and economically sustainable price point. In livestock production two important drivers of profitable production are genetics and reproductive success. ARTs are tools that enable livestock producers to improve average herd performance by leveraging the most elite breeding animals and with some ARTs increase gain from selection by increasing selection intensity and shortening generation intervals (Faber et al. 2003).

When Dolly arrived in 1996, livestock producers were routinely using artificial insemination (AI) and embryo transfer (ET). The value proposition for AI is based on the relative ease of achieving high rates of genetic progress in males (in comparison to rates of gain in females) and through the use of AI extending the genetic impact of high-merit sires across a large number of females. The majority of dairy producers in Europe and the United States routinely use AI to leverage the best male genetics available and improve average herd performance (Foote 2002; Khanal and Gillespie 2013). The adoption of AI by the swine industry, first in Europe and later in the United States, radically changed the nature of pork production (Dominiek et al. 2011).

While AI is an efficient method for leveraging the best male genetics, extending the value of high-merit females requires the use of ET or variations of ET including multiple ovulation and embryo transfer (MOET) and in vitro fertilization (IVF). The commercial use of these ARTs is driven by the high value of breeding animals and the much lower rate of gain achievable in female lines, the consequence of long gestation times and (predominantly) single offspring (Granleese et al. 2015; Hansen 2006; Taylor-Robinson et al. 2014).

Commercial use of bovine ET began in the United States in the late 1970s. In 1987 the concept of MOET was introduced and made it possible to increase the rate of genetic gain by producing multiple progeny from genetically elite females and reducing the generation interval. The use of MOET to produce multiple progeny from elite females is the functional equivalent of using AI to produce multiple progeny from elite males (Hasler 2003; Mapletoft and Hasler 2005).

The use of ET by the cattle industry was driven in part by breeders desiring a higher rate of return on their investments in exotic cattle breeds (Hasler 2003). Advances in the ability to mature, fertilize, and culture embryos in vitro made it possible to reduce generation intervals by collecting oocytes from prepubertal females, extending the breeding life of high-merit females by collecting and storing embryos for later use and realizing the genetic potential of high-merit females that were reproductively challenged (Faber et al. 2003; Hasler 2003).

In the last half of the 1980s, the early successes with nuclear transfer (NT) in frogs were replicated in sheep, cattle, and pigs (Prather et al. 1989; Willadsen 1986, 1989). The scientific advances in NT were key steps along the path to Dolly and of significant interest in the livestock community, especially in cattle. The successful production of sheep, cattle, and pigs cloned from embryonic tissues in the late 1980s along with progress toward repeated cycles of cloning embryos ("multigenerational cloning") (Stice and Keefer 1993) stirred the imagination of cattle producers. In a paper given at a conference on reproductive strategies in cattle, Westhusin et al. described the view of the cattle industry on the potential of NT this way: "With this approach, thousands of genetically identical embryos could be produced that when transferred into recipient females would result in thousands of genetically identical calves. The idea spawned visions of large herds of cloned bulls, cloned feedlot steers and cloned dairy cows" (Westhusin et al. 2005).

Livestock producers and genetics companies viewed NT as a tool analogous to AI, i.e., as a way to use the best genetics more broadly and improve the average genetic value of herds. Beef producers envisioned the same top performing bull in every pasture; dairy producers of having herds of uniformly high-merit milking cows and swine genetics companies could see the potential to clone the top few percent of their genetic pyramids and use the clones as commercial boars and multiplier sows. Consistency of animal performance is a highly valuable attribute in livestock production, and NT offered a way to reduce some of the genetic variability that is introduced through conventional mating.

Although there was great interest in NT among livestock producers and several genetics companies added NT to their genetic toolbox, the high cost of cloning embryos, the low pregnancy and calving rates associated with the practice and the unsure genetic outcomes arising from cloning an embryo of unproven value, proved to be significant barriers to widespread commercial use of cloned embryos. By the time of Dolly's arrival in 1996, interest in cloned embryos was in decline (Stice et al. 1998; Westhusin et al. 2005).

2.2.2   The Arrival of Dolly and SCNT

The ability to produce a clone from an adult cell line reignited interest in cloning in the livestock industry. The same vision of herds of cloned animals that attracted producers and genetics companies to NT was stimulated again by the announcement of Dolly's birth and fueled by two important differences between NT and SCNT: the ability to clone animals of proven genetic merit and the animals produced through SCNT would be exact genetic copies of the original.

A 2003 article appearing in the *Louisiana Agriculture* magazine described the potential of SCNT in livestock this way: "Cloning would provide the cattle producer an opportunity to reproduce genetically valuable seedstock animals, clone animals that have suffered a severe injury such as a fractured leg and can no longer reproduce, or clone males that had been prematurely castrated, such as a prize-winning show steer." The authors went on to add "Cloning technology would also provide livestock producers with ready access to production-tested breeding stock, thus increasing the accuracy of selection in their breeding herds. It has been proposed that cloning F1 terminal-breed males to produce males for market steers might be the ultimate beef production management system" (Godke et al. 2003).

Although the cost of producing cloned animals is high and precludes the use of cloning for producing commercial animals, using SCNT to reproduce animals with proven breeding values can be economically advantageous when cloned animals are used as semen or embryo donors (Kinghorn 2000). Cloning has been used to replicate genetically elite bulls and high-merit dairy cows, rare and endangered breeds, to maintain genetic diversity, and to increase the frequency of rare alleles associated with desirable traits (Loi et al. 2001; Stice et al. 1998).

A unique example of the potential of SCNT to capture the value of rare alleles that occur in production herds can be seen in the work underway by a team of scientists in Texas who have produced a cloned bull from cells collected from the carcass of a slaughtered steer (castrated male used for beef production) after the quality and yield grade of the carcass were known. Tissue samples were collected from carcasses that graded prime (highest quality grade) and yield grade 1 (most muscle with least amount of waste fat), a combination of quality and yield obtained by only 3 of every 1000 fed cattle. The collected tissues were used to establish cell lines and for DNA evaluation and cell lines from carcasses that also carried evidence of genetic attributes associated with the carcass attributes were cloned (Hawkins and Lawrence 2013).

A frequently asked question is how many farm animals have been cloned and the answer is that no one knows for sure. Many farm animal species have been cloned commercially including cattle, pigs, sheep, goats, buffalo, and red deer (Oback 2008). While there is no official tally of the number of animals cloned, it is safe to say that in cattle, which is undoubtedly the species most frequently cloned by the commercial sector, the total number of cloned cattle is quite small when compared to the total number of cattle. Most estimates put the total number of animals cloned thus far in the several thousands and continuing to grow (Michel 2014; Plume 2009).

Outside of farm animals, horses are cloned more frequently than many other species. The first horse was cloned in 2003 (Galli et al. 2003), and high-value sporting horses are being cloned both for breeding and competition (Cohen 2015; Mander 2013; Williams 2015). Companion animals have been and are being cloned commercially including dogs by Sooam Biotech and ViaGen. ViaGen also clones cats as part of its companion animal business.

2.2.3 Cloning Challenges

Widespread adoption of SCNT by commercial livestock interests has been hindered by three important challenges: overcoming technical issues that keep cloning expensive, having a clear regulatory pathway to the market, and gaining public acceptance for the use of cloned animals.

2.2.4 Technical Challenges of SCNT

The production of cloned animals using SCNT has never been efficient. The embryo that became Dolly was 1 of more than 250 implanted by the Roslin team, and while the efficiency of cloning has improved significantly since Dolly, it is still relatively low. In animal species that normally produce single offspring, e.g., cattle, sheep, and horses, efficiency of cloning is quantified as the proportion of all embryos transferred into recipients that produce live offspring. Cattle have been the primary focus of cloning research, and cloning efficiency in cattle has improved from 1 to 3% to 5% to 10% on average with efficiencies as high as 45% reported (Faber et al. 2004; Kasinathan et al. 2015; Long et al. 2014; Wells 2005). Efficiency differences can be affected by the tissue from which cell lines are derived and by the stage of cycle of the donor and recipient cells and can vary significantly among different cell lines derived from the same individual (Kato et al. 1998; Liu et al. 2013; Wells et al. 2003). Cloning efficiency in horse has been reported to be 10% (Walton 2013).

The efficiency of cloning pigs is quantified differently because pigs are litter-bearing animals. In cattle and horses, the common practice is to transfer one SCNT embryo into each recipient. In pigs the common practice is to transfer 100 or more early- to late-stage blastocysts into a surrogate mother and quantify efficiency as the proportion of transfers that produce pregnancy and the number of live, healthy piglets per litter (Petersen et al. 2008). In recent projects involving transgenic and gene-edited pigs, 50–60% of transfers resulted in pregnancies; nearly 100% of pregnancies advanced to parturition, and on average pregnancies produced an average of five piglets per litter (Walton 2016).

Cloned animals can experience a wide range of pre- and postnatal developmental challenges including excess fluid in the allantoin of the surrogate dam, abnormal placentation, extended gestation periods, large offspring, cardiovascular and respiratory distress, problems with tendons, and increased susceptibility to infection (Faber et al. 2004; Niemann and Lucas-Hahn 2012; Panarace et al. 2007; Wells 2005). These challenges are most often seen in ruminant animals and are not as prevalent in non-ruminants (Walton 2013).

Epigenetic factors impacting nuclear programming are generally considered to be the primary causes of the problems seen in the cloning process and early life of the cloned animals (Chavatte-Palmer et al. 2012; Lee and Prather 2013; Long et al. 2014; Niemann and Lucas-Hahn 2012). However, cloned animals that survive the perinatal period are generally healthy as evidenced in a recent paper by Polejaeva et al. in which they followed 96 cloned cows and their offspring (Polejaeva et al. 2013). The longitudinal study compared the reproductive performance of the cloned cows and their offspring to conventionally bred comparators and found that reproductive performance of the clones and their offspring was within normal parameters and not different than the reproductive outcomes of conventionally bred comparators.

2.2.5   Regulation of SCNT

Dolly's birth generated a great deal of discussion about the ethics of cloning with most of the discussion driven by fears that scientists would take the lessons learned in producing Dolly and apply them to cloning humans. In the United States, that fear led to a 1997 presidential ban on the use of federal funds for research on human cloning (NIH 1997).

Although the primary focus of the ethics debate was human cloning, the concept of clones and the use of cloned animals in the food supply created unease among the public, and in 2001 the US Food and Drug Administration (FDA) asked livestock producers to voluntarily refrain from putting meat and milk from cloned animals or their offspring into the food supply, while the agency assessed the safety of meat and milk from clones and progeny of clones. In 2007 the FDA issued a draft risk assessment in which the agency concluded that the meat and milk from cloned cattle, pigs, and goats were as safe as meat and milk from non-cloned animals. The final risk assessment was released in January 2008 with the FDA again stating that "Meat and milk from clones of cattle, swine, goats, and the offspring of all clones, are as safe to eat as food from conventionally bred animals" (FDA 2008; Rudenko and Matheson 2007).

In addition to the FDA, the European Food Safety Authority (EFSA) and the Japan Food Safety Commission (JFSC) conducted cloning risk assessments and like the United States found no evidence that meat and milk from clones and the offspring of clones are less safe than meat and milk from non-cloned animals (EFSA 2009; JFSC 2009).

The completion of three independent risk assessments led many countries, including the United States, Argentina, Brazil, Australia, New Zealand, and China, to remove (if they existed) any requirement for pre-market assessments of cloned animals. In the United States, the USDA continues to ask livestock producers to refrain from placing meat or milk from cloned animals into the food supply, ostensibly to provide sufficient time for a "smooth and seamless transition into the market" (U.S. Department of Agriculture 2008). Given that clones are used as breeding animals, the voluntary moratorium poses no great barrier to the use of cloning. In those countries where cloning is in use, there is no requirement to label food from clones or their progeny as such.

Meat and milk from cloned animals and their offspring are considered novel foods in the European Union (EU) and Canada and as such would require regulatory review and approval before they can enter the market place. Research and use of cloning in food animals is essentially nonexistent in either Canada or the EU, and in 2015 the European Parliament proposed a complete ban on farm animal cloning (Vogel 2015).

2.2.6 Public Acceptance

To quote *The New York Times*, Dolly's birth "created a ruckus," primarily over concerns that the next step in cloning would be to clone humans (Nicholas 2013). In 1998 the controversy grew along with the first report of deriving pluripotent cell lines from human blastocysts (Thomson et al. 1998). Given this volatile mix of controversial issues and the general unease that cloning created in many people, it was impossible for livestock cloning to avoid becoming part of the discussion.

In addition to the early concerns about animal cloning leading to human cloning, as research finding on losses during pregnancy, reduced survival rates of cloned animals, and the physical and physiological issues affecting cloned animals were published in the scientific literature, they were also being reported in newspapers and popular magazines. These reports generated questions about animal welfare and the ethics of cloning became the focus of the livestock cloning debate (Fiester 2005). In the year between publication of the FDA draft risk assessment and the delivery of the final risk assessment, both sides of the cloning argument intensified their efforts to either ensure the risk assessment was published or derail it and delay or prevent cloning from becoming officially declared safe.

The debate about cloning did not end after the FDA, EFSA, and the JFSC risk assessments were completed and communicated, but they did die down. In the United States and South America, several companies provide cloning services for livestock producers with very little public discussion. Only in Europe is cloning still a contentious topic and that is almost exclusively in the context of cloning farm animals.

Although the public angst about cloning has subsided, getting to that condition has had a lasting effect on the use of cloning in the livestock industry. Two segments of the livestock industry that could benefit from the use of cloning, dairy and swine, have largely eschewed the technology. The dairy industry at one point was actively interested in cloning, and more than one clone of prize winning dairy cows have gone on to win prizes in their own right. However, the global dairy genetics industry has never used cloning to any great extent, and in some cases, cloning activities were stopped altogether, usually because customers in Europe were not willing to risk public outcry by using semen from cloned animals. These companies have chosen to forego the benefits of cloning rather than deal with the business and public relations challenges that would be required to use clones in only part of their program.

The pig genetics industry has never used cloning to any significant extent. Beyond a few research efforts relatively early in the history of livestock cloning, the pig industry has not attempted to introduce cloning into their operations. The decision to avoid the cloning debate made sense in light of the intense public scrutiny that was occurring, and the completion of the various cloning risk assessments was not sufficient to overcome concerns the companies had about public reactions to the use of cloning.

2.2.7   Commercial Delivery of ARTs

Livestock producers have long used castration and early weaning as tools for managing the flow of genetics in their operations. These methods of reproductive control fall within the realm of standard animal husbandry practices and are implemented on the farm without the assistance of reproductive specialists. ARTs, however, require varying degrees of skill, expertise, and infrastructure that often fall outside the range of the livestock producer's daily activities. The need for specialized expertise and infrastructure to obtain the benefits of using ARTs has created a commercial sector focused on providing ARTs to livestock producers.

The business of ART began with AI in dairy production. Danish dairy producers established the first AI cooperative in 1933 and the United States followed suit in 1937 (Foote 2002). By the time Dolly was born, 60–70% of American dairies used AI, over 90% of dairies in Europe employed the practice, and the first AI cooperatives had transformed into commercial businesses. Today over 90% of dairy cattle in Europe and North America are impregnated using AI, and bovine semen is collected, sold, and distributed by regional and global genetics companies (Department of Animal Science 2000; Faber et al. 2003; Khanal and Gillespie 2013).

European pig producers began to adopt AI for commercial production of pork in the 1960s, and by 1996 AI was widely used in Europe (Dominiek et al. 2011). The adoption of AI by pork producers changed the face of the pig genetics industry from a focus on pure breeds to one based on sophisticated breeding programs working with proprietary genetic lines. European breeders led the way in this change, and although the United States was slow to change, when it did, in less than a decade the US industry changed from fewer than 10% of sows being impregnated through AI to 70% in 2000 (Estienne and Harper 2009).

By the time of Dolly's birth in 1996, ET was a well-established practice with hundreds if not thousands of ET practitioners in Europe, the United States, and South America. The development of nonsurgical embryo collection and transfer methods, cryopreservation of embryos, and various embryo manipulation techniques including embryo splitting and embryo sexing helped to drive the use of IVF in dairy and high-value beef breeds (Hasler 2003). The International Embryo Transfer Society, founded in 1974, reported that more than 360,000 bovine embryos, including both in vivo and in vitro produced embryos, were transferred in 1996 as were smaller numbers of ovine, caprine, equine, porcine, and cervid embryos (Thibier 1998). In 2016, over 965,000 bovine embryos were transferred of which 53% were in vivo derived embryos and the rest IVF embryos (Perry 2017).

Commercial use of NT began shortly after Willadsen's 1986 report of producing sheep from embryogenic cell lines when Granada Genetics began using NT in its cattle breeding program. Other cattle genetics companies including Alta Genetics, American Breeders Services, and GenMark also incorporated NT into their breeding activities. By 1996 the commercial use of NT was almost nonexistent, a result of the high cost of producing clone embryos, the inefficiency of producing cloned animals, and highly variable outcomes (Westhusin et al. 2005).

The interest of the livestock industry in the use of SCNT was matched by commercial interests ready to deliver cloned animals. Two years after the notice of Dolly's birth, one of the first cloning companies, Infigen, had cloned Holstein heifers on display at the World Dairy Expo in Madison, Wisconsin (Infigen 2001). In 2001 three companies, Infigen, Advanced Cell Therapeutics (ACT), and ProLinia, were offering to clone cattle at prices around $20,000 per calf, and PPL Therapeutics, the Scotland-based company that had licensed the intellectual property generated by the Roslin Institute scientists working in SCNT, was actively cloning pigs for biomedical applications.

Not unexpectedly, by the time the final FDA risk assessment was released, the commercial cloning landscape had changed and two companies, ViaGen Inc. and Cyagra, had become the primary livestock cloning companies in North America. Small-scale commercial activities were getting underway in Argentina and Brazil, and there was some commercial cloning being done in Australia. The only company actively cloning in Europe at that time was Cryozootech, a French company that clones only horses.

In 2016 ViaGen, now a wholly owned subsidiary of Trans Ova Genetics, continues to provide cloning services in several species. Cryozootech and Crestview Genetics, a company that started in the United States but moved to Argentina, are cloning high-value sporting horses. Cabaña Milenium, another Argentine ART company, clones cattle and goats, and InVitro Brasil Clonagem Animal is active in Brazil.

Sooam Biotech is a South Korean cloning company best known for its work in dogs that has cloned a diverse array of animal species (Baer 2015). Sooam and Boyalife, a Chinese biotechnology firm, have announced plans to build the largest animal cloning facility in the world and have announced their intent to produce one million cloned beef cattle per year for slaughter (Phys.org 2015).

2.2.8   SCNT and Genetic Modification of Livestock

Transgenic technology has been used to introduce genes of interest into livestock for over 30 years. After Gordon et al. successfully introduced a transgene into mouse by injecting DNA into a single-cell zygote, pronuclear injection became the primary method for generating transgenic animals (Gordon and Ruddle 1981; Murray and Anderson 2000; Stice et al. 1998; Wheeler 2003). Although pronuclear transformation was effective, it was not efficient, and one of the drivers behind the development of SCNT was a search for more efficient methods of producing transgenic animals.

The efficiency of pronuclear injection was hindered by the difficulty of securing a sufficient number of embryos and the need to produce progeny of the putative transgenic animals before knowing if the transgene was both functioning and stable. SCNT provided an unlimited supply of cells for transformation and made it possible to screen the transformed cells before NT so that every animal produced is transgenic. SCNT-mediated transformation offered the additional advantage of being able to produce additional copies of the transgenic animal should that be necessary by cryopreserving the transgenic cell line and using it to produce new animals (Hodges and Stice 2003; Lai and Prather 2003; Stice et al. 1998).

The low pregnancy rates, low cloning efficiency, and physical challenges experienced by some cloned animals that impact the utility of SCNT in livestock production are much less important in the production of transgenic animals as only a small number of transgenic are needed to launch a transgenic program. When the transgenic animal is proven to be of value, it can be used as a breeding animal, and its traits passed to the next generation through conventional breeding.

The first transgenic animal produced using SCNT approved by the FDA was a transgenic goat in which a human therapeutic protein is produced (U.S. FDA 2009). In March 2016 the FDA granted authority through its enforcement discretion powers for transgenic pig models of cystic fibrosis and atherosclerosis to be sold commercially (Intrexon 2016).

The value of SCNT as a tool for producing genetically modified animals will be even greater when it is combined with the gene technologies such as TALEN and CRISPR. These methods are being used to make very precise genetic changes, including gene deactivation (knockout) and allele substitution. Gene-edited technologies can also be used to insert a transgene into a specific location in the genome, and the combination of SCNT and site-directed insertion will further enhance the efficiency of producing transgenic animals (Carlson et al. 2012; Tan et al. 2012, 2013).

2.2.9   The Future of SCNT

In the 20 years that have elapsed since Dolly was born, there have been many significant advances in livestock genetics, genomics, and reproduction. Animal breeders in many species can use sexed semen or sexed embryos to preselect the sex of animal being produced; within days of birth, a Holstein calf can be genotyped and an accurate estimate of the calf's breeding value known; and gene-editing methods are becoming practical tools for use by livestock breeders. These technologies, along with a number of other technological advances in genetics, genomics, and reproduction, will change livestock breeding and have a significant impact on livestock production. Does SCNT have a role to play in the future and if so what is that role? Two recent papers, one involving SCNT and genomic selection (GS) and the other a gene-editing paper, provide examples of how SCNT might be used in the future.

Kasinathan et al. have demonstrated a breeding approach that combines genome-wide selection with MOET, ovum pickup, IVF using sex-sorted and non-sorted semen, and SCNT to reduce generation interval in the Jersey breed of dairy cattle, one of the two factors that determine gain from selection (Kasinathan et al. 2015). In the study elite Jersey dairy cows were treated to stimulate oocyte production which was collected and used to produce IVF embryos. Multiple IVF embryos were transferred into surrogate mothers and allowed to develop for 21 days then flushed and collected. Cell lines were generated from the collected fetuses and sent for DNA analysis. The cell lines identified as having high genetic merit were used to produce cloned calves using SCNT. 45% of the pregnancies produced healthy calves which were genotyped to confirm the genetic merit of the animal. This approach utilizes not just SCNT but several ARTs and genomic selection to reduce the generation interval in cattle, a major factor limiting gain from selection. The authors claim the method is scalable and provides a realistic way to overcome a biological barrier to more rapid genetic improvement.

In the second example, Carlson et al. combined gene editing with SCNT to substitute the *polled* allele from Angus cattle into a Holstein cell line (Carlson et al. 2016). The *polled* allele is a variant that results in the absence of horns and is desirable because of the dangers to humans and other animals posed by the horns. The methods used to remove horns can induce stress, reduce performance, and are detrimental to the welfare of the animals. Most dairy breeds develop horns, and although the preferred *polled* variant is present in some dairy breeds, it occurs at low frequency and was introduced by crossing with beef breeds. The use of SCNT made it possible to introduce the Angus allele into a Holstein cell line and select those cells in which the desired allele was present to clone. SCNT can be used to introduce the *polled* allele into high-merit dairy lines and significantly reduce the time it would take to introduce the trait through conventional breeding alone.

These examples provide some idea of the utility and value of SCNT in livestock. Until and unless cloning efficiency improves beyond the current levels and the health and welfare of cloned animals is equivalent to the welfare status obtained with other IVF and related ARTs, it will be difficult for livestock producers to utilize

SCNT for applications other than production of breeding animals. However, as shown in the two examples provided, SCNT does have an important role to play in livestock genetics and breeding today.

References

Baer D (2015) Inside the Korean lab that has cloned more than 600 dogs. Business Insider Australia

Baguisi A, Behboodi E, Melican DT, Pollock JS, Destrempes MM, Cammuso C et al (1999) Production of goats by somatic cell nuclear transfer. Nat Biotechnol 17(5):456–461. https://doi.org/10.1038/8632

Barnes FL, Westhusin ME, & Looney CR (1990) Embryo cloning: principles and progress. Edinburgh

Briggs R, King TJ (1952) Transplantation of living nuclei from blastula cells into enculeated frogs eggs. Proc Natl Acad Sci U S A 38(5):455–463

Carlson DF, Lancto CA, Zang B, Kim ES, Walton M, Oldeschulte D et al (2016) Production of hornless dairy cattle from genome-edited cell lines. Nat Biotechnol 34(5):479–481. https://doi.org/10.1038/nbt.3560

Carlson DF, Tan W, Lillico SG, Stverakova D, Proudfoot C, Christian M et al (2012) Efficient TALEN-mediated gene knockout in livestock. Proc Natl Acad Sci U S A 109(43):17382–17387. https://doi.org/10.1073/pnas.1211446109

Chavatte-Palmer P, Camous S, Jammes H, Le Cleac'h N, Guillomot M, Lee RS (2012) Review: placental perturbations induce the developmental abnormalities often observed in bovine somatic cell nuclear transfer. Placenta 33(Suppl):S99–s104. https://doi.org/10.1016/j.placenta.2011.09.012

Cohen H (2015) How champion-pony clones have transformed the game of Polo. Vanity Fair

Department of Animal Science, U. o. W (2000) HIstory of artificial insemination. Retrieved from http://www.ansci.wisc.edu/jjp1/ansci_repro/lec/handouts/hd8.html

Dominiek M, Alfonso LR, Tom R, Philip V, Ann VS (2011) Artificial insemination in pigs, artificial insemination in farm animals milad manafi. IntechOpen, London. https://doi.org/10.5772/16592 Available from: https://www.intechopen.com/books/artificial-insemination-in-farm-animals/artificial-insemination-in-pigs

EFSA (2009) Statement of EFSA prepared by the scientific committee and advisory forum unit on further advice on the implications of animal cloning (SCNT). EFSA J 319:1–15

Estienne MJ, & Harper AF (2009) Using artificial insemination in swine production: detecting and synchronizing estrus and using proper insemination technique, 8. Retrieved from www.ext.vt.edu

Faber D, Ferre LB, Metzger J, Robl J, Kasinathan P (2004) Agro-economic impact of cattle cloning. Cloning Stem Cells 6(2):198–206

Faber DC, Molina JA, Ohlrichs CL, Zwaag DFV, Ferre LB (2003) Commercialization of animal biotechnology. Theriogenology 59:125–138

FDA. (2008). Animal cloning: a risk assessment. Center for Veterinary Medicine, U.S. Food and Drug Administration. Rockville, MD

FDA, U. S. (2009) FDA approves orphan drug atryn to treat rare clotting disorder [Press release]. Retrieved from http://www.fda.gov/NewsEvents/Newsroom/PressAnnouncements/ucm109074.htm

Fiester A (2005) Ethical issues in animal cloning. Perspect Biol Med 48(2):328–343

Foote RH (2002) The history of artificial insemination: selected notes and notables. J Anim Sci 80(E-Suppl_2):1–10. https://doi.org/10.2134/animalsci2002.80E-Suppl_21a

Galli C, Lagutina I, Crotti G, Colleoni S, Turini P, Ponderato N et al (2003) Pregnancy: a cloned horse born to its dam twin. Nature 424(6949):635–635. https://doi.org/10.1038/424635a

Godke RA, Denniston RS, Reggio B (2003) Animal biotechnology and the future. Louisiana Agric 46(4):10–14

Gordon JW, Ruddle FH (1981) Integration and stable germ line transmission of genes injected into mouse pronuclei. Science 214(4526):1244–1246

Granleese T, Clark SA, Swan AA, van der Werf JH (2015) Increased genetic gains in sheep, beef and dairy breeding programs from using female reproductive technologies combined with optimal contribution selection and genomic breeding values. Genet Sel Evol 47:70. https://doi.org/10.1186/s12711-015-0151-3

Gurdon JB (1962) The developmental capacity of nuclei taken from intestinal epithelium cells of feeding tadpoles. J Embryol Exp Morphol 10(4):622–640

Hansen PJ (2006) Realizing the promise of IVF in cattle–an overview. Theriogenology 65(1):119–125. https://doi.org/10.1016/j.theriogenology.2005.09.019

Hasler JF (2003) The current status and future of commercial embryo transfer in cattle. Anim Reprod Sci 79(3–4):245–264

Hawkins D, & Lawrence T (2013) From imagination to reality: using DNA from an exceptional carcass to produce a sire or donor cow. paper presented at the range beef cow symposium, Rapid City, SD. http://digitalcommons.unl.edu/rangebeefcowsymp/309

Hodges CA, Stice SL (2003) Generation of bovine transgenics using somatic cell nuclear transfer. Reprod Biol Endocrinol 1:81. https://doi.org/10.1186/1477-7827-1-81

Infigen (2001) World's first herd of cloned dairy cows in production at infigen [Press release]

Intrexon (2016) Game-changing animal research models offer superior translational research and better predictive efficacy [Press release]. Retrieved from http://investors.dna.com/2016-04-27-Game-Changing-Animal-Research-Models-Offer-Superior-Translational-Research-and-Better-Predictive-Efficacy

JFSC (2009) Risk assessment report on foods derived from cloned cattle and pigs produced by somatic cell nuclear transfer (SCNT) and their offspring (Novel Foods)

Kasinathan, P., Wei, H., Xiang, T., Molina, J. A., Metzger, J., Broek, D., . . . Allan, M. F. (2015). Acceleration of genetic gain in cattle by reduction of generation interval. Sci Rep, 5, 8674. doi:https://doi.org/10.1038/srep08674

Kato Y, Tani T, Sotomaru Y, Kurokawa K, Kato J-y, Doguchi H et al (1998) Eight calves cloned from somatic cells of a single adult. Science 282(5396):2095–2098. https://doi.org/10.1126/science.282.5396.2095

Khanal AR, Gillespie J (2013) Adoption and productivity of breeding technologies: evidence from U.S. dairy farms. AgBioforum 16(1):53–65

Kinghorn B (2000) Animal production and breeding systems to exploit cloning technology. In: Kinghorn BP, Van der Werf JH, Ryan M (eds) Animal breeding: use of new technologies. University of Sydney: Post Graduate Foundation in Veterinary Science, University of Sydney, Sydney

Lai L, Prather RS (2003) Creating genetically modified pigs by using nuclear transfer. Reprod Biol Endocrinol 1:82–82. https://doi.org/10.1186/1477-7827-1-82

Lee K, Prather RS (2013) Advancements in somatic cell nuclear transfer and future perspectives. Anim Front 3(4):56–61. https://doi.org/10.2527/af.2013-0034

Liu J, Wang Y, Su J, Luo Y, Quan F, Zhang Y (2013) Nuclear donor cell lines considerably influence cloning efficiency and the incidence of large offspring syndrome in bovine somatic cell nuclear transfer. Reprod Domest Anim 48(4):660–664. https://doi.org/10.1111/rda.12140

Loi P, Ptak G, Barboni B, Fulka J Jr, Cappai P, Clinton M (2001) Genetic rescue of an endangered mammal by cross-species nuclear transfer using post-mortem somatic cells. Nat Biotechnol 19(10):962–964. https://doi.org/10.1038/nbt1001-962

Long CR, Westhusin ME, Golding MC (2014) Reshaping the transcriptional frontier: epigenetics and somatic cell nuclear transfer. Mol Reprod Dev 81(2):183–193. https://doi.org/10.1002/mrd.22271

Mander B (2013) Polo players look twice at cloned ponies. Financial Times. Retrieved from http://www.ft.com

Mapletoft RJ, Hasler JF (2005) Assisted reproductive technologies in cattle: a review. Rev Sci Tech 24(1):393–403

Michel J (2014) US company in Iowa churns out 100 cloned cows a year. The Tico Times. Retrieved from http://www.ticotimes.net/?s=iowa+company+churns+out+100+cloned+cows+a+year

Murray JD, Anderson GG (2000) Genetic engineering and cloning may improve milk, livestock production. Calif Agric 54(4):57–65

Nicholas W (2013) The clone named dolly. The New York Times. Retrieved from http://www.nytimes.com

Niemann H, Lucas-Hahn A (2012) Somatic cell nuclear transfer cloning: practical applications and current legislation. Reprod Domest Anim 47(Suppl 5):2–10. https://doi.org/10.1111/j.1439-0531.2012.02121.x

NIH (1997) Prohibition on federal funding for cloning of human beings. NIH Grants. Retrieved from https://grants.nih.gov/grants/policy/cloning_directive.htm

Oback B (2008) Climbing mount efficiency–small steps, not giant leaps towards higher cloning success in farm animals. Reprod Domest Anim 43(Suppl 2):407–416. https://doi.org/10.1111/j.1439-0531.2008.01192.x

Palca J (2009) Science Diction: The Origin of the Word 'Clone'. Accessed at https://www.npr.org/2011/03/11/134459358/Science-Diction-The-Origin-Of-The-Word-Clone

Panarace M, Agüero JI, Garrote M, Jauregui G, Segovia A, Cané L et al (2007) How healthy are clones and their progeny: 5 years of field experience. Theriogenology 67(1):142–151. https://doi.org/10.1016/j.theriogenology.2006.09.036

Perry G (2017) 2016 statistics of embryo collection and transfer in domestic farm animals. Retrieved from http://www.iets.org/pdf/comm_data/IETS_Data_Retrieval_Report_2016_v2.pdf

Petersen B, Lucas-Hahn A, Oropeza M, Hornen N, Lemme E, Hassel P et al (2008) Development and validation of a highly efficient protocol of porcine somatic cloning using preovulatory embryo transfer in peripubertal gilts. Cloning Stem Cells 10(3):355–362. https://doi.org/10.1089/clo.2008.0026

Phys.org (2015) World's biggest clone factory raises fears in China. Retrieved from http://phys.org/news/2015-11-world-biggest-animal-clone-factory.html

Plume K (2009) Welcome to the clone farm. Reuters. Retrieved from http://www.reuters.com/article/usfoodcloningidUSTRE5AC07V20091 1 13

Polejaeva IA, Broek DM, Walker SC, Zhou W, Walton M, Benninghoff AD, Faber DC (2013) Longitudinal study of reproductive performance of female cattle produced by somatic cell nuclear transfer. PLoS One 8(12):e84283. https://doi.org/10.1371/journal.pone.0084283

Polejaeva IA, Chen S-H, Vaught TD, Page RL, Mullins J, Ball S et al (2000) Cloned pigs produced by nuclear transfer from adult somatic cells. Nature 407:86–90

Prather RS, Sims MM, First NL (1989) Nuclear transplantation in early pig embryos. Biol Reprod 41:414–418

Rudenko L, Matheson JC (2007) The US FDA and animal cloning: risk and regulatory approach. Theriogenology 67(1):198–206. https://doi.org/10.1016/j.theriogenology.2006.09.033

Spemann H (1938) Embryonic development and induction. Yale University Press, New Haven

Stice SL, Keefer CL (1993) Multiple generational bovine embryo cloning. Biol Reprod 48:715–719

Stice SL, Robl J, Ponce de Leon FA, Jerry J, Golueke PG, Cibelli JB, Kane JJ (1998) Cloning: new breakthroughs leading to commercial opportunities. Theriogenology 49:129–138

Sunderland ME (2012) Hans Adolf Eduard Driesch (1867–1941) Embryo Project Encyclopedia

Tan W, Carlson DF, Lancto CA, Garbe JR, Webster DA, Hackett PB, Fahrenkrug SC (2013) Efficient nonmeiotic allele introgression in livestock using custom endonucleases. Proc Natl Acad Sci U S A 110(41):16526–16531. https://doi.org/10.1073/pnas.1310478110

Tan WS, Carlson DF, Walton MW, Fahrenkrug SC, Hackett PB (2012) Precision editing of large animal genomes. Adv Genet 80:37–97. https://doi.org/10.1016/B978-0-12-404742-6.00002-8

Taylor-Robinson AW, Walton S, Swain DL, Walsh KB, Vajta G (2014) The potential for modification in cloning and vitrification technology to enhance genetic progress in beef cattle in northern Australia. Anim Reprod Sci 148(3–4):91–96. https://doi.org/10.1016/j.anireprosci.2014.06.004

Thibier M (1998) The 1997 Embryo transfer statistics from around the World. Data Retrieval Committee Report, International Embryo Transfer Society. https://www.iets.org/pdf/comm_data/December1998.pdf. Accessed May 2016

Thomas I (2012) Should scientists pursue cloning? Heinemann-Raintree, Chicago

Thomson JA, Itskovitz-Eldor J, Shapiro SS, Waknitz MA, Swiergiel JJ, Marshall VS, Jones JM (1998) Embryonic stem cell lines derived from human blastocysts. Science 282(5391):1145–1147

U.S. Department of Agriculture, O. o. C (2008) USDA statement on FDA risk assessment on animal clones. USDA. Retrieved from http://www.usda.gov/wps/portal/usda/usdahome?contentid only=true&contentid=2008/01/0012.xml

Vogel G (2015) E.U. parliament votes to ban cloning of farm animals. Science News. Retrieved from http://www.sciencemag.org/news/2015/09/eu-parliament-votes-ban-cloning-farm-animals

Walton M (2013) Cloning and animal improvement. Paper presented at the impact of biotechnology on future animal breeding, Berlin, Germany

Walton MF (2016) Use of SCNT in production of biomedical pigs. Unpublished Raw Data

Wells DN (2005) Animal cloning: problems and prospects. Rev Sci Tech 24(1):251–264

Wells DN, Laible G, Tucker FC, Miller AL, Oliver JE, Xiang T et al (2003) Coordination between donor cell type and cell cycle stage improves nuclear cloning efficiency in cattle. Theriogenology 59(1):45–59. https://doi.org/10.1016/S0093-691X(02)01273-6

Wells DN, Misica PM, Tervit RH (1999) Production of cloned calves following nuclear transfer with cultured adult mural granulosa cells. Biol Reprod 60(4):996–1005. https://doi.org/10.1095/biolreprod60.4.996

Westhusin ME, Stroud BK, Kraemer DC, & Long CR 2005) Cloning bovine embryos: current status and future applications. Paper presented at the Proceedings, Applied Reproductive Strategies in Beef Cattle, Texas A&M University, College Station, Texas

Wheeler MB (2003) Production of transgenic livestock: promise fulfilled. J Anim Sci 81(Suppl. 3):32–37

Willadsen SM (1986) Nuclear transplantation in sheep embryos. Nature 320(6057):63–65. https://doi.org/10.1038/320063a0

Willadsen SM (1989) Cloning of sheep and cow embryos. Genome 31(2):956–962. https://doi.org/10.1139/g89-167

Williams O (2015) Battle of the clones: when will a replica horse win olympic gold? Retrieved from http://edition.cnn.com/2015/02/20/equestrian/horse-cloning-olympics/index.html

Wilmut I, Schnieke AE, McWhir J, Kind AJ, Campbell KH (1997) Viable offspring derived from fetal and adult mammalian cells. Nature 385(6619):810–813. https://doi.org/10.1038/385810a0

Epigenetic Features of Animal Biotechnologies

3

Nathalie Beaujean

Abstract

Epigenetic mechanisms play a crucial role in many biological processes, such as regulation of gene expression especially after fertilization and during early embryonic development. Indeed, the parental genomes that carry special epigenetic signatures undergo important chromatin remodelling through epigenetic modifications during the first embryonic cleavages, some of which are crucial for the production of healthy embryos.

It is therefore very important for breeders and embryologists to understand how parentally inherited genomes may be epigenetically altered by animal biotechnologies as it could affect embryo quality and further development. This chapter introduces some of the basic epigenetic parameters underpinning early embryonic development and how they could be affected during the processes of embryo in vitro production, somatic cell nuclear transfer or stem cells derivation.

3.1 Introduction

Epigenetics include heritable changes of the phenotype, which do not involve changes in the DNA sequence itself (Kouzarides 2007; Greally 2018). Epigenetic control is usually based on chemical modifications, which can be transmitted to daughter cells through mitosis and sometimes through meiosis. Epigenetic modifications alter the chromatin structure and nuclear architecture to enable control of the "accessibility" to the DNA (genes) (Schneider and Grosschedl 2007). The major epigenetic modifications of the genome include histone posttranslational

N. Beaujean
Univ. Lyon, Université Claude Bernard Lyon 1, INSERM, INRA, Stem Cell and Brain
Research Institute U1208, USC1361, Bron, France
e-mail: nathalie.beaujean@inserm.fr

© Springer International Publishing AG, part of Springer Nature 2018
H. Niemann, C. Wrenzycki (eds.), *Animal Biotechnology 2*,
https://doi.org/10.1007/978-3-319-92348-2_3

modifications, DNA methylation and remodelling of the chromatin. These modifications, by causing structural changes of the chromatin, affect the expression or silencing of genes and thereby the access of transcription factors controlling gene expression (Kouzarides 2007). This kind of epigenetic regulation takes place in all types of cells, including gametes and embryos. It is now well established that epigenetic reprogramming of gametes and preimplantation embryos is crucial for normal development into a new organism (Beaujean 2014; Beaujean 2015; Sepulveda-Rincon et al. 2016). Moreover, epigenetic modifications seem to be important factors in driving cell fate in mammalian embryos (Graham and Zernicka-Goetz 2016; Wu and Belmonte 2016). Due to the intensity of epigenetic reprogramming that occurs during preimplantation development, this period in particular is very sensitive (El Hajj and Haaf 2013; Anckaert and Fair 2015). Disruption of these control mechanisms can cause aberrant gene expression or silencing which could lead to epigenetic-related diseases. In the case of animal biotechnologies, such as embryo in vitro production (IVP), somatic cell nuclear transfer (SCNT) or stem cells derivation, several steps of the procedures may exert environmental stress on these epigenetic controls. Importantly, it is known that these modifications are dynamic and are rapidly changing within minutes when responding to an arriving stimulus (Kouzarides 2007). In this chapter, the potential epigenetic alterations that may/have been encountered in animal biotechnologies will be discussed.

3.2   Chromatin Compaction and Epigenetic Mechanisms

In eukaryotes, the DNA is packaged with histone proteins to form the chromatin (Margueron and Reinberg 2010). Originally described largely as a method of compaction, the structure of chromatin is now understood to direct and respond to gene expression patterns in a highly dynamic manner. Chromatin can be broadly divided into two categories. Euchromatin is the one with a more relaxed structure and therefore more accessible for transcription mechanisms. This permissive chromatin is usually related with a high gene expression activity although not all genes are necessarily expressed. On the other hand, heterochromatin is considered as repressive because it has a very compact structure which is hard to access and few genes are expressed (Grewal and Jia 2007; Jost et al. 2012; Saksouk et al. 2015).

The histone protein family consists of the core histones H2A, H2B, H3 and H4 and the linker histone H1. Histones are organized in octamers to form the so-called nucleosome, a structure around which linear DNA is wrapped. The different core histones have a similar structure consisting of a globular, hydrophobic internal region and the N-terminal histone tail. Histone tails are protruded from the nucleosome core particle and can be targeted by various enzymes to modify the histone characteristics of particular residues (Kouzarides 2007). These posttranslational modifications (PTM) include acetylation, methylation, ubiquitination and phosphorylation. Histone PTMs may exert their effects on chromatin by changing the relationship between DNA strands and the nucleosome. Acetylation, for instance,

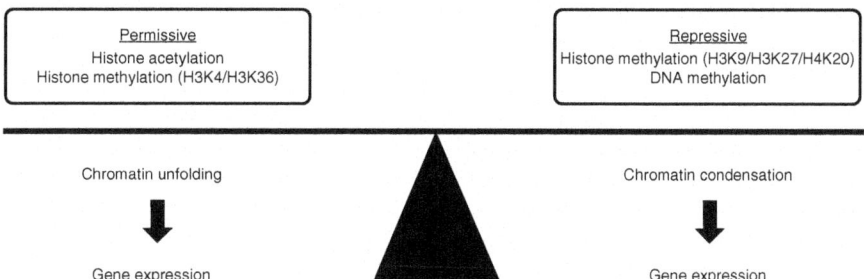

Fig. 3.1   The impact of permissive and repressive epigenetic modifications will influence the equilibrium between chromatin unfolding and chromatin condensation thereby leading either to gene expression or gene repression

decreases the affinity of the histones for DNA by neutralizing the basic charge of the lysine residues and is generally associated with chromatin "unfolding" (Hasan and Hottiger 2002). This open chromatin conformation is often more accessible for transcriptional factors (in other words "permissive") leading to gene expression (Fig. 3.1). On the other hand, histone methylation usually acts on chromatin through the recruitment of other proteins, such as the chromatin-containing CBX-protein family, that are capable of condensing the chromatin structure (Beaujean 2014). Histone methylation is therefore frequently found in condensed chromatin areas together with hypoacetylated histones (Schneider and Grosschedl 2007; Woodcock and Ghosh 2010). However, histone methylation may also promote gene expression depending on which residues are methylated (Kouzarides 2007). Histone methylation has indeed been identified on multiple lysine (K) and arginine (R) residues of histones H1b, H3 and H4; depending on the position of the modified residue, lysine methylation can be associated with transcriptional repression (H3K9, H3K27, H4K20) or activation (H3K4, H3K36 and H3K79). Arginine (R) methylation is also associated with regulation of transcription. Additional complexity is introduced by the possibility of mono-, di- or trimethylation at lysine residues and mono- or dimethylation at arginine residues. Correlation of these histone PTMs with the gene expression status led to the proposal of a "histone code", with different combinations of PTMs leading to the recruitment of specific cofactors, with distinct downstream effects in various biological tasks (Turner and Group 2002).

Each one of these histone PTMs is catalysed by specific enzymes, capable of "writing" or "erasing" the modification. The acetylation status, for instance, is determined by a balance of enzymatic activity between histone acetyltransferases (HATs) and histone deacetylases (HDACs). More than a dozen HDACs have been discovered and several classes of HDACs can be distinguished: class I (HDAC 1–3 and 8), class IIa (HDAC 4, 5, 7 and 9), class IIb (HDAC 6 and 10), class III (SIRT 1–7) and class IV (HDAC 11) (Narlikar et al. 2002). Their importance has been reported in several studies, in particular in cell proliferation control. For this reason many researchers started to be interested in their mechanism of action and in the use of HDAC inhibitors (HDACi), such as trichostatin A (TSA) that induce hyperacetylation of histones by blocking the access of HDAC to their active site (Kretsovali et al. 2012). Similarly, histone methyltransferases (HMTs) are predominantly catalysing methylation of histones H3 and H4: histone lysine methyltransferases (KMTs) target lysine residues, whereas protein arginine methyltransferases (PRMTs) target arginine residues. There are two main families of KMTs: with a SET domain (Su(var)3-9, enhancer of zeste, Trithorax) or without (Dot1 lysine methyltransferases). In all cases, cofactor S-adenosyl methionine (SAM) serves as a cofactor and methyl donor group. On the other hand, the enzymes that promote histone demethylation can be broadly classified into the jmjC (jumonji) demethylases, the LSD1-type KDMs and the PAD-type arginine demethylases (Cyr and Domann 2011).

Apart from histone PTMs, DNA methylation is the other most studied and well-characterized epigenetic modification. Methylation has been observed on cytosine, adenine and guanine, with cytosine methylation being the most abundant and widely studied modification (Auclair and Weber 2012). Cytosine methylation at CpG nucleotides has been well studied, but it is now clear that methylation also occurs in other contexts (Ramsahoye et al. 2000), including at CpA at specific loci on the embryonic genome (Haines et al. 2001). DNA methylation inhibits transcription initiation and removes the engaged transcriptional machinery from active templates (Auclair and Weber 2012). Transcriptional repression depends on methylation density; and even 7% of methylation on CpG sites can cause dramatic transcriptional repression. Low levels of methylation were found to inhibit gene expression by 67–90%, whereas higher levels of methylation were able to completely abolish gene expression. In one specific case, called "imprinting", only one parental allele is silenced, usually by hypermethylation, leading to expression from the other non-methylated parental allele (Reik and Walter 2001).

The enzymes responsible for DNA methylation, the DNA methyltransferases (DNMT), include DNMT1 that is mainly involved in the maintenance of DNA methylation through replication, DNMT3A and DNMT3B as well as the cofactor DNMT3L known as the de novo methyltransferases. Mechanisms leading to active demethylation have been extensively studied recently (Ficz 2015). One of them involves the oxidation of methylated DNA by enzymes of the ten-eleven translocation (TET) family. During this reaction, a hydroxymethyl group replaces the hydrogen atom at the C5 position in cytosine, thereby transforming the 5-methylcytosine (5mC) into to 5-hydroxymethylcytosine (5hmC) (Tahiliani et al. 2009).

3.3   Epigenetic Changes and Development: The Mouse Example

Prior to fertilization, the gametes carry the parental germline epigenetic signature. Interestingly, the spermatozoa nucleus exhibits a highly compacted chromatin stemming from the replacement of histones by protamines and the high acetylation of residual histones associated with methylation of lysine residues (H4K20, H3K27 and H3K9me3; Beaujean 2014) plus eventually with other new posttranslational modifications such as crotonylation (Tan et al. 2011) and high DNA methylation

(Reik and Walter 2001). During follicular maturation, the growing oocytes show a dynamic profile of the methylation with a clear increase of methylation level (from 0.5% in young oocytes to 15.3% in ovulated metaphase II oocytes) that remains lower than in sperm (24.9%) (Smallwood et al. 2011). Histone deacetylation (H4K12), removal of arginine methylation from histones H3/H4 and methylation of H3K9 are also important for the epigenome of female gametes (Bonnet-Garnier et al. 2012; Beaujean 2014).

From fertilization, both the incoming paternal DNA complement and that of the oocyte itself are reprogrammed in a number of steps, resetting chromatin to the embryonic form capable of undergoing further changes required during development (Fig. 3.2) (Beaujean 2014; Beaujean 2015; Sepulveda-Rincon et al. 2016). This results in a series of epigenetic modifications that start during the formation of the paternal and maternal pronuclei (PN) at the 1-cell stage (or zygote). These steps are particularly dramatic for the paternally inherited genome. The protamines associated with the haploid paternal genome in the sperm head are rapidly replaced with histones, including histones H3 and H4 forms that are more acetylated than those associated with the maternal genome (Adenot et al. 1997). Within few hours, there is a widespread intense demethylation of both the paternal and the maternal genome, although both genomes are not similarly affected (Fig. 3.3): methylation is retained on intergenic regions such as constitutive heterochromatin in the paternal genome vs. intergenic regions in the maternal genome (Mayer et al. 2000; Salvaing et al. 2012; Okamoto et al. 2016; Guo et al. 2017). The mechanism responsible for this rapid removal remains unclear: it is suggested that the dioxygenase Tet3 catalyses the oxidation of 5mC to 5hmC in this context (Gu et al. 2011), but other recent data suggest that the initial loss of paternal 5mC does not require 5hmC formation (Amouroux et al. 2016). Imprinted genes are not affected during this wave of

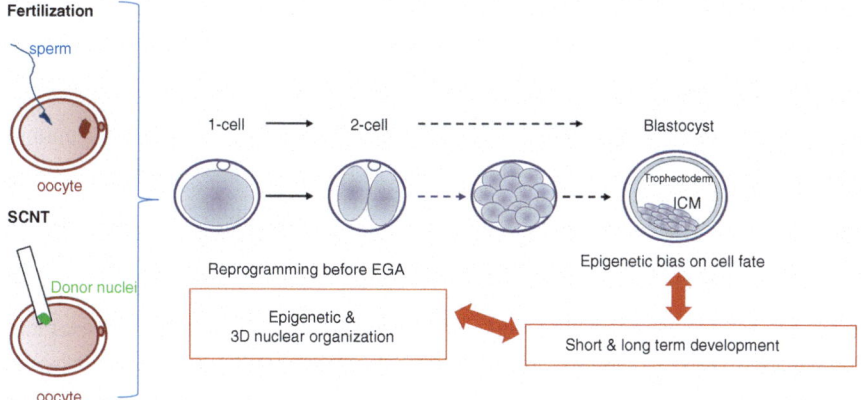

Fig. 3.2 After fertilization of the oocyte by the sperm as well as after somatic cell nuclear transfer, important epigenetic reprogramming events will take place before EGA, involving epigenetic modifications as well as nuclear 3D reorganization. Further epigenetic changes will also occur at the blastocyst stage, either in the ICM or in the trophectoderm leading to cell fate bias. All these events will influence the potential of later development of the embryo

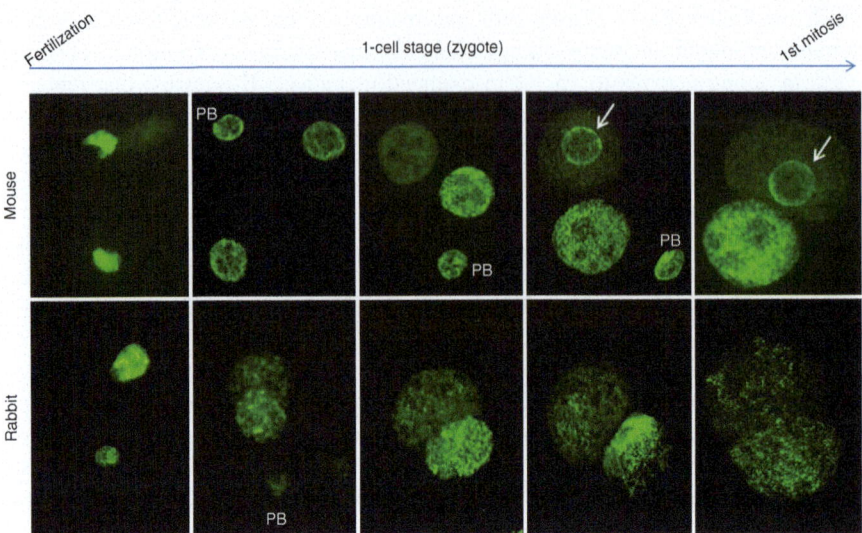

Fig. 3.3 DNA methylation immunofluorescent detection at the 1-cell (zygote) stage in mouse and rabbit embryos showing the paternal and maternal genome in separated pronuclei. Due to the rapid replacement of histones by protamines, the paternal genome is always more decondensed and the pronucleus bigger than the maternal one. The remnant chromosomes expulsed from the oocyte at the end of the meiosis for the so-called polar body (PB) that can often be observed. In both species DNA demethylation is observed on both the paternal and maternal genomes although not to the same extent. In particular, pericentromeric constitutive heterochromatin that surrounds the nucleolar precursor forming rings (white arrows) clearly maintains DNA methylation even in the paternal genome just before the first mitosis

genome-wide DNA demethylation, and parental imprints are maintained in the somatic tissues of the fets throughout life (Reik and Walter 2001).

In addition to the aforementioned asymmetry in histone acetylation, the profiles of several other histone modifications differ between maternally and paternally inherited genomes. In mouse, the paternal genome is initially associated with monomethylation of H3 and H4 (H3K9me1 and H4K20me), while the maternal genome is characterized by di- and trimethylation marks including H3K9me3, H3K27me2/3, H3K4me3 and H4K20me3. After fertilization, H3K9me3 is one of the modifications characterizing the maternal genome, with little H3K9me3 detected on the paternal genome. This asymmetry persists until the 4-cell stage (Liu et al. 2004). All these histones/DNA methylation changes are believed to participate in the onset of the embryonic genome activation (EGA). Indeed, EGA is characterized by a typical chromatin organization that is not observed in somatic cells or at later stages of development. It is believed that these special epigenetic features of the embryo lead to specific changes of gene expression during preimplantation development.

Trimethylation of lysine 27 of histone H3 (H3K27me3) is another histone modification associated with gene repression, but that does not seem to play a major role prior to EGA. In mouse embryos, H3K27me3 is found only on the maternal genome at fertilization, with accumulation on the paternal genome occurring by the late pronuclear stage (Albert and Peters 2009). H3K27me3 is maintained until the morula stage, when a significant drop in levels has been observed (Zhang et al. 2009). This is followed by an increase at the blastocyst stage, in the inner cell mass (ICM) cells only (Bogliotti and Ross 2012). Differential H3K27me3 levels at the promoters of several key transcription factors have recently been shown to be functionally significant in determining whether cells assume an ICM or trophectoderm fate (Saha et al. 2013). Immunostaining has shown H3K27me3 at the inactive X chromosome in extraembryonic and embryonic cells (Plath et al. 2003), and whole genome analyses have indicated that its accumulation at the promoters of key developmental genes coincides with their repression during differentiation (Pan et al. 2007; Hawkins et al. 2010). Another epigenetic mark seems to play a key role in late preimplantation stages: analysis of the expression of Carm1, an enzyme mediating methylation of arginine residues on H3, has shown overexpression in the ICM of mouse blastocysts (Torres-padilla et al. 2007), suggesting an epigenetic bias of the cell fate at the blastocyst stage (Parfitt 2010; Wu and Belmonte 2016). Indeed H3R26 methylation regulates Sox2 transcription factor binding to DNA as early as the 4-cell stage thereby modulating the balance between pluripotency (ICM cells) and differentation (trophectoderm cells) (Goolam et al. 2016; White et al. 2016). Similarly, high levels of 5meC colocalize with the ICM in blastocysts (Ruzov et al. 2011), and knockdown of Tet1 in preimplantation embryos results in a bias towards trophectoderm differentiation, suggesting a role of TET1 protein in ICM specification (Ito et al. 2010). Knockdown of Set1a, one of the H3K4 methyltransferase, also demonstrated that H3K4 methylation is not only required for early embryonic development but also for the emergence of the ICM (Fang et al. 2016).

Such knockdowns of enzymes catalysing histone PTMs confirm that these histone modifications are essential regulators of chromatin remodeling after fertilization that modulate the maternal-to-embryonic transcriptional transition. Both histone methyl-transferases and demethylases are crucial for preimplantation development. For example, without Setdb1, an HMT that controls H3K9me2, embryos do not develop properly and exhibit severe cell cycle defects (Eymery et al. 2016; Kim et al. 2016). However, embryonic development is also disrupted after knockdown of KDMs targeting H3K9me2/me3 such as KDM4A or KDM1A (Ancelin et al. 2016; Sankar et al. 2017) underlying the requirement of a good epigenetic balance.

Remarkably, the same changes in histone PTMs/DNA methylation have been shown to regulate embryonic development by knockdowns or overexpression of regulating enzymes in other species such as pig (Cao et al. 2017; Huan et al. 2015a, b; Ding et al. 2017) and bovine (Li et al. 2015; Chung et al. 2017; Fu et al. 2017). However, in these cases, embryos are often produced after superovulation, in vitro maturation and in vitro culture (Fig. 3.4). All these biotechnologies have been shown to affect the epigenetic profile of the derived embryos/offspring that will be described hereafter.

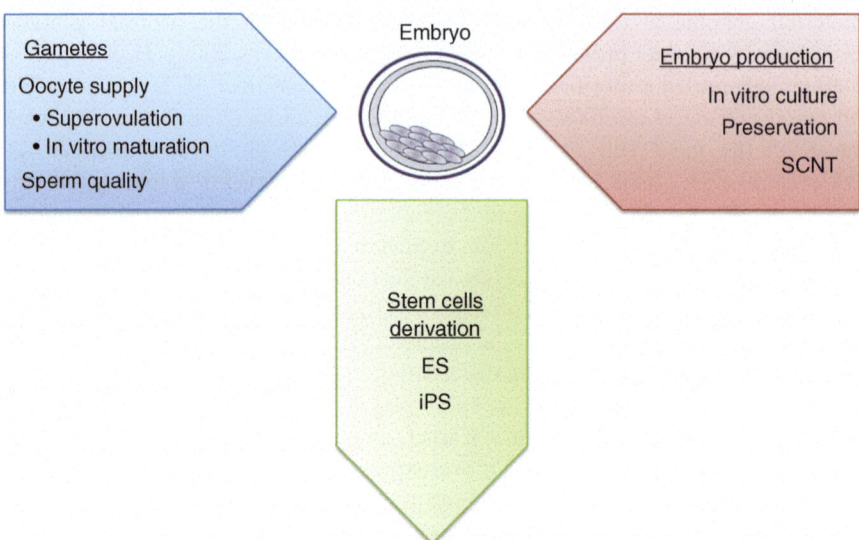

Fig. 3.4 Animal biotechnologies are mostly based on the quality and the development of embryos that may be affected either at the level of the gametes used (oocyte and sperm) or during embryo production by various protocols. On the other hand, once obtained, these embryos may then be used to produce stem cells that are also very interesting both for biomedical and veterinarian applications

3.4 Embryo In Vitro Production (IVP)

3.4.1 Oocyte Supply

In the female germ line, methylation patterns are established predominantly during the last steps of oocyte development. Hormonal stimulation of the follicles (superovulation) during this period to obtain oocytes used for embryo IVP has been suggested to affect the completion of this epigenetic event. Superovulation seems associated with reduced oocyte quality and delayed embryonic development, but it has been also shown in the mouse to affect DNA methylation remodelling in the resulting embryo (Shi and Haaf 2002), that would correlate with altered gene expression at later stages. Superovulation has been shown to be associated with aberrant imprinted methylation profiles, especially at high hormone dosage: loss of methylation was observed in maternally imprinted genes *Snrpn*, *Peg3* and *Kcnq1ot1*, and gain of methylation at the maternal allele in paternally imprinted *H19* gene altered its expression (Denomme and Mann 2012). The transcriptomic profile divergences between bovine oocytes collected from stimulated vs. non-stimulated donors suggests that epigenetic changes may contribute to the reduced developmental competence of oocytes under certain conditions (high dosages of gonadotropins, prepubertal cattle), but only minor changes have been observed, and further studies are still required (Urrego et al. 2014).

In livestock species, in vitro maturation (IVM) is often used to obtain oocytes prior to embryo IVP. In the mouse, IVM has been shown to downregulate the expression of key epigenetic modifiers such as HDAC1 both in the oocyte and in the 2-cell embryo (Wang et al. 2010). Imprinted genes also are affected: IVM resulted under some conditions in a loss of methylation at the Igf2r and Mest loci and a gain of methylation at the H19 imprinted region. However, improved culture conditions (e.g., with different metabolic compounds present in the media) yielded fewer epigenetic abnormalities. Similarly, oocyte culture in bovine and ovine species does not seem to have significant effects on imprinted genes (Anckaert et al. 2012). However, other epigenetic mechanisms may be affected (e.g., histone methylation or acetylation) that could contribute to differences in the transcriptomic profile of those in vitro matured oocytes with various developmental competence.

3.4.2 Sperm Quality

Spermatogenesis is a long process of cellular differentiation requiring a well-orchestrated series of epigenetic modifications to obtain a stably packed chromatin, facilitating all sperm functions (motility, energy production, capacitation, acrosomal reaction, etc.) and thus ultimately fertilization. DNA methylation, histone/protamine exchange, histone PTMs and noncoding RNAs have important, but so far underestimated, roles in the production of fertile sperm and for subsequent embryo development (Boissonnas et al. 2013). Any aberrations in the sperm epigenetic landscape may indeed have detrimental consequences for early embryo development. Literature suggests that stochastic, environmentally and genetically induced deviation in the genome-wide epigenetic reprogramming process during germ cell development may lead to epigenetic aberrations impacting spermatogenesis, semen quality and the male fertility as a whole (Laurentino et al. 2016). Identifying subfertile males is thus a major issue that has stimulated research on sperm characteristics associated with poor or successful fertilization events. Much progress has been made in this area, especially in dairy cattle due to the extensive use of artificial insemination. Current factors such as motility, chromatin structure or seminal proteins help in identifying defects mainly associated with fertilization. Recently, the epigenetic impact of DNA methylation on embryonic development and pregnancy has been addressed in bovine (Kropp et al. 2017), showing a link between spermatozoa DNA methylation and fertility, as in human.

3.4.3 In Vitro Culture of Preimplantation Embryos

It was shown already 20 years ago that embryo culture can alter the imprinted H19 locus in mouse (Anckaert et al. 2012). The impact of embryo culture, especially the presence of foetal calf serum in the media, on imprinted gene methylation and expression was then confirmed in several studies, both in mouse and in other species. In general, imprinting errors appeared to occur during the preimplantation period in culture and persisted during gestation, most probably because the trophec-toderm is directly in contact with the culture medium, which can potentially affect placental development. The "large offspring syndrome" (LOS) was the first obvious incidence of abnormal development following transfer of in vitro culture (IVC) ruminant embryos into surrogate mothers. This syndrome is characterized by an overgrowth phenotype and appeared to be linked to culture conditions (serum con-taining; coculture) (Van Soom et al. 2014). Indeed, LOS sheep foetuses exhibited aberrant hypomethylation and reduced expression of the imprinted IGF2R gene (insulin-like growth factor receptor), a gene known to influence body size and car-cass traits (Young et al. 2001).

In 2002, W. Shi and T. Haaf demonstrated by 5mC staining of mouse 2-cell embryos cultured in vitro that suboptimal culture media can lead to disturbances of the genome-wide DNA demethylation process and concomitantly to developmental arrest. Similar results were obtained in the rabbit although timing and the degree of DNA demethylation differ between these species (Fig. 3.5) emphasizing the impor-tance of DNA methylation in preimplantation development (e Silva et al. 2012; Salvaing et al. 2016). Using a similar immunostaining approach for histone PTMs such as H3K9 methylation, H4 acetylation or H3S10 phosphorylation no differ-ences were observed between in vitro fertilized/IVC mouse embryos and their in vivo counterparts (Beaujean 2015), only H3K4me3 levels were significantly lower in the IVC embryos (Wu et al. 2012). A gene candidate approach based on chromatin immunoprecipitation (ChIP) has also enabled studies of several PTMs in embryos after IVC and confirmed few chromatin configuration changes and histone PTMs alterations (Urrego et al. 2014).

The importance of this IVC step has led (and is still leading) numerous laborato-ries worldwide to improve culture conditions in order to reduce the deficiencies that

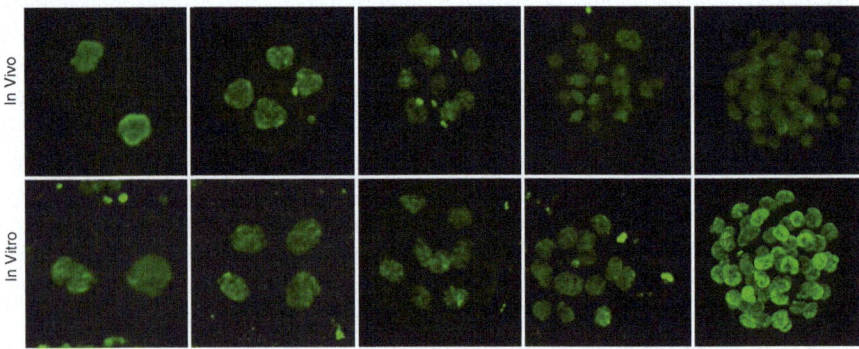

Fig. 3.5 DNA methylation immunofluorescent detection in rabbit embryos from the 2-cell to the morula stages. To evaluate the impact of in vitro culture, embryos were fertilized in vivo but were either collected at 2-cell and cultured or left in vivo and collected just before being processing. DNA demethylation during preimplantation development was observed in both cases, but with different kinetics (quantification was performed to confirm this observation; Salvaing et al. 2012). Unexpectedly, the level of DNA methylation increased between the 16-cell and morula stages after in vitro culture

might lead to epigenetic alterations and changes in gene expression. Some culture media have been tested in mouse and bovine that seem to have lower impact on the DNA methylation and histone acetylation levels, resembling quite closely the in vivo situation (El Hajj and Haaf 2013; Rollo et al. 2017; Canovas et al. 2017).

3.5   Cryopreservation and Vitrification of Gametes and Embryos

Storage of gametes and embryos is a routine procedure in all animal biotechnologies, but few studies analysed the impact on epigenetic marks. In mouse, it was shown that epigenetic reprogramming of H3K4me3 and H4K12ac up to the blastocyst stage was similar in embryos derived from frozen sperm (at -20 °C) and in fertilized embryos derived from fresh sperm (Chao et al. 2012). Reversely, oocytes can be preserved, either by cryopreservation or vitrification. In the mouse, histone PTMs alterations have been observed after vitrification with vitrified oocytes showing abnormally high levels of H4K12ac and low HDAC1 expression (Suo et al. 2010) that correlated with lower developmental rates. Similar results were obtained in ovine and bovine vitrified oocytes (Chen et al. 2016; Shirazi et al. 2016). In cryopreserved bovine embryos, DNMT3 expression and global DNA methylation levels also seem to be affected. However, such studies are quite rare in livestock animals, with no evaluation of the potential impact on later development.

3.6   Somatic Cell Nuclear Transfer (SCNT or "Cloning")

Differentiated cells can be reprogrammed back to a pluripotent state through somatic cell nuclear transfer (SCNT). This technique consists of a single cell nucleus being injected into an enucleated oocyte in order to produce a viable embryo (Ogura et al. 2013). However, the efficiency of SCNT is very low. The first evidence for reprogramming was shown by Briggs and King when tadpoles were produced from transplanting the nuclei of blastula cells into enucleated frog oocytes in 1952 (Gurdon and Wilmut 2011). In 1997, Wilmut et al. were able to produce a healthy cloned sheep, the first mammal to be cloned from an adult cell (Wilmut et al. 1997). Since then, features of cloning techniques have been improved in order to get more accurate and reproducible procedures for generating live mammalian offspring. The importance of the type and age of the nucleus donor cell, as well as the strain and genotype of the oocyte that will receive this nucleus, has been demonstrated. Reasons to explain why some cells are more suitable for cloning than others include epigenetic reprogramming problems. Nuclear epigenetic reprogramming is known to be an essential mechanism needed for embryonic development, both after fertilization (as described earlier) but even more after nuclear transfer (Fig. 3.2). The somatic donor nucleus will need to be reprogrammed—the expression profile of the differentiated donor cell needs to be replaced by an embryo-specific one—in order to give rise to a totipotent embryo (Beaujean 2015; Sepulveda-Rincon et al. 2016). It is now clear that reprogramming errors may accumulate during this process and that reprogramming of DNA methylation and histone PTMs profiles of the donor nucleus is a critical step. The removal of the epigenetic profile associated with the donor cell type, and its replacement by the epigenetic profile of the embryo, has been widely suggested to be incomplete or incorrect, thereby limiting the success of SCNT.

A comparison of H3K27me3 levels in normal and cloned mouse embryos, for example, failed to detect any H3K27me3 in the ICM of cloned embryos (Zhang et al. 2009). This work indicated a role for H3K27me3 in repressing expression of genes associated with differentiation, allowing cells to maintain pluripotency during development. The authors suggested that in cloned embryos, decreased H3K27me3 levels resulted in faulty expression of such genes, resulting in arrest of development (Zhang et al. 2009). Several other histone PTMs levels, such as H3K9me3 and H3K4me3, are also abnormal in SCNT embryos as compared to in vivo produced embryos. Remarkably, the degree of reprogramming of these levels correlates with the developmental potential of these clones into adult offspring (Maalouf et al. 2009; Kallingappa et al. 2016).

Similarly, DNA methylation reprogramming has been shown to be critical for the success of SCNT (Beaujean et al. 2004). DNA methylation is usually associated with repression of gene expression and is found at higher levels in somatic cells than in embryos. Successful nuclear reprogramming involves genome wide removal of DNA methylation, profiles with observations suggesting that demethylation is necessary for gene expression after reprogramming (Simonsson and Gurdon 2004). Persistent cytosine methylation after nuclear transfer has been suggested as a factor contributing to the low efficiency of cloning and to the high incidence of abnormalities observed in cloned animals (Kang et al. 2001). Indeed, several studies in cloned embryos suggest that most cloned animals do not develop because of hypermethylation changes in their genome, both at a gene-specific level but also genome-wide. This retention of 5mC is thought to be responsible for the so-called persistent cellular memory after SCNT. Genome-wide methylation analyses of cloned mouse embryos revealed that the DNA demethylation process normally observed after fertilization also occurs after SCNT. However, this demethylation was less drastic, and the resulting DNA methylation profile was more similar to the donor fibroblast one than to a fertilized embryo. This aberrant demethylation is concomitant with aberrant nuclear reorganization of the heterochromatin that forms somatic-like clusters usually not observed after fertilization (Fig. 3.6). In ovine, bovine and mouse, a clear correlation was found between the efficiency of nuclear heterochromatin reorganization and the percentage of SCNT embryos surviving the first embryonic cleavages, up to embryonic genome activation (Beaujean et al. 2004; Zink et al. 2006).

Importantly, the kinetics of DNA demethylation/remethylation in SCNT cloned embryos clearly differs from that in fertilized embryos, leading to an awkward situation at the blastocyst stage. Indeed, most SCNT blastocysts do not present the typical asymmetry of DNA methylation between ICM and trophectoderm cells (Beaujean et al. 2004). In fact, most SCNT-derived blastocysts present a hypermethylation of the trophectoderm cells that might potentially be associated with developmental abnormalities in cloned foetus. This hypothesis is particularly sustained by

SCNT   SCNT + scriptaid

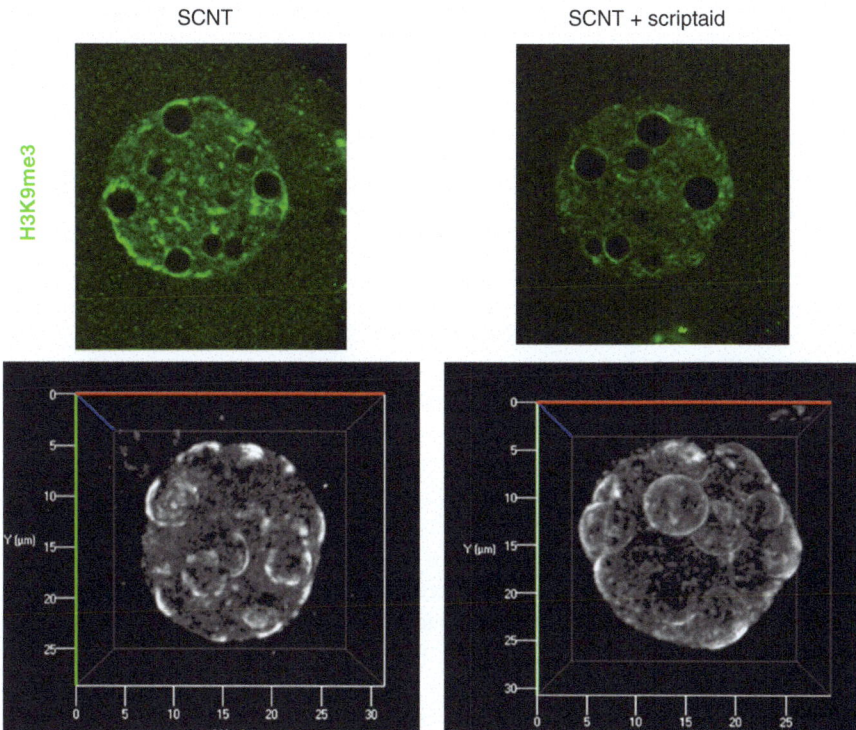

Fig. 3.6 H3K9me3 immunofluorescent detection in mouse SCNT embryos at the 1-cell stage. The upper panel corresponds to a single focal plan through the nucleus and the lower panel to the 3D-reconstructions obtained from the confocal scanning that was performed through the nucleus. Clusters of heterochromatin usually not observed after fertilization actually accumulated at the periphery (arrow) in the nucleus of cloned embryos. However, treatment of the embryos just after nuclear transfer with an HDAC inhibitor (Scriptaid) improves the reprogramming of H3K9me3, and the density as well as the number of heterochromatin clusters clearly decreases. This improvement is correlated with an improved development to term (Maalouf et al. 2009)

the high frequency of the large offspring syndrome after SCNT. It has indeed been confirmed that several imprinted regions of the genome, which should remain methylated, are in fact demethylated after SCNT (Niemann 2016). Several researchers have produced data on SCNT cloned foetuses or offspring showing that the aberrant phenotypes described after SCNT bear strong similarities to abnormalities associated with deletion/mutations in imprinted regions. Aberrant proper expression of genes such as *IGF2* gene and other members of this gene family (*IGF2R* and *H19*) is critical for normal embryonic and foetal development (Young et al. 2001). The maintenance of such high DNA methylation levels in SCNT embryos has been related to the presence of the somatic form of the DNA methyltransferase brought by the somatic donor cell that probably interferes with the genome-wide demethylation process that normally takes place after fertilization. Similarly, the *NDN* (necdin; paternally expressed) and XIST (paternally expressed) imprinted genes were found to be aberrantly expressed in cloned bovine embryos (compared to fertilized ones), and this aberrant expression was at least partially associated with histone aberrant H4K5 acetylation (Wee et al. 2006).

Because these abnormal gene expression profiles observed during preimplantation development often persist throughout foetal development up to birth in some cases (Koike et al. 2016), it suggests that very early epigenetic modifications can have very late impact and that preimplantation development is a very sensitive period in this regard.

3.7 Use of Epigenetic Inhibitors to Improve SCNT

As epigenetic reprogramming clearly plays a key role in the development of SCNT-derived embryos, many researchers have chosen to use HDAC inhibitors (HDACi)—such as TSA—to correct epigenetic errors and improve cloning efficiency (Ogura et al. 2013). They examined the efficiency of SCNT with TSA treatments, demonstrating that TSA enhances the pool of acetylated histones in SCNT embryos as well as DNA demethylation. They also reported that the optimal concentration of TSA to obtain development to the blastocyst stage was 5–50 nM for a few hours after nuclear transfer, during the first embryonic cycle (usually 6–10 h post-activation). Toxicity was shown at 500 nM and when the treatment was extended for more than 10 h. One study using TSA demonstrated that this inhibitor can assist with reprogramming of Oct4, a major factor for pluripotency maintenance: while SCNT embryos often showed aberrant expression of this gene in the trophectoderm cells of mouse blastocysts, TSA treatment helped embryos to express Oct4 in the correct number cells during development (Hai et al. 2011). The nuclear reorganization of heterochromatin was also clearly improved in SCNT embryos treated with HDAC inhibitors (Fig. 3.6) (Maalouf et al. 2009).

In most species studied so far, it has been shown that histone acetylation can be enhanced by TSA and other HDACi (e.g., scriptaid, valproic acid (VPA)), even in inter-species (porcine-bovine) cloned embryos (Opiela et al. 2017). They have now been demonstrated to play a similar beneficial role on cloned embryo development, albeight with different efficiencies and outcomes (Ono et al. 2010). It was demonstrated that scriptaid can indeed increase transcriptional activity in SCNT embryos but could not support full-term development of inbred cloned embryos in contrast to TSA. It also appears that SAHA (suberoylanilide hydroxamic acid) and oxamflatin—but not valproic acid—could reduce apoptosis in SCNT-derived blastocysts and thus improve full-term development of the clones. Since valproic acid inhibits HDAC classes I and IIa and the rest of HDACi's act on HDAC classes IIa/b, this suggests that the inhibition of HDAC class IIb is an important step for reprogramming mouse cloning efficiency. Analyses performed on imprinted genes also showed improved reprogramming after HDACi treatments: in the porcine SCNT protocols, scriptaid treatment of the donor cells rescued disrupted methylation at H19 imprinted region after SCNT and oxamflatin treatment improved the DNA methylation profile at the imprinted XIST locus that lead to higher in vitro developmental rates and offspring rates (Xu et al. 2013; Hou et al. 2014). Similar results were obtained on developmental key genes Oct4, Nanog and Sox2 (Huan et al. 2015a, b; Jin et al. 2017; Sun et al. 2017).

Other histone PTMs abnormal levels have also been targeted: removal of H3K9me3 by overexpression of the lysine demethylase Kdm4d has been proven to restore transcriptional reprogramming in mouse cloned embryos and to efficiently improve blastocysts rates both in mouse and human cloning experiments (Matoba et al. 2014; Chung et al. 2015). Similarly, treatment with a DOT1L inhibitor to reduce H3K79 methylation improved cloning efficiency in pig (Tao et al. 2017), and overexpression of USP21 in cloned mouse embryos improved their transcriptional reprogramming (Jullien et al. 2017). Combinations of several approaches are now being explored. Recently, the very first SCNT monkeys were obtained by a combined treatment with TSA and Kdm4D overexpression (Liu et al. 2018).

Strategies directing DNA methylation have also been reported. In the mouse and pig, the removal of DNA methylation in donor cells by knockout of the *DNMT3l* or Dnmt1s genes, respectively, was associated with significant changes of the epigenetic profile of the cloned embryos and significant improvements of the SCNT process with regard to blastocyst rates, indicating that modulation of DNMTs activity could indeed be beneficial (Liao et al. 2015; Song et al. 2017). Similarly, helping the cloned embryos to remove DNA methylation by overexpression of the AID (activation-induced cytidine deaminase, one of the demethylation mechanism) significantly improves the cleavage and blastocyst rates in the bovine (Ao et al. 2016).

These findings provide clues towards more efficient cloning protocols by modulating epigenetic reprogramming. However, it remains to be shown whether this yields to more healthy live cloned offspring as only few studies have been performed so far on this subject.

3.8 Stem Cells Derivation

Embryonic stem (ES) cells can be derived from the ICM of mammalian blastocysts. They are pluripotent and have the ability to differentiate into cells of all three germ layers—mesoderm, endoderm and ectoderm (Rossant 2008). All specialized cells are derived from one of these three germ layers, which implies that pluripotent cells can differentiate into any type of cell in the body. ES cells also have a high capacity for renewal and can be expanded indefinitely in culture. These properties have led to expectations that ES cells could be useful for research into understanding disease mechanisms, screening for safe and effective drugs and treating diseases and injuries. Pluripotent cells have indeed the potential to be cultured and differentiated in the lab to make specialized cell types that could be used to replace damaged cells and tissues in the organisms. However, ES cells production is limited by embryo availability.

On the other hand, induced pluripotent stem (iPS) cells are somatic cells that can be reprogrammed to a pluripotent state by overexpressing certain transcription factors (initially Oct4, Sox2, Myc and Klf4) (Takahashi and Yamanaka 2006). They are able to function similarly to ES cells and are more readily available for use in research and therapeutics than ES cells. When mouse iPS cells are transplanted into blastocysts, they are able to give rise to adult chimaeras, which are competent for germline transmission. This has introduced the opportunity to develop custom-made cells in order to study and treat many diseases.

Animal stem cells play an important role as a research model in testing research theories before human clinical trials and clinical use. Animal pluripotent stem cells could also serve for agricultural purposes: such stem cells could be used for genetic engineering to generate livestock with superior genes that are important for economic and disease-resistant traits; they could also be used for studying functional genomics in those mammals. As a result, monkey, bovine, pig, sheep, goat, horse and rabbit iPS cells have been successfully established although there is a lack of research on stem cells from farm animals (Nowak-Imialek and Niemann 2012; Ogorevc et al. 2016). However, those non-human/non-murine stem cells are often unable to colonize embryos after injection into blastocysts; they do not show all the attributes of pluripotent stem cells and lacked some of the key features of pluripotency compared to mouse stem cells. Consequently, it has been proposed that most of these stem cells are not in a "naïve" pluripotency status like in mouse ES cells but rather in a more "primed" status with less flexibility (Piedrahita and Olby 2010).

Studies have shown that mouse iPS cells have a higher ratio of euchromatin to heterochromatin when compared to differentiated cells such as somatic cells (Mattout et al. 2011). This is similarly seen in ES cells, which display an even more opened and hyperdynamic chromatin structure in comparison to somatic cells. These characteristics resemble the ones observed in vivo in the ICM of mouse blastocysts, which are the source of ES cells (Ahmed et al. 2010). It has been hypothesized that an open chromatin structure allows rapid switching of transcriptional programmes when differentiation is induced (Meshorer and Misteli 2006). As these stem cells have the ability to differentiate into many different cells, a broad spectrum of differentiation opportunities is indeed necessary. In agreement with this hypothesis, the ES cell genome is transcriptionally globally hyperactive, a hallmark that seems to be characteristic of ES cell pluripotency and contributes to their plasticity. On the other hand, it has been proposed that reduction of pluripotency potential in primed pluripotent cells correlates with a reduction of the actively transcribed portion of the genome and that, similarly to the situation observed after SCNT, abnormal reprogramming of DNA and histones involved in the removal of the original somatic cell epigenetic landscape (Savatier et al. 2017).

DNA demethylation and cytosine hydroxymethylation appeared to be key elements in the reactivation of pluripotency genes in mice stem cells as they are hypermethylated and silenced in somatic cells (Mikkelsen et al. 2008). Indeed, the TET family of enzymes, which promote the conversion of 5mC to 5hmC, are highly expressed in embryonic stem cells (Tahiliani et al. 2009; Ito et al. 2010). Moreover, DNA methylation deficient ESCs are unable to differentiate (Jackson et al. 2004). In humans, it has been shown that DNA methylation patterns in iPS and ES cells were similar but exhibited different methylated regions in which 55% of methylated regions in iPS cells are not found in the somatic cell of origin or in ES cells. Finally, it has been found that iPS cells retain a methylation signature of their tissue of origin (Kim et al. 2010). Histone PTMs also seem involved: histone acetylation (particularly H3K9ac) is increased in undifferentiated human ES cells (Krejcí et al. 2009), and H3K9me3/H3K27me3 was shown to increase from 4% in ES cells to 12–16% coverage in differentiated cells (Hawkins et al. 2010). Moreover, H3K9me is associated with Oct4 inactivation suggesting that this mark could indeed act as a barrier for reprogramming (Feldman et al. 2006). Those observations have been confirmed by genome-wide analysis of H3K9ac, H3K27me3 and H4K3me3 (ChIP-sequencing technology) (Bernstein et al. 2006; Azuara et al. 2006). In rabbbits, clear histone PTMS/DNA methylation differences can be observed between highly "primed" iPS cells and more "naïve" ones (Tapponnier et al. 2017).

There have been studies conducted on iPS cells that focus on particular chromatin treatments in order to decondense and open the chromatin, similar to the chromatin structure of ES cells. Treatment with agents that promote chromatin decondensation has been shown to increase the efficiency of iPS cell generation (Szablowska-Gadomska et al. 2012). It was shown that even at low doses, the histone deacetylase inhibitor TSA was able to stabilize expression of pluripotent genes. Similarly, VPA was found to stabilize histone acetylation and the expression of pluripotent gene in bovine iPS cells (Mahapatra et al. 2015). It was also shown that the use of such chemicals to increase reprogramming efficiency could also replace one or more of the key factors found to induce reprogramming (Ma et al. 2017). Removal of the barrier imposed by H3K9me also leads to better reprogrammed iPS cells (Chen et al. 2013; Wei et al. 2017). Interestingly, studies decreasing H3K9me2 via inhibition of the histone methyltransferase G9a can replace Oct4 during pluripotency induction (Shi et al. 2008). Similarly, studies have shown that it is possible to decrease the retention of DNA methylation and increase pluripotency using chromatin-modifying compounds (Kim et al. 2010).

3.9   Conclusions and Perspectives

Epigenetic reprogramming naturally occurs in gametes and embryos and is essential criteria for undisturbed development. However, during this vulnerable time window of the lifespan, environmentally induced epigenetic defects may occur, some of which may have long-term effects leading to subsequent changes in gene expression. In this chapter, several stress factors leading to such defects in animal biotechnologies have been mentioned, but other factors such as environmental exposures to toxic compounds, nutrients or infectious agents may also interfere and should be taken into account (Doherty et al. 2014).

Some of these phenotypic modifications can be seen during preimplantation development but also during gestation, at birth or shortly after, or even during adult life. This concept called DOHaD (Developmental Origin of Health and Diseases) has attracted much attention recently, and numerous reviews can be found in the literature (Dupont et al. 2012; Lucas and Watkins 2017). Finally, it should be remembered that many studies have shown that such environmental changes alter epigenetic modification not only in the animals studied but also in the descendants as epigenetics can be passed from one generation to the next (Jammes et al. 2011; Feil and Fraga 2011; Feeney et al. 2014).

References

Adenot P, Mercier Y, Renard J, Thompson E (1997) Differential H4 acetylation of paternal and maternal chromatin precedes DNA replication and differential transcriptional activity in pronuclei of 1-cell mouse embryos. Development 124:4615–4625

Ahmed K, Dehghani H, Rugg-Gunn P et al (2010) Global chromatin architecture reflects pluripotency and lineage commitment in the early mouse embryo. PLoS One 5:e10531. https://doi.org/10.1371/journal.pone.0010531

Albert M, Peters AHFM (2009) Genetic and epigenetic control of early mouse development. Curr Opin Genet Dev 19:113–121. https://doi.org/10.1016/j.gde.2009.03.004

Amouroux R, Nashun B, Shirane K et al (2016) De novo DNA methylation drives 5hmC accumulation in mouse zygotes. Nat Cell Biol 18:225–233. https://doi.org/10.1038/ncb3296

Ancelin K, Syx L, Borensztein M, Ranisavljevic N, Vassilev I, Briseño-Roa L, Liu T, Metzger E, Servant N, Barillot E, Chen CJ, Schüle R, Heard E (2016) Maternal LSD1/KDM1A is an essential regulator of chromatin and transcription landscapes during zygotic genome activation. Elife 5:pii: e08851. https://doi.org/10.7554/eLife.08851

Anckaert E, Fair T (2015) DNA methylation reprogramming during oogenesis and interference by reproductive technologies: studies in mouse and bovine models. Reprod Fertil Dev 27:739. https://doi.org/10.1071/RD14333

Anckaert E, De Rycke M, Smitz J (2012) Culture of oocytes and risk of imprinting defects. Hum Reprod Update 0:1–15. https://doi.org/10.1093/humupd/dms042

Ao X, Sa R, Wang J et al (2016) Activation-induced cytidine deaminase selectively catalyzed active DNA demethylation in pluripotency gene and improved cell reprogramming in bovine SCNT embryo. Cytotechnology 68:2637–2648. https://doi.org/10.1007/s10616-016-9988-8

Auclair G, Weber M (2012) Mechanisms of DNA methylation and demethylation in mammals. Biochimie 94:2202–2211. https://doi.org/10.1016/j.biochi.2012.05.016

Azuara V, Perry P, Sauer S et al (2006) Chromatin signatures of pluripotent cell lines. Nat Cell Biol 8:532–538. https://doi.org/10.1038/ncb1403

Beaujean N (2014) Histone post-translational modifications in preimplantation mouse embryos and their role in nuclear architecture. Mol Reprod Dev 81:100–112. https://doi.org/10.1002/mrd.22268

Beaujean N (2015) Epigenetics, embryo quality and developmental potential. Reprod Fertil Dev 27:53–62. https://doi.org/10.1071/RD14309

Beaujean N, Taylor J, Gardner J et al (2004) Effect of limited DNA methylation reprogramming in the normal sheep embryo on somatic cell nuclear transfer. Biol Reprod 71:185–193. https://doi.org/10.1095/biolreprod.103.026559

Bernstein BE, Mikkelsen TS et al (2006) A bivalent chromatin structure marks key developmental genes in embryonic stem cells. Cell 125:315–326. https://doi.org/10.1016/j.cell.2006.02.041

Bogliotti YS, Ross PJ (2012) Mechanisms of histone 3 lysine 27 trimethylation remodeling during early mammalian development. Epigenetics 7:976–981

Boissonnas CC, Jouannet P, Jammes H (2013) Epigenetic disorders and male subfertility. Fertil Steril 99:624–631. https://doi.org/10.1016/j.fertnstert.2013.01.124

Bonnet-Garnier A, Feuerstein P, Chebrout M et al (2012) Genome organization and epigenetic marks in mouse germinal vesicle oocytes. Int J Dev Biol 887:877–887. https://doi.org/10.1387/ijdb.120149ab

Canovas S, Ivanova E, Romar R et al (2017) DNA methylation and gene expression changes derived from assisted reproductive technologies can be decreased by reproductive fluids. elife 6:e23670. https://doi.org/10.7554/eLife.23670

Cao Z, Hong R, Ding B, Zuo X, Li H, Ding J, Li Y, Huang W, Zhang Y (2017) TSA and BIX-01294 induced normal DNA and histone methylation and increased protein expression in porcine somatic cell nuclear transfer embryos. PLoS One 12(1):e0169092. https://doi.org/10.1371/journal.pone.0169092

Chao S, Li J, Jin X et al (2012) Epigenetic reprogramming of embryos derived from sperm frozen at −20°C. Sci China Life Sci 55:349–357. https://doi.org/10.1007/s11427-012-4309-8

Chen JJ, Liu H, Liu J et al (2013) H3K9 methylation is a barrier during somatic cell reprogramming into iPSCs. Nat Genet 45:34–42. https://doi.org/10.1038/ng.2491

Chen H, Zhang L, Deng T et al (2016) Effects of oocyte vitrification on epigenetic status in early bovine embryos. Theriogenology 86:868–878. https://doi.org/10.1016/j.theriogenology.2016.03.008

Chung YG, Matoba S, Liu Y et al (2015) Histone demethylase expression enhances human somatic cell nuclear transfer efficiency and promotes derivation of pluripotent stem cells. Cell Stem Cell 17:758–766. https://doi.org/10.1016/j.stem.2015.10.001

Chung N, Bogliotti YS, Ding W, Vilarino M, Takahashi K, Chitwood JL, Schultz RM, Ross PJ (2017) Active H3K27me3 demethylation by KDM6B is required for normal development of bovine preimplantation embryos. Epigenetics 12(12):1048–1056. https://doi.org/10.1080/15592294.2017

Cyr AR, Domann FE (2011) The redox basis of epigenetic modifications: from mechanisms to functional consequences. Antioxid Redox Signal 15:551–589. https://doi.org/10.1089/ars.2010.3492

Denomme MM, Mann MRW (2012) Genomic imprints as a model for the analysis of epigenetic stability during assisted reproductive technologies. Reproduction 144:393–409. https://doi.org/10.1530/REP-12-0237

Ding B, Cao Z, Hong R, Li H, Zuo X, Luo L, Li Y, Huang W, Li W, Zhang K, Zhang Y (2017) WDR5 in porcine preimplantation embryos: expression, regulation of epigenetic modifications and requirement for early development. Biol Reprod 96(4):758–771. https://doi.org/10.1093/biolre/iox020

Doherty R, Farrelly CO, Meade KG (2014) Comparative epigenetics: relevance to the regulation of production and health traits in cattle. Anim Genet 45:3–14. https://doi.org/10.1111/age.12140

Dupont C, Cordier AG, Junien C et al (2012) Maternal environment and the reproductive function of the offspring. Theriogenology 78:1405–1414. https://doi.org/10.1016/j.theriogenology.2012.06.016

e Silva ARR, Bruno C, Fleurot R et al (2012) Alteration of DNA demethylation dynamics by in vitro culture conditions in rabbit pre-implantation embryos. Epigenetics 7:440–446. https://doi.org/10.4161/epi.19563

El Hajj N, Haaf T (2013) Epigenetic disturbances in in vitro cultured gametes and embryos: implications for human assisted reproduction. Fertil Steril 99:632–641. https://doi.org/10.1016/j.fertnstert.2012.12.044

Eymery A, Liu Z, Ozonov EA, Stadler MB, Peters AH (2016) The methyltransferase Setdb1 is essential for meiosis and mitosis in mouse oocytes and early embryos. Development 143(15):2767–2779. https://doi.org/10.1242/dev.132746

Fang L, Zhang J, Zhang H et al (2016) H3K4 methyltransferase set1a is a key Oct4 coactivator essential for generation of Oct4 positive inner cell mass. Stem Cells 34:565–580. https://doi.org/10.1002/stem.2250

Feeney A, Nilsson E, Skinner M (2014) Epigenetics and transgenerational inheritance in domesticated farm animals. J Anim Sci Biotechnol 5:48. https://doi.org/10.1186/2049-1891-5-48

Feil R, Fraga MF (2011) Epigenetics and the environment: emerging patterns and implications. Nat Rev Genet 13:97–109. https://doi.org/10.1038/nrg3142

Feldman N, Gerson A, Fang J et al (2006) G9a-mediated irreversible epigenetic inactivation of Oct-3/4 during early embryogenesis. Nat Cell Biol 8:188–194. https://doi.org/10.1038/ncb1353

Ficz G (2015) New insights into mechanisms that regulate DNA methylation patterning. J Exp Biol 218:14–20. https://doi.org/10.1242/jeb.107961

Fu Y, Xu JJ, Sun XL, Jiang H, Han DX, Liu C, Gao Y, Yuan B, Zhang JB (2017) Function of JARID2 in bovines during early embryonic development. PeerJ 5:e4189 https://doi.org/10.7717/peerj.4189

Goolam M, Scialdone A, Graham SJL et al (2016) Heterogeneity in Oct4 and Sox2 targets biases cell fate in 4-cell mouse embryos. Cell 165:61–74. https://doi.org/10.1016/j.cell.2016.01.047

Graham SJL, Zernicka-Goetz M (2016) The acquisition of cell fate in mouse development: how do cells first become heterogeneous? Curr Top Dev Biol 117:671–695. https://doi.org/10.1016/bs.ctdb.2015.11.021

Greally JM (2018) A user's guide to the ambiguous word epigenetics. Nat Rev Mol Cell Biol 19(4):207–208. https://doi.org/10.1038/nrm.2017.135

Grewal SIS, Jia S (2007) Heterochromatin revisited. Nat Rev Genet 8:35–46. https://doi.org/10.1038/nrg2008

Gu T, Guo F, Yang H et al (2011) The role of Tet3 DNA dioxygenase in epigenetic reprogramming by oocytes. Nature 477:606–610. https://doi.org/10.1038/nature10443

Guo F, Li L, Li J et al (2017) Single-cell multi-omics sequencing of mouse early embryos and embryonic stem cells. Cell Res 27:967–988. https://doi.org/10.1038/cr.2017.82

Gurdon JB, Wilmut I (2011) Nuclear transfer to eggs and oocytes. Cold Spring Harb Perspect Biol 3:1–14. https://doi.org/10.1101/cshperspect.a002659

Hai T, Hao J, Wang L et al (2011) Pluripotency maintenance in mouse somatic cell nuclear transfer embryos and its improvement by treatment with the histone deacetylase inhibitor TSA. Cell Reprogram 13:47–56. https://doi.org/10.1089/cell.2010.0042

Haines TR, Rodenhiser DI, Ainsworth PJ (2001) Allele-specific non-CpG methylation of the Nf1 gene during early mouse development. Dev Biol 240:585–598. https://doi.org/10.1006/dbio.2001.0504

Hasan S, Hottiger MO (2002) Histone acetyl transferases: a role in DNA repair and DNA replication. J Mol Med (Berl) 80:463–474. https://doi.org/10.1007/s00109-002-0341-7

Hawkins RD, Hon GC, Lee LK et al (2010) Distinct epigenomic landscapes of pluripotent and lineage-committed human cells. Cell Stem Cell 6:479–491. https://doi.org/10.1016/j.stem.2010.03.018

Hou L, Ma F, Yang J et al (2014) Effects of histone deacetylase inhibitor oxamflatin on in vitro porcine somatic cell nuclear transfer embryos. Cell Reprogram 16:253–265. https://doi.org/10.1089/cell.2013.0058

Huan Y, Zhu J, Huang B et al (2015a) Trichostatin a rescues the disrupted imprinting induced by somatic cell nuclear transfer in pigs. PLoS One 10:e0126607. https://doi.org/10.1371/journal.pone.0126607

Huan Y, Wu Z, Zhang J, Zhu J, Liu Z, Song X (2015b) Epigenetic modification agents improve gene-specific methylation reprogramming in porcine cloned embryos. PLoS One 10(6):e0129803. https://doi.org/10.1371/journal.pone.0129803

Ito S, Alessio ACD, Taranova OV et al (2010) Role of Tet proteins in 5mC to 5hmC conversion, ES-cell self-renewal and inner cell mass specification. Nature 466:1129–1133. https://doi.org/10.1038/nature09303

Jackson M, Krassowska A, Gilbert N et al (2004) Severe global DNA hypomethylation blocks differentiation and induces histone hyperacetylation in embryonic stem cells. Mol Cell Biol 24:8862–8871. https://doi.org/10.1128/MCB.24.20.8862-8871.2004

Jammes H, Junien C, Chavatte-Palmer P (2011) Epigenetic control of development and expression of quantitative traits. Reprod Fertil Dev 23:64–74. https://doi.org/10.1071/RD10259

Jin L, Guo Q, Zhu H-Y et al (2017) Quisinostat treatment improves histone acetylation and developmental competence of porcine somatic cell nuclear transfer embryos. Mol Reprod Dev 84:340–346. https://doi.org/10.1002/mrd.22787

Jost KL, Bertulat B, Cardoso MC (2012) Heterochromatin and gene positioning: inside, outside, any side? Chromosoma 121(6):555–563. https://doi.org/10.1007/s00412-012-0389-2

Jullien J, Vodnala M, Pasque V et al (2017) Gene resistance to transcriptional reprogramming following nuclear transfer is directly mediated by multiple chromatin-repressive pathways. Mol Cell 65:873–884.e8. https://doi.org/10.1016/j.molcel.2017.01.030

Kallingappa PK, Turner PM, Eichenlaub MP et al (2016) Quiescence loosens epigenetic constraints in bovine somatic cells and improves their reprogramming into totipotency. Biol Reprod 95:16–16. https://doi.org/10.1095/biolreprod.115.137109

Kang YK, Koo DB, Park JS et al (2001) Aberrant methylation of donor genome in cloned bovine embryos. Nat Genet 28:173–177. https://doi.org/10.1038/88903

Kim K, Doi A, Wen B et al (2010) Epigenetic memory in induced pluripotent stem cells. Nature 467:285–292. https://doi.org/10.1038/nature09342

Kim J, Zhao H, Dan J, Kim S, Hardikar S, Hollowell D, Lin K, Lu Y, Takata Y, Shen J, Chen T (2016) Maternal Setdb1 is required for meiotic progression and preimplantation development in mouse. PLoS Genet 12(4):e1005970. https://doi.org/10.1371/journal.pgen.1005970

Koike T, Wakai T, Jincho Y et al (2016) DNA methylation errors in cloned mouse sperm by germ line barrier evasion. Biol Reprod 94:128. https://doi.org/10.1095/biolreprod.116.138677

Kouzarides T (2007) Chromatin modifications and their function. Cell 128:693–705. https://doi.org/10.1016/j.cell.2007.02.005

Krejcí J, Uhlírová R, Galiová G et al (2009) Genome-wide reduction in H3K9 acetylation during human embryonic stem cell differentiation. J Cell Physiol 219:677–687. https://doi.org/10.1002/jcp.21714

Kretsovali A, Hadjimichael C, Charmpilas N (2012) Histone deacetylase inhibitors in cell pluripotency, differentiation, and reprogramming. Stem Cells Int 2012:184154. https://doi.org/10.1155/2012/184154

Kropp J, Carrillo JA, Namous H et al (2017) Male fertility status is associated with DNA methylation signatures in sperm and transcriptomic profiles of bovine preimplantation embryos. BMC Genomics 18:280. https://doi.org/10.1186/s12864-017-3673-y

Laurentino S, Borgmann J, Gromoll J (2016) On the origin of sperm epigenetic heterogeneity. Reproduction 151:R71–R78. https://doi.org/10.1530/REP-15-0436

Li CH, Gao Y, Wang S, Xu FF, Dai LS, Jiang H, Yu XF, Chen CZ, Yuan B, Zhang JB (2015) Expression pattern of JMJD1C in oocytes and its impact on early embryonic development. Genet Mol Res 14(4):18249–18258. https://doi.org/10.4238/2015.December.23.12

Liao H-F, Mo C-F, Wu S-C et al (2015) Dnmt3l-knockout donor cells improve somatic cell nuclear transfer reprogramming efficiency. Reproduction 150:245–256. https://doi.org/10.1530/REP-15-0031

Liu H, Kim J-MM, Aoki F (2004) Regulation of histone H3 lysine 9 methylation in oocytes and early pre-implantation embryos. Development 131:2269–2280. https://doi.org/10.1242/dev.01116

Liu Z, Cai Y, Wang Y et al (2018) Cloning of macaque monkeys by somatic cell nuclear transfer. Cell 172:881–887.e7. https://doi.org/10.1016/j.cell.2018.01.020

Lucas ES, Watkins AJ (2017) The long-term effects of the periconceptional period on embryo epigenetic profile and phenotype; the paternal role and his contribution, and how males can affect offspring's phenotype/epigenetic profile. In: Advances in experimental medicine and biology. Springer, Cham, pp 137–154

Ma X, Kong L, Zhu S (2017) Reprogramming cell fates by small molecules. Protein Cell 8:328–348. https://doi.org/10.1007/s13238-016-0362-6

Maalouf WE, Liu Z, Brochard V et al (2009) Trichostatin a treatment of cloned mouse embryos improves constitutive heterochromatin remodeling as well as developmental potential to term. BMC Dev Biol 9:11. https://doi.org/10.1186/1471-213X-9-11

Mahapatra PS, Singh R, Kumar K et al (2015) Valproic acid assisted reprogramming of fibroblasts for generation of pluripotent stem cellsin buffalo (Bubalus bubalis). Int J Dev Biol 61(1-2):81–88. https://doi.org/10.1387/ijdb.160006sb

Margueron R, Reinberg D (2010) Chromatin structure and the inheritance of epigenetic information. Nat Rev Genet 11:285–296. https://doi.org/10.1038/nrg2752

Matoba S, Liu Y, Lu F et al (2014) Embryonic development following somatic cell nuclear transfer impeded by persisting histone methylation. Cell 159(4):884–895. https://doi.org/10.1016/j.cell.2014.09.055

Mattout A, Biran A, Meshorer E (2011) Global epigenetic changes during somatic cell reprogramming to iPS cells. J Mol Cell Biol 3:341–350. https://doi.org/10.1093/jmcb/mjr028

Mayer W, Niveleau A, Walter J et al (2000) Demethylation of the zygotic paternal genome. Nature 403:501–502. https://doi.org/10.1038/35000654

Meshorer E, Misteli T (2006) Chromatin in pluripotent embryonic stem cells and differentiation. Nat Rev Mol Cell Biol 7:540–546. https://doi.org/10.1038/nrm1938

Mikkelsen TTS, Hanna J, Zhang X et al (2008) Dissecting direct reprogramming through integrative genomic analysis. Nature 454:49–55. https://doi.org/10.1038/nature07056.Dissecting

Narlikar GJ, Fan H-Y, Kingston RE (2002) Cooperation between complexes that regulate chromatin structure and transcription. Cell 108:475–487

Niemann H (2016) Epigenetic reprogramming in mammalian species after somatic cell nuclear transfer based cloning. Theriogenology 86(1):80–90. https://doi.org/10.1016/j.theriogenology.2016.04.021

Nowak-Imialek M, Niemann H (2012) Pluripotent cells in farm animals: state of the art and future perspectives. Reprod Fertil Dev 25:103–128. https://doi.org/10.1071/RD12265

Ogorevc J, Orehek S, Dovč P (2016) Cellular reprogramming in farm animals: an overview of iPSC generation in the mammalian farm animal species. J Anim Sci Biotechnol 7:10. https://doi.org/10.1186/s40104-016-0070-3

Ogura A, Inoue K, Wakayama T (2013) Recent advancements in cloning by somatic cell nuclear transfer. Philos Trans R Soc Lond Ser B Biol Sci 368:20110329. https://doi.org/10.1098/rstb.2011.0329

Okamoto Y, Yoshida N, Suzuki T et al (2016) DNA methylation dynamics in mouse preimplantation embryos revealed by mass spectrometry. Sci Rep 6:1–9. https://doi.org/10.1038/srep19134

Ono T, Li C, Mizutani E et al (2010) Inhibition of class IIb histone deacetylase significantly improves cloning efficiency in mice. Biol Reprod 83:929–937. https://doi.org/10.1095/biolreprod.110.085282

Opiela J, Samiec M, Romanek J (2017) In vitro development and cytological quality of interspecies (porcine→bovine) cloned embryos are affected by trichostatin A-dependent epigenomic modulation of adult mesenchymal stem cells. Theriogenology 97:27–33. https://doi.org/10.1016/J.THERIOGENOLOGY.2017.04.022

Pan G, Tian S, Nie J et al (2007) Whole-genome analysis of histone H3 lysine 4 and lysine 27 methylation in human embryonic stem cells. Cell Stem Cell 1:299–312. https://doi.org/10.1016/j.stem.2007.08.003

Parfitt D (2010) Epigenetic modification affecting expression of cell polarity and cell fate genes to regulate lineage specification in the early mouse embryo. Mol Biol Cell 21:2649–2660. https://doi.org/10.1091/mbc.E10

Piedrahita J, Olby N (2010) Perspectives on transgenic livestock in agriculture and biomedicine: an update. Reprod Fertil Dev 23(1):56–63

Plath K, Fang J, Mlynarczyk-Evans SK et al (2003) Role of histone H3 lysine 27 methylation in X inactivation. Science 300:131–135. https://doi.org/10.1126/science.1084274

Ramsahoye BH, Biniszkiewicz D, Lyko F et al (2000) Non-CpG methylation is prevalent in embryonic stem cells and may be mediated by DNA methyltransferase 3a. Proc Natl Acad Sci U S A 97:5237–5242. https://doi.org/10.1073/PNAS.97.10.5237

Reik W, Walter J (2001) Genomic imprinting: parental influence on the genome. Nat Rev Genet 2:21–32. https://doi.org/10.1038/35047554

Rollo C, Li Y, Jin XL, O'Neill C (2017) Histone 3 lysine 9 acetylation is a biomarker of the effects of culture on zygotes. Reproduction 154:375–385. https://doi.org/10.1530/REP-17-0112

Rossant J (2008) Stem cells and early lineage development. Cell 132:527–531. https://doi.org/10.1016/j.cell.2008.01.039

Ruzov A, Tsenkina Y, Serio A, Dudnakova T (2011) Lineage-specific distribution of high levels of genomic 5-hydroxymethylcytosine in mammalian development. Cell Res 21:1332–1342. https://doi.org/10.1038/cr.2011.113

Saha B, Home P, Ray S et al (2013) EED and KDM6B coordinate the first mammalian cell lineage commitment to ensure embryo implantation. Mol Cell Biol 33:2691–2705. https://doi.org/10.1128/MCB.00069-13

Saksouk N, Simboeck E, Déjardin J (2015) Constitutive heterochromatin formation and transcription in mammals. Epigenetics Chromatin 8:3. https://doi.org/10.1186/1756-8935-8-3

Salvaing J, Aguirre-Lavin T, Boulesteix C et al (2012) 5-Methylcytosine and 5-hydroxymethylcytosine spatiotemporal profiles in the mouse zygote. PLoS One 7:e38156. https://doi.org/10.1371/journal.pone.0038156

Salvaing J, Peynot N, Bedhane MNN et al (2016) Assessment of "one-step" versus "sequential" embryo culture conditions through embryonic genome methylation and hydroxymethylation changes. Hum Reprod 31:2471–2483. https://doi.org/10.1093/humrep/dew214

Sankar A, Kooistra SM, Gonzalez JM, Ohlsson C, Poutanen M, Helin K (2017) Maternal expression of the histone demethylase Kdm4a is crucial for pre-implantation development. Development 144(18):3264–3277. https://doi.org/10.1242/dev.155473

Savatier P, Osteil P, Tam PPL (2017) Pluripotency of embryo-derived stem cells from rodents, lagomorphs, and primates: slippery slope, terrace and cliff. Stem Cell Res 19:104–112. https://doi.org/10.1016/j.scr.2017.01.008

Schneider R, Grosschedl R (2007) Dynamics and interplay of nuclear architecture, genome organization, and gene expression. Genes Dev 21:3027–3043. https://doi.org/10.1101/gad.1604607

Sepulveda-Rincon LP, del Llano Solanas E, Serrano-revuelta E et al (2016) Early epigenetic reprogramming in fertilized, cloned, and parthenogenetic embryos. Theriogenology 86:91–98. https://doi.org/10.1016/j.theriogenology.2016.04.022

Shi W, Haaf T (2002) Aberrant methylation patterns at the two-cell stage as an indicator of early developmental failure. Mol Reprod Dev 63:329–334. https://doi.org/10.1002/mrd.90016

Shi Y, Desponts C, Do JT et al (2008) Induction of pluripotent stem cells from mouse embryonic fibroblasts by Oct4 and Klf4 with small-molecule compounds. Cell Stem Cell 3:568–574. https://doi.org/10.1016/j.stem.2008.10.004

Shirazi A, Naderi MM, Hassanpour H et al (2016) The effect of ovine oocyte vitrification on expression of subset of genes involved in epigenetic modifications during oocyte maturation and early embryo development. Theriogenology 86:2136–2146. https://doi.org/10.1016/j.theriogenology.2016.07.005

Simonsson S, Gurdon J (2004) Dna demethylation is necessary for the epigenetic reprogramming of somatic cell nuclei. Nat Cell Biol 6:984–990

Smallwood SA, Tomizawa S, Krueger F et al (2011) Dynamic CpG island methylation landscape in oocytes and preimplantation embryos. Nat Genet 43:811–814. https://doi.org/10.1038/ng.864

Song X, Liu Z, He H et al (2017) Dnmt1s in donor cells is a barrier to SCNT-mediated DNA methylation reprogramming in pigs. Oncotarget 8:34980–34991. https://doi.org/10.18632/oncotarget.16507

Sun JM, Cui KQ, Li ZP et al (2017) Suberoylanilide hydroxamic acid, a novel histone deacetylase inhibitor, improves the development and acetylation level of miniature porcine handmade cloning embryos. Reprod Domest Anim 52:763–774. https://doi.org/10.1111/rda.12977

Suo L, Meng Q, Pei Y et al (2010) Effect of cryopreservation on acetylation patterns of lysine 12 of histone H4 (acH4K12) in mouse oocytes and zygotes. J Assist Reprod Genet 27:735–741. https://doi.org/10.1007/s10815-010-9469-5

Szablowska-Gadomska I, Sypecka J, Zayat V et al (2012) Treatment with small molecules is an important milestone towards the induction of pluripotency in neural stem cells derived from human cord blood. Acta Neurobiol Exp (Wars) 72:337–350

Tahiliani M, Koh KP, Shen Y et al (2009) Conversion of 5-methylcytosine to 5-hydroxymethylcytosine in mammalian DNA by MLL partner TET1. Science 324:930–935. https://doi.org/10.1126/science.1170116

Takahashi K, Yamanaka S (2006) Induction of pluripotent stem cells from mouse embryonic and adult fibroblast cultures by defined factors. Cell 126:663–676

Tan M, Luo H, Lee S et al (2011) Identification of 67 histone marks and histone lysine crotonylation as a new type of histone modification. Cell 146:1016–1028. https://doi.org/10.1016/j.cell.2011.08.008

Tao J, Zhang Y, Zuo X et al (2017) DOT1L inhibitor improves early development of porcine somatic cell nuclear transfer embryos. PLoS One 12(6):e0179436. https://doi.org/10.1371/journal.pone.0179436

Tapponnier Y, Afanassieff M, Aksoy I et al (2017) Reprogramming of rabbit induced pluripotent stem cells toward epiblast and chimeric competency using Krüppel-like factors. Stem Cell Res 24:106–117. https://doi.org/10.1016/j.scr.2017.09.001

Torres-padilla ME, Parfitt DE, Kouzarides T, Zernicka-goetz M (2007) Histone arginine methylation regulates pluripotency in the early mouse embryo. Nature 445:214–218

Turner BM, Group GE (2002) Cellular memory and the histone code. Cell 111:285–291

Urrego R, Rodriguez-Osorio N, Niemann H (2014) Epigenetic disorders and altered gene expression after use of assisted reproductive technologies in domestic cattle. Epigenetics 9:803–815

Van Soom A, Peelman L, Holt W, Fazeli A (2014) An introduction to epigenetics as the link between genotype and environment: a personal view. Reprod Domest Anim 49:2–10

Wang N, Le F, Zhan Q et al (2010) Effects of in vitro maturation on histone acetylation in metaphase II oocytes and early cleavage embryos. Obstet Gynecol Int 2010:989278. https://doi.org/10.1155/2010/989278

Wee G, Koo D-B, Song B-S et al (2006) Inheritable histone H4 acetylation of somatic chromatins in cloned embryos. J Biol Chem 281:6048–6057. https://doi.org/10.1074/jbc.M511340200

Wei J, Antony J, Meng F et al (2017) KDM4B-mediated reduction of H3K9me3 and H3K36me3 levels improves somatic cell reprogramming into pluripotency. Sci Rep 7:1–14. https://doi.org/10.1038/s41598-017-06569-2

White MD, Angiolini JF, Alvarez YD et al (2016) Long-lived binding of Sox2 to DNA predicts cell fate in the four-cell mouse embryo. Cell 165:75–87. https://doi.org/10.1016/j.cell.2016.02.032

Wilmut I, Schnieke AE, McWhir J et al (1997) Viable offspring derived from fetal and adult mammalian cells. Nature 385:810–813. https://doi.org/10.1038/385810a0

Woodcock CL, Ghosh RP (2010) Chromatin higher-order structure and dynamics. Cold Spring Harb Perspect Biol 2:a000596

Wu J, Belmonte JCI (2016) The molecular harbingers of early mammalian embryo patterning. Cell 165:13–15. https://doi.org/10.1016/j.cell.2016.03.005

Wu F-R, Liu Y, Shang M-B et al (2012) Differences in H3K4 trimethylation in in vivo and in vitro fertilization mouse preimplantation embryos. Genet Mol Res 11:1099–1108. https://doi.org/10.4238/2012.April.27.9

Xu W, Li Z, Yu B et al (2013) Effects of DNMT1 and HDAC inhibitors on gene-specific methylation reprogramming during porcine somatic cell nuclear transfer. PLoS One 8:e64705. https://doi.org/10.1371/journal.pone.0064705

Young LE, Fernandes K, McEvoy TG et al (2001) Epigenetic change in IGF2R is associated with fetal overgrowth after sheep embryo culture. Nat Genet 27:153–154. https://doi.org/10.1038/84769

Zhang M, Wang F, Kou Z et al (2009) Defective chromatin structure in somatic cell cloned mouse embryos. J Biol Chem 284:24981–24987. https://doi.org/10.1074/jbc.M109.011973

Zink D, Martin C, Brochard V et al (2006) Architectural reorganization of the nuclei upon transfer into oocytes accompanies genome reprogramming. Mol Reprod Dev 73:1102–1111. https://doi.org/10.1002/mrd.20506

Current Status of Genomic Maps: Genomic Selection/GBV in Livestock

4

Agustin Blasco and R. N. Pena

Abstract

Our understanding on how the genome is structured has improved substantially since the human genome was first sequenced in 2001. The sequencing of livestock and other model animals, in addition to other organisms, has also helped to identify common genomic patterns and features, which can now be summarised in genome maps. The annotation of sequence variation in the livestock genomes has opened up the possibility of using its genomic information for improving the prediction accuracy of its genetic merit. This chapter will give a general view on the main features annotated to the livestock genomes and outline the application of molecular information in the prediction of the genetic breeding value of the animals. The advantages and limitations of implementing this methodology in distinct production systems are also discussed.

4.1 The Evolution of Genetic Maps

Before the sequence of the genome was available for most livestock and model animals, researchers used genetic maps to orderly map genes and markers in the genome. A genetic map is simply a representation of the distribution of genes and other genetic features within the genome of one species. Specific techniques were developed to respond to questions such as in which chromosome a certain gene (or

A. Blasco (✉)
Institute for Animal Science and Technology, Universitat Politècnica de València,
Valencia, Spain
e-mail: ablasco@dca.upv.es

R. N. Pena
Department of Animal Science, University of Lleida – Agrotecnio Centre, Lleida, Spain
e-mail: romi.pena@prodan.udl.cat

© Springer International Publishing AG, part of Springer Nature 2018 61
H. Niemann, C. Wrenzycki (eds.), *Animal Biotechnology 2*,
https://doi.org/10.1007/978-3-319-92348-2_4

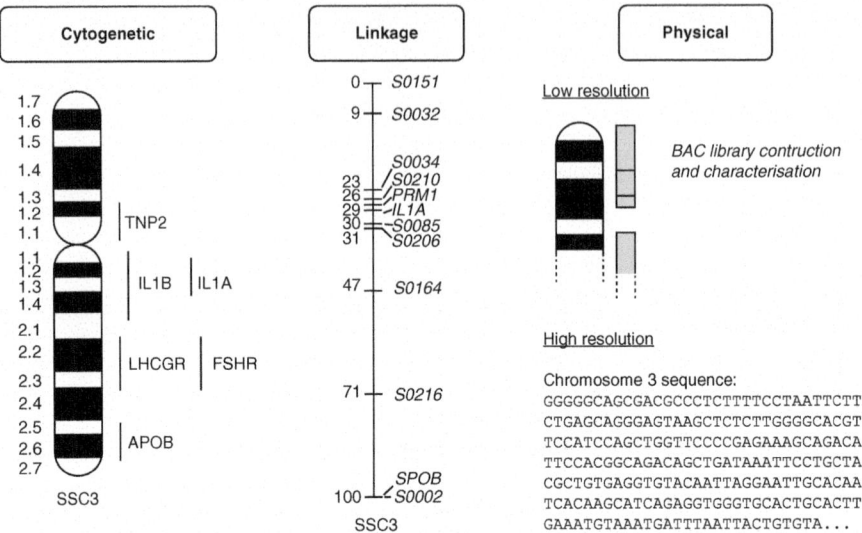

Fig. 4.1 Comparison of cytogenetic, linkage and physical maps. The three types of maps are shown for loci in pig chromosome 3 (SSC3). The cytogenetic and linkage maps extend over the entire chromosome, and loci are position in relative order and distance. Chromosome bands are indicated in the long (q) and short (p) chromosomal arms in the cytogenetic map. Linkage units are centiMorgans. The three maps can be connected to allow complementing the information from one another. Genetic maps from different species can also be compared by matching blocks of synteny (comparative mapping). Data source: Archibald et al. (1995), Groenen et al. (2011), Yerle et al. (1995)

marker) is mapped, or which were its closest genes/markers, or even in which particular order a small number of loci were mapped in a given chromosome. Thus, three distinct types of genetic maps—cytogenetic, linkage and physical—were developed to answer the questions above complementary (Fig. 4.1). Silver (1995) includes an excellent revision on genetic maps. A brief summary is presented here.

Cytogenetic maps relied on the hybridisation of a fluorescently labelled gene-specific probe (a synthetic DNA fragment) to its target gene in condensed whole-chromosome preparations (such as in karyotypes). The resolution of this type of mapping was low, but it allowed mapping a limited number of genes to the telomeric, centromeric or short (named 'p') or long (named 'q') arms of chromosomes. Complementary to these efforts, other researchers developed *linkage maps*, which were based on the frequency of recombination between two or more heterozygous loci (markers or genes) over generations. Loci that are close together in the same chromosome tend to be inherited together more often than loci that are apart. Linkage maps are generated by counting the number of offspring that receive either parental or recombinant allele combinations from a heterozygous parent. The frequency of recombination between two loci is directly related to the distance between them, measured in centiMorgans (1 cM equals a crossover rate of 1%). This measure of the linkage disequilibrium between loci allowed establishing their relative order and distance, a critical information in the pre-genomic era. Finally, the *physical maps* analysed the genomic DNA directly, usually by subcloning large DNA fragments into

DNA vectors such as BACs (bacteria artificial chromosomes) or YACs (yeast artificial chromosomes), which could be easily propagated in the lab using standard microbiology methods. These DNA fragments were usually generated by restricting targeted fractions of chromosomal DNA with several restriction enzymes to obtain overlapping fragments. By comparing the structure of these fragments, the relative position of each gene and their upstream and downstream flanking sequences could be identified. At its highest resolution, a physical map will give us the full sequence of the whole genome. Consequently, physical maps are measured in base pairs (bp) or its derived units (kbp, Mbp, Gbp). Nowadays, the genome of the main livestock species (chicken, cow, sheep, pig, horse and rabbit) has been sequenced, and efforts are being made to update and improve the information annotated to them. A summary and comparison of this information are given in the following sections.

4.2   Current State of the Livestock Genomes

While the first draft of the human genome sequence was delivered in 2001, we had to wait a number of years for the first sequence of the cow (2004), chicken (2005), horse (2007), pig (2010), rabbit (2014) and sheep (2014) genomes. Although whole genomes can be sequenced by different methods, in practice all of them result in a pool of millions of short (75–150 bp) or long (>500 bp) sequence reads. The first hurdle in describing a genome is to identify and assemble overlapping sequences into larger fragments (called contigs) to eventually reconstitute the sequence of whole chromosomes. For this, new bioinformatic programmes able to deal with these massive data had to be developed and implemented. In all species, the first genome drafts had a large number of gaps rendering incomplete chromosomes. However, these have progressively been filled in as newer versions were released. The exception is the chicken genome, which is structured in 38 autosomes, many of which are relatively small and uniform in size, often termed microchromosomes. Several properties (e.g. %GC content, gene and repeat density) contribute to the fact that some of them are not yet assembled (or only partially) even in the latest version of the genome (Warren et al. 2017). In this species, linkage groups estimated from linkage maps are still of use to study genes located in these missing regions.

The most updated version of farm animal genomes is available at www.ensembl. org. The importance of these updated versions is double: first, it is a precious material for researchers to study the structure of the genome and to investigate genes related to production traits or disease. They also provide a scaffold to assemble new whole-genome sequencing (WGS) data from other animals of the same species in a much faster and accurate way. As the costs of WGS have become more affordable, it is now feasible to describe genetic variability in a population by sequencing key genetic contributors. Sound scaffolds are critical to identify, map and compare sequence variants across these individuals.

As a result of the genome sequencing projects, we have been able to measure the *total size* of the genome, which is specific to each species. In the five farm animals analysed here, it ranges from 1.2 Gbp in chicken to 2.7 Gbp in rabbit (Table 4.1). As a reference, the human genome is slightly longer (3.1 Gbp), but the longest genome

Table 4.1 Genome size and annotation of genes, transcripts and sequence variation features in the latest assemblies available for the main livestock species

	Chicken	Genome Cow	Sheep	Pig	Horse	Rabbit	Human
Assembly	Gga5.0	UMD_3.1	Oar_v3.1	Sscrofa11.1	Equ Cab 2	OryCun2.0	GRCh38.p10
Length (bp)	1,230,258,557	2,670,422,299	2,619,054,388	2,501,912,388	2,474,929,062	2,737,490,501	3,096,649,726
Genes							
Protein-coding genes	18,346	19,994	20,921	22,452	20,449	19,293	20,338
Non-coding genes	6491	3825	5843	3250	2142	3375	22,521
– Small nc-genes	1705	3650	3624	2503	1967	3059	5363
– Long nc-genes	4643	175	1858	361	–	–	14,720
– Miscellaneous nc-genes	144	797	361	386	175	316	2222
Pseudogenes	43	26,740	290	178	4400	1001	14,638
Gene transcripts	38,118	19,994	29,118	49,573	29,196	24,964	200,310
Sequence variation							
Short variants	23,873,479	102,499,615	60,323,418	64,310,125	5,217,806	–	329,465,985
Structural variants	–	10,462	2	224,038	193,747	–	5,864,995

Data from the human genome are also included as a reference. Source: www.ensembl.org

so far sequenced is the loblolly pine tree (*Pinus taeda*) which spans 23.2 Gbp (Neale et al. 2014). As we will see below, there is no linear correlation between the size of a genome and the number of genes it contains.

4.3 Gene Annotation in the Livestock Genomes

Once the sequence is established, the next step in order to build a genomic map is to annotate the genetic elements underlying each genome. This annotation step is constantly evolving as new elements are still being discovered. The first features to be mapped to the genomes were the *protein-coding genes*. By doing so, researchers realised that animal genomes were, at once, simpler and more complex than expected. Humans, farm animals, mice and simpler animals such as the earthworm *Caenorhabditis elegans* have all approximately the same number of genes, around 20,000 (Fig. 4.2 and Table 4.1). This number of genes seemed too low to explain the complexity of larger mammals. Moreover, the coding sequences only spanned a very small percentage of the total genomic sequence of farm animals, about 1.5–2%. This means 98% of the genome does not encode for proteins, the ultimate effectors of cellular functions. About a quarter of this non-coding (nc) DNA are intron sequences, that is, gene sequences that are transcribed by the RNA polymerases but that are spliced out of the mature mRNA by the spliceosome. Half of the 70% remaining genomic DNA contains repetitive DNA elements such as micro-/minisatellites or transposon-derived sequences (LINEs, SINEs, Alu, LTRs, etc.).

Strikingly, the proportion of ncDNA in the genome, unlike the total number of protein-coding genes, increases in parallel with evolutionary complexity (Fig. 4.2). Thus, in simple organisms such as prokaryotes or yeasts, 70–85% of the genome encodes proteins, while in invertebrates (earthworm, fruitful), this figure drops to 20–25% and reaches the overwhelming 1.5–2% value in humans and farm animals. The presence of ncDNA has been explained by several mechanisms. Initially, all this additional ncDNA of unknown (and unpredictable) function was thought to be an evolutionary artefact, a carry-over of non-functional (and non-damaging) DNA that had accumulated over evolution without adding any specific advantage to the species. Moreover, although there was a degree of sequence conservation in the protein-coding DNA, sequences were much more divergent in ncDNA, reinforcing the hypothesis of lack of function. In consequence, the ncDNA was often called 'junk DNA' to designate its lack of purpose. However, as it became more and more obvious that the number of protein-coding genes was not the main drive of biological evolution, the attention was turned into ncDNA.

In this context, the ENCODE project was set up to annotate functional elements in the genome of humans and model organisms. The consortia of research groups participating in this initiative designed two types of experiments: one group aiming at identifying DNA that was being transcribed into RNA and another group targeting chemical labels in the chromatin (epigenome). One of the first results reported by the ENCODE consortia was that more than 80% of the genome was being transcribed into RNA. This phenomenon was called pervasive transcription to express

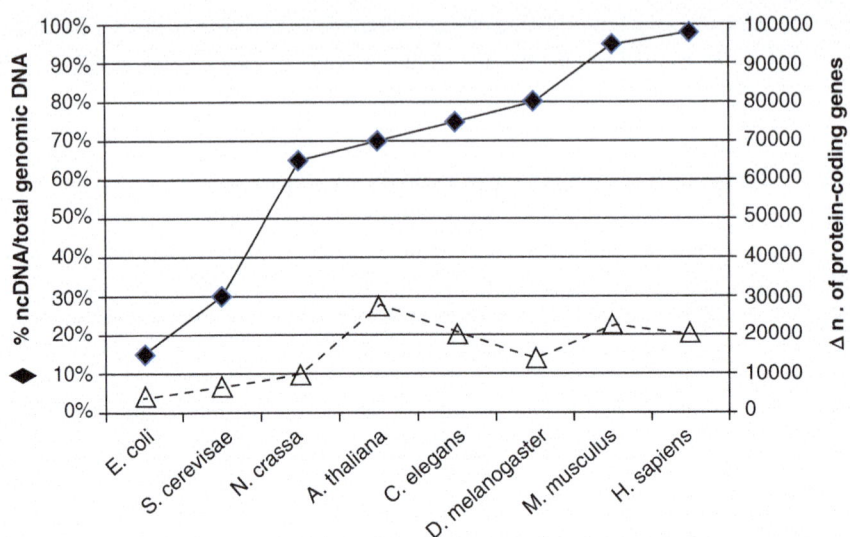

Fig. 4.2 The proportion of non-coding DNA (ncDNA) in the genome increases with developmental complexity (adapted from Mattick 2011), but the number of protein-coding genes does not (source www.ensembl.org)

the permanent state of transcription of most of the genome (Libri 2015). As protein-coding genes only span 25% of the genome (adding exons and introns together), that means most of the transcribed RNA was in fact non-coding RNA (ncRNA) molecules. Thus, a new category of *non-coding genes* was defined (Table 4.1), which, like coding genes, are also organised in exons and introns and have regulatory elements that control expression.

Currently, the annotation of protein-coding genes is almost complete in most animal genomes. The exception is again the chicken genome, where 360 genes are still missing in the current annotation, which most likely map to unassembled microchromosomes (Warren et al. 2017). In contrast, the mapping of nc-genes is still at its initial stages, particularly in farm animals. As a reference, in humans there are similar numbers of protein-coding and non-coding genes, indicating that annotation in farm animals is probably underestimated (Table 4.1). Based on the length of the transcripts, ncRNAs can be classified into small ncRNAs (usually <200 nt-long) and long ncRNA (>200 nt-long) molecules. Small ncRNA can be divided into further categories (Wright 2014), although probably the best characterised are the family of microRNA (*miRNA*) genes. These represent a group of genes that, once transcribed and processed, generate short structures of double-stranded (ds) RNA, usually ~21 nt-long. About 80% of miRNA genes map to intronic DNA, usually in polycistronic clusters from which up to ten miRNAs are co-expressed (Hausser and Zavolan 2014). This has facilitated their mapping, and they are probably the best annotated class of nc-genes in the farm animal genomes. miRNA are strong regulators of the translation rate of protein-coding mRNAs. By binding usually to the 3′ untranslated regions (3′UTR) of the mRNA, the miRNA are able to put the translation of that miRNA on hold. This represents an additional layer of regulation of the expression of protein-coding genes, from DNA to proteins. On the other hand, long intergenic ncRNA (*lincRNA*) represents a new class of ncRNA that has brought much excitement, even though few data are yet available for most of them. In general, these genes are transcribed at very low levels (about 100- to 1000-fold lower than the average protein-coding gene) from >60% of the genome. Most lincRNA genes are active only in some cell types or at certain developmental stages and are thought to be one of the key organisers of development and probably a main evolutionary drive (Hangauer et al. 2013).

The third type of genes mapped to the genomes is *pseudogenes* (Table 4.1). These represent 'dead genes', relics from former protein-coding genes, usually generated by gene duplication, that have been inactivated in the course of evolution through accumulation of mutations. The gene graveyard is extensive in the human and cow genome (14,638 and 26,740 pseudogenes, respectively) but is probably underrepresented in chicken, pig, sheep and horse. It is not unusual for a pseudogenes to be transcribed into mRNA, but they very rarely get translated into proteins, due to unstable messengers or to accumulation of premature STOP codons (Xu and Zhang 2016).

Altogether, protein-coding genes, non-coding genes and pseudogenes generate a large number of transcripts (around 20,000–50,000 in farm animals but close to 200,000 in humans). The tenfold higher number of transcripts in humans is explained mainly by alternative splicing of exons and introns, which takes place in ~94% of the human (protein-coding and non-coding) genes. This is a process that also takes place in the animal transcripts but to a lower extent (for instance, if has been estimated to affect 21% of cow genes). Current genomic maps also include information on alternative transcripts and predicted proteins generated by each gene. Beyond question, this is a major source of functional variation that can explain the larger biological complexity of livestock animals and certainly that of humans.

4.4   Annotation of Regulatory Elements

The second set of experiments carried out in the frame of the ENCODE project had the aim to identify the regulatory elements of the genome, that is, stretches of genomic DNA that regulate (activate/inactivate) the expression of genes. The two main types of regulatory elements are promoters and enhancers (Fig. 4.3). *Promoters* are DNA sequences around the transcription start site of a gene where the proteins of the transcription machinery assemble. The transcription complex represents a runway for the RNA polymerase II to land and start transcription. *Enhancers*, on the other hand, are usually located remotely from gene promoters. They physically interact with promoters stabilising or disassembling the transcription complex. Enhancers are essential for the correct spatio-temporal activation of gene expression (Andersson 2015). For instance, an enhancer may act to increase the transcription of a gene with a possibly weak promoter or may provide essential, additional information not encoded in the gene promoter itself. Enhancer function is highly specific to cell type and state

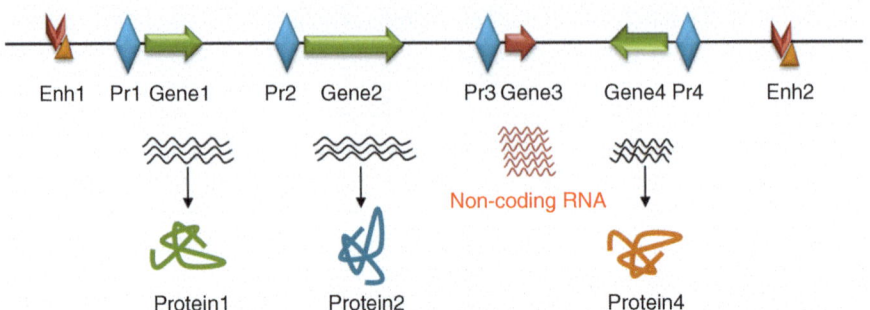

Fig. 4.3 Spatial relationship between enhancers (Enh), promoters (Pr) and genes. Promoter elements are positioned close to the transcriptional start site of both protein-coding and non-coding genes. Both types of genes are transcribed into RNA, but only the protein-encoding genes are translated into proteins. Enhancers can be located upstream, downstream and even inside the genes they are regulating

compared to protein-coding genes. Hence, a gene may be regulated by different enhancers in different cell types, at different developmental stages and in response to different signals. Enhancers can be hundreds of kbp away from the regulated genes, and it is not unusual to find several (untargeted) genes between them (Fig. 4.4). Hence, to put enhancers proximal to the correct target gene promoters in three-dimensional space, the DNA must be structured into chromatin loops (Fig. 4.5). A current hype is the elaboration of 3D dynamic genomic maps of how these loops evolve during cellular differentiation according to the required change in gene expression.

These functional elements are currently being annotated to the livestock genomes thanks to the efforts of the FAANG (Functional Annotation of the Animal Genomes) initiative. Annotations are much more advanced in humans and model animals. As a reference, there are 70,292 promoters and 399,124 enhancers in the human genome (ENCODE Project Consortium 2012), and about half of each are active in any given cell (Won et al. 2013). Regulatory elements are difficult to identify by computational analysis of the genome sequence as in general they lack evolutionary constraint, which means their sequence is not conserved across species despite having the same function. A combination of wet-lab techniques is needed to position epigenetic labels that are characteristic of silent, poised or active regulatory elements (ENCODE Project Consortium 2012). A second common feature of promoters and enhancers is that they are bidirectionally transcribed; that is, RNA is synthesised from both strands flanking the element, producing relatively short non-polyadenylated enhancer RNAs (eRNAs). Synthesis of eRNA is essential for full enhancer functionality. This explains a large part of the non-coding RNA pervasively transcribed in the genome and indicates there is an extensive overlap between transcription and regulation. Overall, the results from the ENCODE project claim a shift from the gene-centric vision of the genome to a more dynamic and holistic interpretation of genomic function.

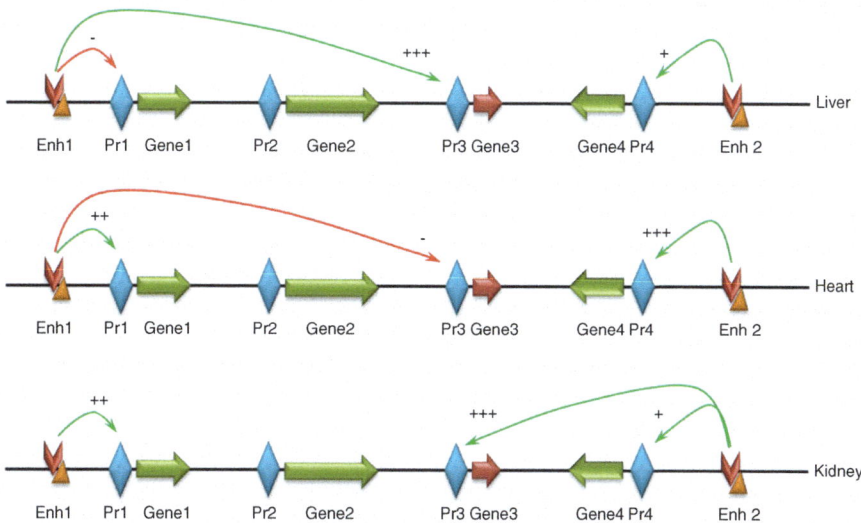

Fig. 4.4 Enhancers (Enh) regulate gene expression by interacting with gene promoters (Pr), which might be several genes away from the enhancer site. To bring enhancers and promoters together, genomic DNA needs to bend in a 3D loop. Enhancers can activate or silence transcription depending on the gene, the tissue and the stage of development. Enhancer engagement and displacement are very dynamic events as they regulate more than one gene at a time

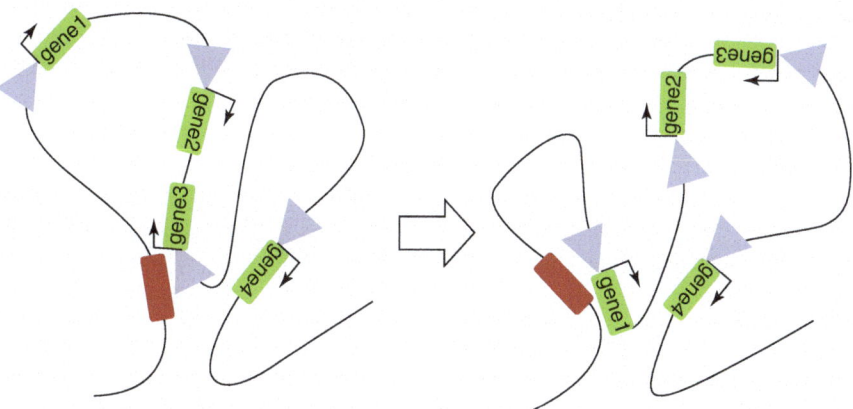

Fig. 4.5 Three-dimensional chromatin looping is necessary for the correct regulation of gene expression by distal enhancers. On the left, the enhancer (dark block) interacts only with the promoter (triangle) of gene 3. In order to regulate the expression of gene 1, a different loop needs to be formed (right). This is a dynamic process that can change rapidly in response to cellular signals

4.5   Mapping the Genomic Sequence Variation

Another objective of the genome annotation initiatives has been to catalogue the new mutations described in each genome and to map them in the genomic context. For simplification, in the annotated genomes, mutations are classified as either short variants or structural variants. *Short variants* include single nucleotide polymorphisms (SNPs) and insertion/deletions (indels) of short nucleotide runs. By large, most short variants have two possible versions, called alleles. For instance, for a given SNP, an adenine can change into a guanidine, so A and G constitute the two alternative alleles. The data to annotate these variants come from specialised databases such as the dbSNP (www.ncbi.nlm.nih.gov/snp) and from sequencing centres (e.g. the Broad Institute). Short variants are very common. Taking figures in Table 4.1, it is estimated that in farm animals there is one short variant per 30–50 bp (even more frequent in humans; 1 in 15 bp). However, although the amount of intra-species sequence variation is disturbingly high, the numbers are expected to be much lower within a given breed or commercial line. In genomic maps, short variants are annotated over other genomic features such as genes. For variants overlapping protein-coding genes, an estimation of the effects on the final protein is also calculated and annotated on the map. Protein structure predictor programmes such as SWIFT (www.bioinfo.org.cn/swift) are routinely used for this purpose. Under *structural variants*, repetitions in larger regions of DNA, of at least 1 kb in size, are gathered. It can include inversions, balanced translocations or genomic imbalances (insertions and deletions), commonly referred to as copy number variants (CNVs). This is an area of uneven annotation across the genomes, with total numbers ranging from ~200,000 in horse and pig to none in the chicken genome. The number of possible alleles per structural variants is more variable and can go from complete deletion (zero copies) up to three to four copies of the fragment.

Variation data, particularly SNP information, have been used to build for each species dense panels of markers evenly distributed across the genome. Novel biotechnological tools have been developed to genotype these panels. Currently, two companies lead the market of genotyping platforms for livestock animals. They provide a range of SNP-based arrays (also known as *SNP-chips*) to genotype at variable densities (Table 4.2). These chips are currently used to improve the accuracy of predictions of breeding values in several species, as we will see in the following sections.

4.6   Genomic Selection

The use of genetic markers to improve the efficiency of current selection programmes was proposed 40 years ago by Moses Soller (1978). At that time, few markers were available, and the expectation was to find some gene with a substantial effect linked to the marker and increase its frequency in the population. Unfortunately, most production traits in livestock species are determined by a large number of genes with small effect, and consequently the method was inefficient.

Table 4.2 Summary of the commercial high-density genotyping chips currently available for the main livestock species

Species	Chip name	No. of SNPs	Average interval between SNPs (kb)	Average MAF across tested populations	Supplier
Chicken	Axiom Chicken array	580,961			Affymetrix
Cattle	BovineSNP50	54,001	50.6	0.26	Illumina
	BovineHD	777,000	3.43	0.25	Illumina
	BovineLDv2	7931	383	0.31	Illumina
	Bovine3K[a]	2900			Illumina
	Axiom BOS1 array	648,855			Affymetrix
Horse	Axiom Equine Array	650,000			Affymetrix
Pig	PorcineSNP60	64,232	43.4	0.28	Illumina
	GGP Porcine HD	>65,000	43.0		GeneSeek
	Axiom Porcine 650K	658.692	3.34	0.32	Affymetrix
Sheep	OvineSNP50	54,241	50.9	0.28	Illumina
	OvineHD	603,350	5	0.30	Illumina
Rabbit	Axiom OricunSNP	200,000	15–20	0.20	Affymetrix

[a]Subset of SNP50 panel for prediction of milk yield, protein % and fertility

When QTL detection started in the 1990s and it was feasible to use more markers, it seemed that alleles of medium effect could increase their frequencies by marker-assisted selection, and higher responses to selection could be obtained (Lande and Thompson 1990). However, as Blasco (2008) noticed, there was a notorious discrepancy between simulation results, relatively optimistic, and practical application of marker-assisted selection, which gave deceptively small improvements. The problem was, as Smith and Smith (1993) stressed, the lack of enough markers to cover the whole genome and capture the signals of genes with small effect on the traits. When the genome sequences of livestock species were published, first in 2004 in cattle and later in the other species, chips with a large amount of SNPs became available at an affordable cost, and its use in selection programmes was examined. The first chips of 10,000 SNPs were not well distributed along the genome and were not efficient, but in December 2007 a well-distributed 57,000 SNPs chip was for the first time commercialised. In 2008 the first genetic evaluations for dairy cattle using genomics started in several countries, and in 2009, the USDA published the first official dairy cattle genomic evaluations. The impact in dairy cattle selection programmes was dramatic, doubling the rate of improvement of total genetic merit (Wiggans et al. 2017); thus the use of genomic selection was rapidly investigated for the other livestock species. Dairy cattle has some special characteristics that permit an efficient use of genomic selection, as we will see later, but the use of genomic selection in other species is not as straightforward (Blasco and Toro 2014; Jonas and de Koning 2015). Nevertheless, genomic selection can contribute to the efficiency of current selection programmes, if the strategies of implementation are carefully studied and the cost of genotyping is low enough. It seems that very high-density SNP chips do not lead to a much higher accuracy of prediction; for example, Van Raden et al. (2011) obtained a gain in accuracy of only 1.6% when using 500,000 markers instead of 50,000. Even SNP chips of 3000 markers, using imputation techniques that we will comment later, give good results (Berry and Kearney 2011, in cattle; Cleveland and Hickey 2013, in pigs), which permits examining scenarios less favourable than the dairy cattle one.

4.7   Predicting Breeding Values with Genomic Selection

The methods for predicting breeding values with genomic selection were developed in a seminal paper by Meuwissen et al. (2001) before we had access to the SNP chips. Predicting breeding values has two steps. First, we collect data form a set of animals, for example, 4000 animals, the 'reference population', and genotype all of them with a high-density chip with, for example, 50,000 SNPs. Now we need to prepare the prediction equation. To do this, we generate one variable z_i per SNP having an arbitrary value indicating whether the SNP 'i' is 'homozygous' for one base, 'heterozygous' (i.e. has different bases) or 'homozygous' for the other base. Calling 'M' and 'm' the two positions of the bases of one SNP, we have for each SNP.

SNP$_i$	M$_i$M$_i$	M$_i$m$_i$	m$_i$m$_i$
z_i	1	0	−1

The values 1, 0 and −1 are arbitrary and can be substituted by other values (e.g. 2, 1, 0). The coding is additive; it is related to the number of copies of one reference allele, 'M' in this example. The use of capital letter 'M' does not mean that we are considering dominance effects, although models that are more complex can include this possibility (Vitezica et al. 2016). We will consider in this simple example that the data are pre-corrected to make the formula simpler. The regression equation is

$$y = a_0 + a_1 z_1 + a_2 z_2 + a_3 z_3 + \cdots + a_{50,000} z_{50,000} + e$$

where $a_1, a_2, a_3 \cdots a_{50,000}$ are the coefficients of regression and $z_1, z_2, z_3 \cdots z_{50,000}$ are the variables associated to each SNP. The genetic value of the animal is

$$a = a_1 z_1 + a_2 z_2 + a_3 z_3 + \cdots + a_{50,000} z_{50,000}$$

(we can add, if we like, the intercept a_0 of the regression equation). Now we have to estimate the coefficients, and we are faced with the problem that there are much more unknowns than the equations that we have; the equation system cannot be solved by classical procedures. However, the system has a solution using Bayesian statistics under some assumptions about the prior information on the SNPs we have (see Blasco 2017, for details). We can thus obtain the estimates of the regression coefficients $\hat{a}_1, \hat{a}_2, \hat{a}_3 \cdots \hat{a}_{50,000}$, and we are ready to predict the breeding value of new individuals.

In the second step, we will predict the genetic value of animals that may have or not have their own data. Suppose first that the animal has no phenotypic data, but it has been genotyped, and we know the values of each of the variables z_i for this animal. By substituting in the equation, we can predict its genetic value:

$$\hat{a} = \hat{a}_1 z_1 + \hat{a}_2 z_2 + \hat{a}_3 z_3 + \cdots + \hat{a}_{50,000} z_{50,000}$$

The genetic value of each new animal will be predicted using the same coefficients $\hat{a}_1, \hat{a}_2, \hat{a}_3 \cdots \hat{a}_{50,000}$ with the variables z_i of the new animal provided by its SNPs. It is important to notice that the coefficients do not indicate the importance of each SNP, since the variables z_1, z_2, $z_3 \cdots z_{50,000}$ are correlated. We have said before that SNPs close to each other are frequently associated in its genetic transmission; even if they are in different chromosomes, they can be associated in its transmission, for example, due to selection. The equation is useful to predict the whole genetic value '$\hat{a}$' of an animal, but not to detect single genes. The coefficient of a SNP in a multiple regression is not the same as the coefficient that can be found when this SNP is fit in isolation. The coefficients of the equation will also change depending on the number of SNPs considered, because the sum of all of the terms in the multiple regression equation should give the same genetic value $\hat{a}$ of the animal; thus when many SNPs are considered, they have smaller individual effects.

The different Bayesian statistical methods for solving the equations depend on different prior assumptions about the genetic determination of the traits; for example, the trait can be determined by many genes of small effect each one or by some major genes, some intermediate ones and many genes with small effects. The success of each method depends on whether the actual genetic determination of the trait reflects well what the prior information assumes, although Fernando and Garrick (2013) have noticed that in real applications, the simplest model that considers the traits determined by many genes with small effects works just as well as the more complex models and sometimes even better. This occurs probably because in practice, even if there are genes with medium-large effects, they are not in close association with only few markers, but their effect is captured by many markers.

When the animal has no data, its breeding value can be estimated by weighing the information of its relatives appropriately, a technique called selection index, in which several traits can be simultaneously used for selection weighed according to their economic importance (Falconer and Mackay 1996). Now, if the animal is genotyped, the estimated breeding value from the genomic equation can also be appropriately weighed and integrated with the breeding value provided by the selection index. The information given by the SNP chips can be used to better assess the actual relationships between individuals. For example, we know that on average full sibs share half of their genomic information, but by crossing two heterozygotes Aa × Aa, we could produce full sibs that are more similar than others. If we have three full sibs AA, AA and aa coming from this cross, the two first full sibs are more similar than the first and the third or the second and the third sib. Taking into account all SNPs, we can have a more accurate idea about the actual correlation between relatives. This allows being more accurate in the genetic evaluation.

In current breeding programmes, the correction of environmental effects (parity, season, herd, batch, etc.) is done at the same time as the genetic evaluation, using a technique called best linear unbiased prediction (BLUP, see Blasco 2017, for details). When all genomic relationships are used in the evaluation, this procedure is known as 'genomic-BLUP' (G-BLUP; see e.g. Clark and Van der Werf 2013). It can be shown that this procedure is equivalent to solving the genomic equations under a model assuming that the genetic determination of the trait depends on many genes with small effects each one (Habier et al. 2007). Nowadays there is a wide consensus about the reasons of the success of genomic selection; rather than a better assessment of the 'genetic architecture' of the trait, it is mainly related to a better determination of the actual relationships between relatives. Genomic information can be integrated with BLUP, and the evaluation is made with all data of all animals and all important traits, integrating the information provided by the genomic equations, a procedure that is called 'single step' (Legarra et al. 2009; Misztal et al. 2009).

4.8   Difficulties in Implementing Genomic Selection

Blasco and Toro (2014) and Jonas and de Koning (2015) have detailed some of the difficulties of implementing genomic selection in current breeding programmes. First, to create the equations to be used in prediction, we need a large 'reference population' of several thousand animals. This is not a problem in dairy cattle, but in other species, it can be a serious problem. In prolific species, for example, selection is performed in small nucleuses, sometimes with few hundred females. Some alternatives can be considered, for example, using sibs from multiplication farms or using animals from several generations (Chen et al. 2011) or crossbred animals (Knol et al. 2016), but the efficiency of the equations rapidly decays, thus alternative strategies should be examined with care. A second problem is the need of generating new equations every three or four generations, because due to recombination, the associations between SNPs and causal genes are lost with time. Ibañez and Blasco (2011) have shown that the accuracy of the equations is rapidly lost generation after generation, which means that new large reference populations are needed from time to time. In practice, instead of having large reference populations every few generations, phenotypes are collected every generation to update the equations. This is not a problem for routinely recorded traits (e.g. litter size), but it can be a problem for more expensive traits.

Another major problem of genomic selection is the cost of genotyping. This cost has been dramatically reduced in the last years, but it is still important for species in which the individual value of the animal is small and the generation interval is short (pigs, poultry, rabbits), which implies frequent genotyping with high cost with respect to the value of the animal. A way of facing this problem is to use low-density chips with only few hundreds or thousands SNPs, inferring the missing SNPs from high-density chips. This technique, called 'imputation', is based on that recombination which is low in a single generation and has produced efficient results (Huang et al. 2012; Cleveland and Hickey 2013). Imputation from high-density chips should be repeated every three or four generations, because recombination leads to errors of imputation. Nowadays, instead of having a reference population, some high-density chips are used every generation for repeating imputation.

Genomic selection was considered as a possible procedure for improving 'difficult' traits. For example, meat quality traits were considered natural candidates for genomic selection, hoping that after collecting the data in a reference population, many animals could be evaluated using their genomic data without the need of collecting their phenotypic data or data from relatives. However, as the equations have to be reformulated after three or four generations, there is the need of continuous data collection to avoid reconstituting reference populations every few generations; thus genomic selection became less attractive, at least for short generation interval species such as pigs, poultry or rabbit. In dairy cattle, the index of conversion of food for milk is economically attractive but difficult to be recorded, but by collecting these records in some specialised farms, a reference population and the equations needed for genomics could be prepared. As dairy cattle, particularly the Holstein breed, constitutes a global population in which most farmers use the same bulls, genomics could be used to estimate the genetic value of animals that have not been measured for this trait. Here the problem comes from the genotype per environment interactions. Farms measuring food efficiency for milk production are good farms having the cows under a good environment. It is not clear that the best genetic animals in these farms will be the best in common farms under other environments. This has happened yet with another 'new trait' in pigs, residual feed intake, where the relationship between the breeding values of the animals in the nucleus of selection and the commercial farms was null (Knap and Wang 2012).

The difficulties in the implementation of genomic selection do not invalidate genomics for selection programmes, since genomics is a tool and how to use it efficiently is a matter of research. As we will see below, genomic selection has proved to be extremely useful in dairy cattle, but the cost of genotyping prevents its use in rabbit breeding programmes and complicates its application in pigs, lamb or poultry. Nevertheless, in pigs, poultry, lamb and beef cattle, genomic selections is, or can be, a useful complement to current selection programmes.

4.9   The Use of Genomic Selection in Breeding Programmes

Genomic selection has been applied with success in breeding programmes, with spectacular results in dairy cattle and with more modest results in other species. Nowadays there is no doubt that genomics is a useful tool for selection, but careful strategies for its implementation should be developed in most species to ensure its profitability.

Dairy cattle. Genomic selection has revolutionised the dairy cattle breeding programmes. As Schaeffer (2006) predicted before the first SNP chip was available, dairy cattle is particularly suitable for genomic selection. It has a long generation interval (6 years) due to the need of progeny test, the traits of interest (milk production and quality) cannot be measured in the sire, selection pressure has to be applied essentially in sires because the average parities of dams is about 2.7, and the dissemination of the genetic progress is cheap and easy via artificial insemination. It was not a problem to create large reference populations and maintain a continuous recording system to update the equations, since a single bull can have many daughters and all farms constitute a global nucleus linked by artificial insemination. Moreover, sires have a high price; thus genomic cost is not an impediment for developing genomic selection compared to other species. Genomic selection was implemented in 2008 in several countries and nowadays is widely used for sire evaluation. Nowadays, as using imputation with 3000 SNPs chips has a high accuracy (up to 99%, Van Eenennaam et al. 2013), dairy cows are also being genotyped. As a result of this wide implementation, generation interval has been halved and genetic progress doubled (Wiggans et al. 2017). It is interesting to notice that other efficient programmes based on reducing the generation interval were proposed in the past, for example, MOET (multiple ovulation and embryo transfer, Nicholas and Smith 1983). However, in addition to difficulties in implementation MOET (Simianier 2016), as the accuracy of bulls evaluation was lower compared to proven bulls with 100 daughters (0.45 versus 0.95), farmers were reluctant to use them. Now, genomic bulls have still lower accuracy (around 0.8), but farmers accept the loss of accuracy and use several genomic bulls to lower the risk. Obviously, this loss of accuracy is compensated by far by the reduction of the generation interval, but the fascination for the new technique may have played a role in its rapid acceptance.

Beef cattle. The success of genomic selection in dairy cattle has moved the whole industry to consider the introduction of genomics in current breeding programmes. However, beef cattle is organised in many breeding associations, with a much lower size than the dairy cattle breeds. Moreover, beef cattle are not always well connected by artificial insemination. Because of this, it is not feasible for most beef cattle associations to have a 'training population' and a continuous recording as large as in dairy cattle. This has led to the proposal of using multibreed training populations for predictions, but the problem is that effectiveness of genomic breeding value prediction is higher when training populations are close to the animals to be predicted, otherwise the prediction is poor (Lund et al. 2014); thus the use of multibreed populations is now under discussion. Another problem is the cost of genotyping. In beef cattle, the most commonly measured traits are weights at a given age. Usually these traits have relatively high heritabilities (about 0.40), which means that the accuracy of the individual phenotype is about 0.6–0.7, and it can become higher by adding information from relatives. Therefore, genomics should improve accuracy over 0.7 when the trait of interest can be measured just using a scale, although in some extensive systems collecting samples for genomics may be easier than using a scale. Imputation may be a solution, but imputation is precise only when the low-density chip is used in animals closely related to the ones used for imputation (Rolf et al. 2014); thus multibreed low-density chips may be of little utility. Although it is true that genomics has been used by commercial companies as a marketing tool (Rolf et al. 2014), genomics could improve the accuracy for traits not directly measured, for example, when the objective is weight at slaughter but only weight at weaning is measured, or for carcass traits. Even in all these cases, a careful study should be made taking into account the large training populations needed and the permanent cost of genotyping in relation to the benefits expected.

Sheep and goat. Lambs and goats bred for milk production have the same scheme as in dairy cattle at a much lower scale, which limits the application of genomic selection. Meat sheep shares with beef cattle most of its problems for the efficient use of genomic selection. In both cases, the low price of the animals limits the application of genomics due to the relatively high cost of genotyping. Rupp et al. (2016) have recently reviewed the application of genomics in sheep and goats. Gains in accuracy when applying genomic selection were rather modest, around 10–20%, even for milk production traits. Considering costs of genotyping, Shumbusho et al. (2016) estimated the economic advantages of using genomic selection in sheep meat to be only 15% in the best scenario. Similar results were found in Australian merino breed by Horton et al. (2015). Multibreed SNP chips have also been proposed, but they share the same problems as in beef cattle.

Pigs. In pigs, progeny test is not performed, and generation interval is consequently short (around 1 year). Selection objectives are traits expressed in males and females with the exception of litter size, dissemination of genetic progress is made through a pyramidal structure of nucleus-multiplier-commercial farms (where genetic improvement is performed only in the nucleus), and selection can be applied on dams because they are prolific animals. There is no global nucleus but several companies competing in a free market, having small nucleuses of around 25–50 males and 300–2000 females per line, and the price of selected animals is much lower than in dairy cattle. Moreover, as pigs are normally produced in a three-way cross scheme, the costs of genotyping are three times higher than when a single breed is used in production. With all of these constrains, the application of genomics has had less spectacular results than in dairy cattle; nevertheless the increment in profit when using genomic selection has been evaluated from 10% (Lillehammer et al. 2013) to 50% (Knol et al. 2016), depending on the implementation. Litter size is an obvious candidate for genomic selection because the trait is not expressed in the female when it should be selected, but heritability of litter size is very low, so large reference populations are needed, and the strategy for obtaining them is not evident; for example, information from multipliers can be used or even information from crossbred commercial females, as we mentioned before. The success of genomic selection in pigs comes from a careful study of the strategies for implementing genomic selection (see Ibáñez et al. 2014 and Knol et al. 2016 for a detailed description of some strategies). Imputation is important because genotyping is still economically relevant relatively to the price of selected animals, and the accuracy of imputation is high (around 97%, Cleveland and Hickey 2013). The success of genomic selection comes, again, from a better estimation of the relationships between animals.

Poultry. Similar constraints to pigs arise in poultry, in which four-way cross schemes are common; nucleuses are also small, although it can be found large nucleuses up to 2000 males and 10,000 females. Generation interval is also very short, females produce a large amount of eggs, and the relevant traits are expressed mainly in females in layers and in both sexes in broilers. Genomics was implemented in 2013 in both production systems. Careful imputation procedures have obtained very good results in both layers and broilers, with accuracies of around 97% with respect to the high-density SNP chip, and due to the continuous decreasing in genotyping cost, medium density SNP chips are being used, removing the need of imputation (Wolc et al. 2016). A selection experiment in layers has evaluated the response to selection using genomics when compared with a line in which the same sort of selection was performed without genomics. The results were variable depending on the trait used in the selection index; traits like egg production number showed little advantage, but for some traits like egg weight, the use of genomic selection was much more efficient (Wolc et al. 2015). Efficiency of genomic selection in broilers has been evaluated by comparing the increasing in precision when evaluating the genetic merit of some traits (Chen et al. 2011). In the sire line, selected mainly for growth rate, the increment of precision for body weight when using genomic selection was 20%, and for ultrasound measurements of the breast, it was 17%; the dam line had better results for the same traits, but it was selected mainly for reproductive traits. In general, the best advantage of the use of genomic selection, as in the other species, comes from traits that are not available at the moment of selection (Wolc et al. 2016).

Rabbits. Genomics has not been implemented in rabbits yet, mainly due to the cost of genotyping. The rabbit chip of 200,000 SNPs appeared recently (October of 2015), and no low-density chips have been produced yet. Rabbit selection schemes are three-way crosses with the same structure as in pigs, and nucleuses are even smaller (from 20 males and 150 females per line); dam lines are mainly selected for litter size and sire lines for growth rate. Generation interval is very short (6–9 months), and the price of the animals is low, which represents the main constraint for the application of genomic selection. Research needs to be done to find the best strategy for implementing genomic selection in rabbits.

References

Andersson R (2015) Promoter or enhancer, what's the difference? Deconstruction of established distinctions and presentation of a unifying model. BioEssays 37:314–323. https://doi.org/10.1002/bies.201400162

Archibald AL et al (1995) The PiGMaP consortium linkage map of the pig (Sus scrofa). Mamm Genome 6:157–175

Berry DP, Kearney JF (2011) Imputation of genotypes from low- to high-density genotyping platforms and implications for genomic selection. Animal 5:1162–1169

Blasco A (2008) The role of genetic engineering in livestock production. Livestock Sci 113:191–201

Blasco A (2017) Bayesian statitics for animal scientists. Springer, New York

Blasco A, Toro MA (2014) A short critical history of the application of genomics to animal breeding. Livstock Sci 166:4–9

Chen CY, Misztal I, Aguilar I, Tsuruta S, Meuwissen THE, Aggrey SE, Wing T, Muir WM (2011) Genome-wide marker-assisted selection combining all pedigree phenotypic information with genotypic data in one step: an example using broiler chickens. J Anim Sci 89:23–28

Clark SA, van der Werf J (2013) Genomic best linear unbiased prediction (gBLUP) for the estimation of genomic breeding values. In: Gondro C, van der Werf J, Hayes B (eds) Genome-wide association studies and genomic prediction. Springer, New York

Cleveland MA, Hickey JM (2013) Practical implementation of cost-effective genomic selection in commercial pig breeding using imputation. J Anim Sci 91:3583–3592

ENCODE Project Consortium (2012) An integrated encyclopedia of DNA elements in the human genome. Nature 489:57–74. https://doi.org/10.1038/nature11247

Falconer D, Mackay TFC (1996) Introduction to quantitative genetics. Longman, Edinburgh

Fernando RL, Garrick D (2013) Bayesian methods applied to GWAS. In: Gondro C, van der Werf J, Hayes B (eds) Genome-wide association studies and genomic prediction. Springer, New York

Groenen MAM, Schook LB, Archibald AL (2011) Pig genomics. In: Rothschild MF, Ruvinsky A (eds) The genetics of the pig, 2nd edn. CAB International, Wallingford, UK, p 496. https://doi.org/10.1079/9781845937560.0000

Habier D, Fernando RL, Dekkers JCM (2007) The impact of genetic relationship information on genome-assisted breeding values. Genetics 177:2389–2397

Hangauer MJ, Vaughn IW, McManus MT (2013) Pervasive transcription of the human genome produces thousands of previously unidentified long intergenic noncoding RNAs. PLoS Genet 9:e1003569. https://doi.org/10.1371/journal.pgen.1003569

Hausser J, Zavolan M (2014) Identification and consequences of miRNA-target interactions-beyond repression of gene expression. Nat Rev Genet 15:599–612. https://doi.org/10.1038/nrg3765

Horton BH, Banks R, Van der Werf JHJ (2015) Industry benefits from using genomic information in two- and three-tier sheep breeding systems. Anim Prod Sci 55:437–446

Huang Y, Hickey JM, Cleveland MA, Maltecca C (2012) Assessment of alternative genotyping strategies to maximize imputation accuracy at minimal cost. Genet Sel Evol 44:25

Ibañez N, Blasco A (2011) Modifying growth curve parameters by multitrait genomic selection. J Anim Sci 89:661–668

Ibáñez-Escriche N, Forni S, Noguera JL, Varona L (2014) Genomic information in pig breeding: science meets industry needs. Livestock Sci 166:94–100

Jonas E, de Koning DJ (2015) Genomic selection needs to be carefully assessed to meet specific requirements in livestock breeding programs. Anim Front 6:1–8

Knap PW, Wang L (2012) Pig breeding for improved feed efficiency. In: Patience JF (ed) Feed efficiency in swine. Wageningen Academic Publishers, Wageningen

Knol EF, Nielsen B, Knap PW (2016) Genomic selection in commercial pig breeding. Anim Front 6:15–22

Lande R, Thompson R (1990) Efficiency of marker-assisted selection in the improvement of quantitative traits. Genetics 124:743–756

Legarra A, Aguilar I, Misztal I (2009) A relationship matrix including full pedigree and genomic information. J Dairy Sci 92:4656–4663

Libri D (2015) Sleeping beauty and the beast (of pervasive transcription). RNA 21:678–679. https://doi.org/10.1261/rna.050948.115

Lillehammer M, Meuwissen THE, Sonesson AK (2013) Genomic selection for two traits in a maternal pig breeding scheme. J Anim Sci 91:3079–3087

Lund MS, Su G, Janss L, Guldbrandtsen B, Brøndum RF (2014) Genomic evaluation of cattle in a multi-breed context. Livestock Sci 166:101–110

Mattick JS (2011) The central role of RNA in human development and cognition. FEBS Lett 585:1600–1616

Meuwissen TH, Hayes BJ, Goddard ME (2001) Prediction of total genetic value using genome-wide dense marker maps. Genetics 157:1819–1829

Misztal I, Legarra A, Aguilar I (2009) Computing procedures for genetic evaluation including phenotypic, full pedigree and genomic information. J Dairy Sci 92:4648–4655

Neale DB et al (2014) Decoding the massive genome of loblolly pine using haploid DNA and novel assembly strategies. Genome Biol 15:R59. https://doi.org/10.1186/gb-2014-15-3-r59

Nicholas FW, Smith C (1983) Increased rates of genetic change in dairy cattle by embryo transfer and splitting. Anim Prod Sci 36:341–353

Rolf MM, Decker JE, Mckay SD, Tizioto PC, Branham KA, Whitacre LK, Hoff JL, Regitano LCA, Taylor JF (2014) Genomics in the United States beef industry. Livestock Sci 166:84–93

Rupp R, Mucha S, Larroque H, McEwan J, Conington J (2016) Genomic application in sheep and goat breeding. Anim Front 6:39–44

Schaeffer LR (2006) Strategy for applying genome-wide selectionin strategy for applying genome-wide selection in dairy cattle. J Anim Breed Genet 123:218–223

Shumbusho F, Raoul J, Astruc JM, Palhiere I, Lemarié S, Fugeray-Scarbel A, Elsen JM (2016) Economic evaluation of genomic selection in small ruminants: a sheep meat breeding program. Animal 6:1033–1041

Silver LM (1995) Mouse genetics. Oxford University Press, Bar Harbor, Maine

Simianier H (2016) Genomic and other revolutions why some technologies are quickly adopted and others are not. Anim Front 6:53–58

Smith C, Smith DJ (1993) The need for close linkages in markers-assisted selection for economic meritin livestock. Anim Breed Abst 61:197–204

Soller M (1978) The use of loci associated with quantitative traits in dairy cattle improvement. Anim Prod 27:133–139

Van Eenennaam AL, Weigel KA, Young AE, Matthew AC, Dekkers JCM (2013) Applied animal genomics: results from the field. Annu Rev Anim Biosci 2:9.1–9.35

Van Raden PM, O'Connell JR, Wiggans GR, Weigel KA (2011) Genomic evaluations with many more genotypes. Genet Sel Evol 43:10

Vitezica ZG, Varona L, Elsen JM, Misztal I, Herring W, Legarra A (2016) Genomic BLUP including additive and dominant variation in purebreds and F1 crossbreds, with an application in pigs. Genet Sel Evol 48:6

Warren WC et al (2017) A new chicken genome assembly provides insight into avian genome structure. G3 7:109–117. https://doi.org/10.1534/g3.116.035923

Wiggans GR, Cole JB, Hubbard SM, Sonstegard TS (2017) Genomic selection in dairy cattle: the USDA experience. Annu Rev Anim Biosci 5:309–327

Wolc A, Zhao HH, Arango J, Settar P, Fulton JE, O'Sullivan NP, Preisinger R, Stricker C, Habier D, Fernando RL, Garrick DJ, Lamont SJ, Dekkers JCM (2015) Response and inbreeding from a genomic selection experiment in layer chickens. Genet Sel Evol 47:59

Wolc A, Kranis A, Arango J, Settar P, Fulton JE, O'Sullivan NP, Avendano A, Watson KA, Hickey JM, De los Campos G, Fernando RL, Garrick DJ, Dekkers JCM (2016) Implementation of genomic selection in the poultry industry. Anim Front 6:23–31

Won KJ et al (2013) Comparative annotation of functional regions in the human genome using epigenomic data. Nucleic Acids Res 41:4423–4432. https://doi.org/10.1093/nar/gkt143

Wright MW (2014) A short guide to long non-coding RNA gene nomenclature. Hum Genomics 8:7. https://doi.org/10.1186/1479-7364-8-7

Xu J, Zhang J (2016) Are human translated pseudogenes functional? Mol Biol Evol 33:755–760. https://doi.org/10.1093/molbev/msv268

Yerle M et al (1995) The PiGMaP consortium cytogenetic map of the domestic pig (Sus scrofa domestica). Mamm Genome 6:176–186

Embryo Biopsies for Genomic Selection

5

Erik Mullaart and David Wells

Abstract

Embryo genomic selection (preimplantation genetic screening) is increasingly being used to select the best embryos within cattle breeding programs. The procedure starts with the collection of a few cells (biopsy) from each of the embryos before they are individually cryopreserved. The biopsy samples are then genotyped, and the genomic estimated breeding value for each embryo is calculated from prediction equations. These are based on algorithms developed from large reference populations of previously genotyped and phenotyped animals. Based on the genomic estimated breeding value, a decision is made whether to thaw and transfer the embryo or not. Due to the recent availability of low-density bovine single-nucleotide polymorphism (SNP) microarrays, this method is now cost effective. The data in this review describe field results and show that the breeding values calculated from the embryo biopsies are reliable enough for selection. Importantly, the embryo manipulation associated with the procedure only has a very limited negative effect on the resulting pregnancy rate. The method can also be used to prevent the transfer of embryos that are carriers of known recessive lethal genetic defects or other chromosomal aberrations. Therefore it can be concluded that embryo genomic selection can be used in breeding programs to accelerate the rate of genetic gain compared to animal-based genomic selection due to an increased selection intensity among full- and half-sib embryos. Although this review only describes results in dairy cattle, embryo genomic selection can also be used in beef cattle and other livestock species where accurate genomic prediction equations exist.

E. Mullaart (✉) · D. Wells
CRV BV, Arnhem, The Netherlands

AgResearch, Ruakura Research Centre, Hamilton, New Zealand
e-mail: Erik.mullaart@crv4all.com; david.wells@agresearch.co.nz

© Springer International Publishing AG, part of Springer Nature 2018   81
H. Niemann, C. Wrenzycki (eds.), *Animal Biotechnology 2*,
https://doi.org/10.1007/978-3-319-92348-2_5

5.1   Introduction

Genomic selection is increasingly being used in dairy cattle breeding programs all over the world. Due to very large reference (training) populations in Europe (~40,000), the United States (~60,000), and New Zealand (~6000), reliabilities of the genomic breeding values approach 65–75% for most traits.

Genomic selection is now commonly used to select young animals just after birth or to identify the best bulls and bull mothers to generate future sires for the artificial insemination industry. These genomic assessments and selections are typically conducted on existing, live animals. However, selection based on genomics can also be performed on the early embryo, before transplantation to a recipient female. The advantage of this approach is that a large number of full- and half-sib embryos can be easily produced and only the best embryos of the desired sex and genotype are selected for transfer. In addition, embryos carrying known recessive lethal genetic defects can also be detected and excluded, thereby lowering the number of carrier animals in the population.

The optimal use of genomic selection in an intensive embryo breeding program will accelerate the rate of genetic progress by further increasing selection intensity and reducing generation interval (especially with embryos produced from juvenile animals). The herd improvement cooperative CRV (Arnhem, the Netherlands) has a significant European Holstein Friesian breeding program and produced around 8000 embryos in 2015 (4000 in vivo-flushed embryos and 4000 in vitro-produced [IVP] embryos). Increasingly CRV and other breeding companies are utilizing embryo genotyping to select those embryos possessing the greatest genetic merit for transfer. This is especially in situations where the numbers of recipients are limited, and selecting only the best embryos based on genomics for transfer offers considerable economic advantages.

5.2   Biopsy Methods

In order to perform a DNA test on an embryo, a sample of cells is required. There is a compromise between taking enough cells to enable an accurate DNA test without reducing the developmental competency of the embryo and thus not decreasing its potential to establish a pregnancy. In principle, preimplantation-stage embryos from the two-cell stage onward can be biopsied. However, for practical reasons in cattle, typically only morula- or blastocyst-stage embryos are biopsied for DNA testing. These stages possess a greater number of cells (32–150) and strike a balance, whereby a biopsy of a few cells can be obtained without overly compromising development of the remaining embryo.

There are different methods available for obtaining biopsies from early embryos. Two of the more common methods are (1) the blade biopsy method to cut a portion of a compacted morula or the polar trophectoderm from blastocyst-stage embryos and (2) the needle biopsy method to aspirate cells from cleavage- or morula-stage embryos. Both methods have their advantages and disadvantages

(Mullaart 2002). Some groups (MasterRind, Personal Communication) have good results with needle biopsies, whereas others (e.g., CRV and Midatest, France, Personal Communication) have better results with the blade biopsy method. In general, the success of the biopsy depends on the training and experience of personnel in specific methods. In addition, especially in cattle breeding where sometimes large numbers of embryos have to be biopsied within a limited time, the practicality and labor intensity of the method are important factors. The ultimate choice of method also depends on the stage of the embryo available to be biopsied. The needle biopsy method is typically more suitable for less-advanced embryos (up to morula stage), where cell adhesion is not as strong. In contrast, the blade biopsy method is better suited for more advanced compacted morula- and blastocyst-stage embryos (Mullaart 2002). Also, the quality of the embryo is an important selection criterion for biopsies. At CRV, only grade 1 quality embryos as categorized by the International Embryo Technology Society (IETS) (Robertson and Nelson 1998) are used for biopsy. With IETS grade 2 embryos, it is commonly observed that the remaining embryo deteriorates after taking the biopsy and has lower viability.

As mentioned above, it is important to collect sufficient cells for DNA testing but not so many that compromises embryo competence. Following nuclear staining with DAPI and counting in the fluorescence microscope, the average blade biopsy obtained from blastocyst-stage embryos possessed about 15 cells, but the variation was large (between 8 and 40 cells). In Fig. 5.1, an example of a nine-cell biopsy is illustrated. There does not appear to be any relationship between the number of cells in the biopsy and the embryo stage, probably indicating operator variation.

There is continued debate whether the cells in the biopsy are a representative sample of the entire embryo. Indeed, it has been reported that a large percentage of IVP embryos, and the cells within the trophectoderm in particular, are mixoploid (Viuff et al. 2002). This is especially relevant for blade biopsies that are obtained from the trophectoderm of blastocysts.

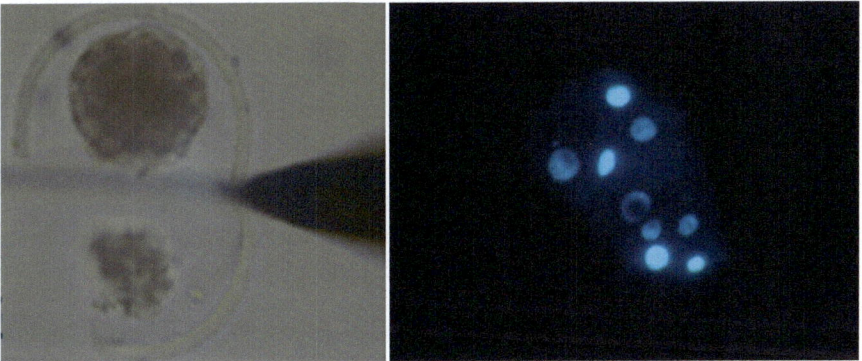

Fig. 5.1 Example of a blade biopsy (left) and DAPI stained biopsy (right) comprising nine trophectodermal cells obtained from a blastocyst

5.3   DNA Amplification

The embryo biopsy can be utilized for various DNA analyses. For instance, it can be used for sex determination, or the identification of specific candidate alleles. But it can also be used for more sophisticated whole-genome analyses, such as genotyping with thousands of single-nucleotide polymorphism (SNP) markers or DNA sequencing. Sex determination by PCR (Bredbacka 1998) is a relatively sensitive assay that only requires a very limited amount of starting material (around ten cells). Blastocyst biopsies therefore provide a sufficient amount of DNA template for the PCR to be immediately performed on the cell sample (and are often done so directly in the field). However, other assays, such as the Illumina SNP chip-based genotyping platform (Illumina, San Diego, CA, USA), require at least 50 ng of DNA. Assuming a cellular DNA content of 4 pg, this corresponds to approximately 12,500 cells. Since the biopsy contains around a thousand-fold less DNA, a "pre-amplification" step is needed before SNP chip-based genotyping can be performed.

This pre-amplification can be done in two general ways: (1) either by culturing the biopsy for several days or (2) by enzymatic pre-amplification. At present, however, only the enzymatic pre-amplification method is used routinely to obtain SNP genotypes from an embryo biopsy. The in vitro culture of embryo biopsies is not yet consistently reliable for cellular amplification in most cases (Ramos-Ibeas et al. 2014; Shojaei Saadi et al. 2014).

For enzymatic pre-amplification, there are various protocols, many of which are based on isothermal multiple displacement amplification (MDA) using Phi29 polymerase. For an overview of different pre-amplification methods, see Shojaei Saadi et al. (2014). It should be noted that the process of taking a small biopsy from an embryo, performing the enzymatic pre-amplification, followed by SNP chip-based genotyping, is a technically challenging process that is error prone. The embryo biopsy (comprising only a few cells) can very occasionally be lost through handling mistakes, but also pre-amplification of the minute amount of template DNA can introduce errors (Ponsart et al. 2013). A major issue with enzymatic pre-amplification is the so-called allele drop-out (ADO) or loss of heterozygosity. This is where only one of the two heterozygous alleles (either paternal or maternal) is pre-amplified and this leads to a false homozygous call at this locus. While ADO tends to be the most common genotyping error, allele drop-ins (gain of heterozygosity) and homozygosity reversal (an erroneous shift from one homozygous genotype to another), although rare, can also occur dependent on the whole-genome amplification method used (Shojaei Saadi et al. 2014).

The whole-genome amplification method used routinely at CRV is the Single-Cell Repli-g Kit (Qiagen, the Netherlands). It has provided the most consistent results, leading to at least a 10,000–20,000-fold amplification of the DNA. This generates sufficient material for further downstream analysis (i.e., SNP chip-based DNA genotyping). However, in a project at AgResearch in New Zealand comparing several different kits, the Illustra GenomiPhi V2 DNA Amplification Kit (GE Healthcare Life Sciences, New Zealand) proved to be superior among those tested. The difference between laboratories was most likely due to a subtle differences in the consistency of the biopsy size (related to sample quality) and experience of the technicians in particular methods.

5.4   DNA Genotyping

After pre-amplification, enough DNA is generated to perform complex DNA analyses, such as on the Illumina genotyping platform with low (7–10K)-, medium (50K)-, or high (777K)-density SNP chips, according to standard protocols. In terms of quality control, the genotyping results are first checked for their call rate. This is a measure for the fraction of the markers on the chip that give a result. So, for instance, a call rate of 0.9 means that genotypes were assigned for 90% of the SNP markers on the chip. As shown in Table 5.1, the call rate and error rate of the genotypes are affected by the number of cells in the biopsy.

The more cells present in the sample at the start of the pre-amplification, the better the genotyping results are. This is most likely caused by fewer errors during the pre-amplification (e.g. less ADO). With more amplification required, a one-cell sample is clearly more sensitive to this than a larger (10–15 cell) sample. With a trophectoderm (blade) biopsy, the average call rate is 0.88 with an error rate of ~1%.

Data from the Fisher et al. (2012) study also indicate that there is a clear inverse correlation between the error rate (measured as the difference between the biopsy compared to the remaining embryo) and the call rate with the 7K density SNP chip (Fig. 5.2).

Table 5.1   Effect of biopsy size on genotype result (modified from Fisher et al. 2012)

Sample	Call rate	Replication error (%)[a]
One-cell biopsy[b]	0.78 ± 0.06	7.8 ± 3.5
Three-cell biopsy[b]	0.86 ± 0.03	2.9 ± 1.7
Trophectoderm biopsy (~10–15 cells)[c]	0.88 ± 0.05	1.1 ± 1.5
Bisected blastocyst[c]	0.94 ± 0.04	0.1 ± 0.1

[a]Based on two or three samples from the same embryo
[b]Based on 50K chip
[c]Based on 7K chip

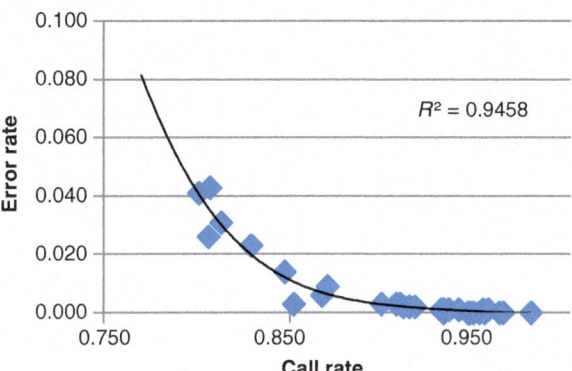

Fig. 5.2   Correlation between call rate vs. error rate

Table 5.2 Mean call rate and percentage of biopsies with a call rate above 0.85 using 10K and 50K SNP chips

Chip type	Number of embryo biopsies	Mean call rate (±SD)	Percentage of biopsies with a call rate >0.85
10K	514	0.84 ± 0.18	80
50K	1378	0.90 ± 0.14	83

From Fig. 5.2 it is also observed that when the call rate is greater than 0.85, the error rate is less than 1%. Comparable results were also obtained using both the 50K and 777K SNP chips (results not shown).

Based on these results, the standard procedure at CRV is to only use genotype results where the call rate is greater than 0.85. A high call rate is a proxy for an inherently lower error rate with SNP genotyping. The effect of call rate on the quality of genotypes and the subsequent calculation of genomic estimated breeding values is also shown in Sect. 5 below.

Over recent years, almost 2000 embryo biopsies have been genotyped by CRV. Initially, the 50K SNP chips were used, but since the cheaper 10K chips became available, they are now used extensively. The results from genotyping these embryos are shown in Table 5.2.

There were no significant differences in the call rates observed between the 10K and 50K SNP chips. As can be seen, the average call rate is between 0.84 and 0.90, and at least 80% of the biopsies gave a call rate above 0.85. Considering that this method is technically challenging (starting with only a few cells), the results are very acceptable and can be used to calculate genomic estimated breeding values for the selection of embryos in commercial breeding programs.

5.5   Breeding Value Estimation

Genotypes from the medium- and high-density SNP chips can be used directly in the calculation of genomic estimated breeding values. However, genotypes from the low-density chip (7–10K) must first be imputed to a reference set consisting of 50K SNPs by using a combination of LinkPHASE, DAGPHASE (for both software packages see Druet and Georges 2010), and Beagle (Browning and Browning 2007). In situations where both parents of the embryo are already genotyped, imputation accuracies are very high (~99%). Genomic evaluation is described by de Roos et al. (2009), where the core of the evaluation is replaced by the method described by Calus et al. (2014). Genomic breeding values are estimated using the EuroGenomics reference database containing more than 35,000 bulls (see also Lund et al. 2011) for 48 different traits including production, health, and fertility, among others.

The genotypes with call rates above 0.85 are used to calculate genomic breeding values, and based on those breeding values, the embryos can be either selected for transfer or discarded. To demonstrate the accuracy of genomic breeding values calculated from embryo biopsies following SNP genotyping, we compared them to the

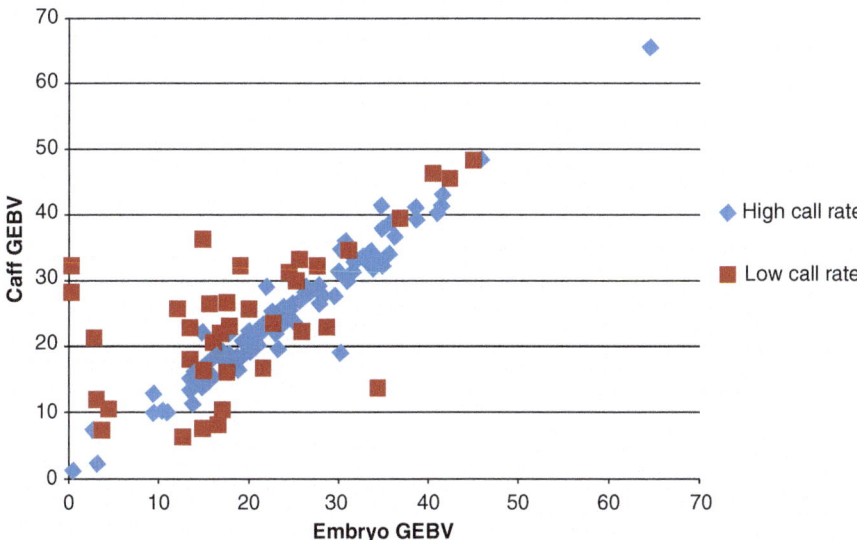

Fig. 5.3 Correlation between the genomic estimated breeding value (GEBV) for kilograms of milk protein (based on 305-day production) in genotyped embryo biopsies with that following genotyping the corresponding calf

corresponding breeding values determined from tissue DNA obtained from each of the resulting calves. In Fig. 5.3, the genomic breeding values for kilograms of milk protein based on genotypes of the embryos is positively correlated with those values obtained by genotyping the corresponding animal after birth. The results show that the correlation is very high ($r^2 = 0.95$) when only genotypes with a call rate above 0.85 are included. If embryo samples with lower call rates are included in the analysis, the correlation is considerably lower ($r^2 = 0.71$). Note that in this case, some embryos with lower call rates were indeed transferred because of their potential value for the breeding program with respect to certain other traits (e.g., polled, red factor, etc.).

The lower correlation for samples with a low call rate was expected, since it was previously shown that lower call rates are associated with a higher error rate and higher ADO (Fig. 5.2). It is also in complete agreement with the results from other groups that demonstrate such correlations to be generally above 0.95 when call rates are high (Ponsart et al. 2013; Shojaei Saadi et al. 2014).

Based on the genomic estimated breeding values, embryos can been ranked in order of superiority and only the highest selected and subsequently transferred. In Fig. 5.4, an example is provided of the genomic breeding values obtained from the different full-sib embryos within a single in vivo flush on Day 7. When no information is available on the genomics of the embryo, all embryos recovered within a single flush have the same "expected" parental average breeding value. However, after genotyping the individual embryos, it became clear that in flush 1, embryo C is predicted to be the most superior for kilogram milk and should be the one selected

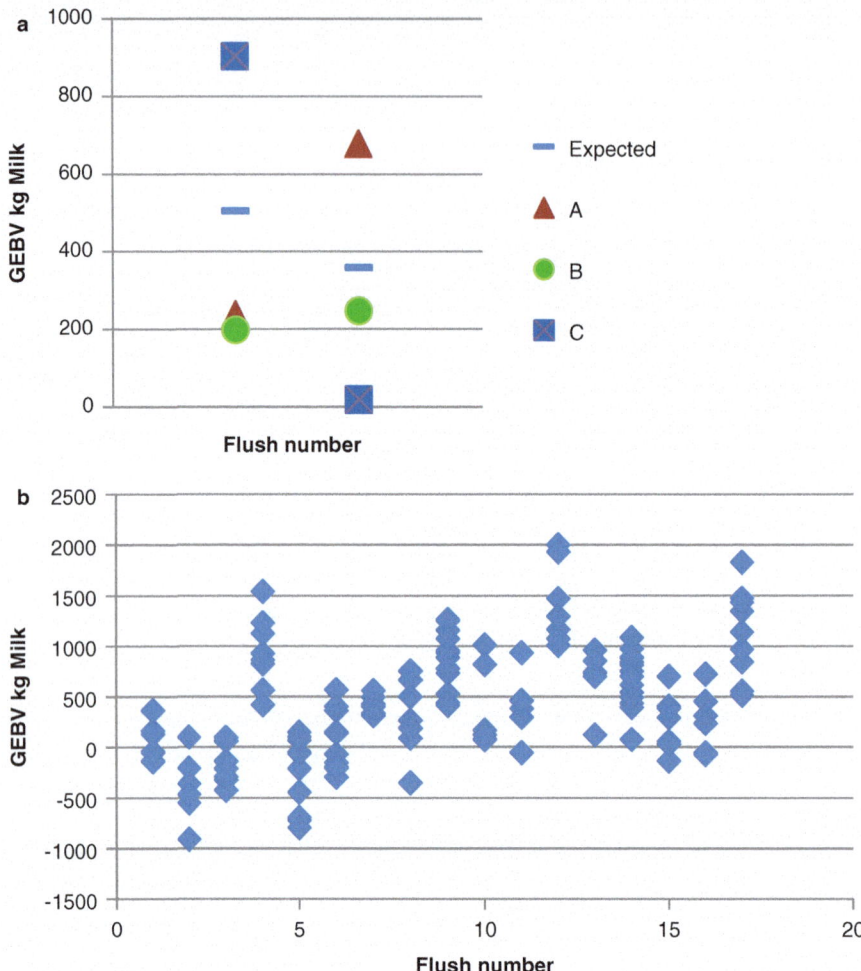

Fig. 5.4 Genomic estimated breeding values (GEBV) for kilogram milk predicted among full-sib in vivo-derived embryos. (**a**) Two flushes, each with three different embryos (A, B, and C), with their expected parental average and GEBV. (**b**) Data from an additional 17 flushes, each with at least six embryos, with their respective estimated GEBV

for transfer (Fig. 5.4a). Likewise in flush 2, embryo A was identified as the best for this particular trait among the three full-sib embryos recovered.

Analysis of additional flushes, with more than six genotyped embryos per flush, showed that there is considerable variation among full-sib embryos (Fig. 5.4b). The average difference within a flush between the embryo with the highest genomic breeding value and the expected breeding value (based on only the parental average) is 446 kg milk (on a yearly production of 8500 kg milk). Within a flush, the average difference between the highest and the lowest is 875 kg milk. This indicates that when all of the embryos cannot be transferred due to limited availability of recipients or financial constraints, selection of embryos based on genomics can significantly improve genetic progress and reduce embryo transfer costs.

Besides calculating a genomic breeding value, the SNP genotyping data can also be used for other applications. The CRV-customized chip also contains SNPs for sex determination, specific traits, several milk proteins, and certain diseases. Based on X- and Y-chromosome-specific markers, the sex of the embryo can be determined. This sex determination can assist decisions in breeding programs to specifically transfer embryos of the desired gender. For instance, the male embryos can be kept within a company's breeding program and transferred at specific recipient farms, while the females can be sold to the farmer. Also, other simple traits including coat color, polled, and certain milk protein variants (e.g., kappa casein A and B, beta-lactoglobulin A and B, etc.) can be determined for each embryo. This can be important for breeding programs focused on specific traits.

In addition, SNPs for various genetic defects (e.g., BLAD, CVM, lethal haplotypes, etc.) are present on the chip. Based on these, the status with regard to several known genetic diseases for each embryo can be determined. This is very important for two reasons. First, it allows the use of top sires or cows that are otherwise carriers of a certain genetic disease. In such cases, the embryos from these matings can be screened, and only those free of the genetic mutation are transferred. Secondly, it will result in higher pregnancy results, since embryos that are homozygous for certain lethal mutations, and will not survive in vivo, are not transferred.

5.6   Pregnancy Rates of Biopsied Embryos

It is obviously very important to not compromise the embryo by removing too many cells in the biopsy (see Sect. 2). Since the introduction of embryo genotyping at CRV, we have transferred, in excess of 1000 biopsied, in vivo-derived embryos following conventional slow freezing. The freezing of the embryo after biopsy is necessary since the current procedure entailing the pre-amplification of DNA from the biopsy, genotyping, and the calculation of genomic estimated breeding values may take several days. Moreover, in some countries there may be a greater requirement for seasonal calving patterns necessitating cryopreservation of biopsied embryos produced over several months and subsequent transfer of selected embryos to generate spring-born calves. The pregnancy results after single transfer of biopsied frozen embryos, compared to normal intact (non-biopsied) embryos, are shown in Table 5.3.

These results indicate a significant decrease in pregnancy rate after biopsy and freezing. Nevertheless, the 46% pregnancy rate for biopsied frozen in vivo embryos is still very acceptable in the field.

In contrast to in vivo embryos, the pregnancy results following conventional slow freezing of biopsied IVP embryos in ethylene glycol plus sucrose are currently less than ideal. Although our results show that there is negligible impact of slow freezing on the development of intact IVP blastocysts to Day 60 of gestation compared to nonfrozen controls (29/64 = 45% vs. 27/56 = 48%, respectively), the

Table 5.3 Pregnancy results of frozen normal and biopsied in vivo embryos

	Number of embryos transferred	Pregnancy rate[a]
Not biopsied	13,067	54%[b]
Biopsied embryos	1190	46%[b]

[a]Pregnancy as determined by scanning 5 months after transfer
[b]$P < 0.05$

cryosurvival of slow-frozen blastocysts following trophectoderm biopsy is typically much lower (9/54 = 17%; Oback et al. 2017).

The ultrarapid cooling and warming rates afforded by various embryo vitrification methods appear to provide a potential solution. Preliminary results with biopsied vitrified IVP blastocysts indicate embryo survival on Day 65 (43/96 = 45%) to be significantly better than following slow freezing and comparable to fresh control IVP embryos (Fisher et al. 2012; Oback et al. 2017). More experiments are necessary to prove this in practice on a large scale, especially with more user-friendly methods for field situations.

In implementing an embryo genomic selection program, it is critical to optimize each of the manipulation steps to maximize subsequent embryo survival. There is additional expense associated with biopsy and genotyping, and although savings are made by not transferring embryos of lower breeding value, it is important to increase the probability of generating a calf from each selected genotype.

5.7 Optimizing Survival of Genomically Selected Embryos

While much research aims to identify noninvasive biomarkers predictive of oocyte and embryo competence (e.g., metabolomics in spent embryo culture media), a physical sample of the embryo enables direct determination of not only its genotype but also its karyotype, epigenotype, and transcriptome that might all be related to developmental outcomes (Orozco-Lucero and Sirard 2014). In this regard, embryonic cell biopsies are superior and more versatile compared to genomic analyses with fragmented DNA collected from blastocoelic fluid, despite this being a less invasive procedure (Zhang et al. 2016).

Based on the genotype profile consisting of thousands of DNA markers spread evenly throughout the genome, SNP microarray analysis, and ultimately methods utilizing next generation sequencing, enables a form of molecular karyotyping from biopsies. The extent of chromosomal anomalies in the blastomeres of early mammalian embryos is now being revealed by SNP-based karyotyping (Destouni et al. 2016; Treff et al. 2016). However, these SNP-based karyotypes are complicated by any heterogeneity within the sample. This is a particular issue in embryos that are commonly mixoploid (Viuff et al. 2002) and sometimes even chimeric (Garcia-Herreros et al. 2010). It has been reported that on average 11% of trophectoderm cells were polyploid in 96% of Day 7–8 bovine blastocysts (Viuff et al. 2002). While the frequency was less in the inner cell mass, the issue of mixoploidy raises the question about how representative a small trophectoderm biopsy is of the whole embryo. Despite this concern, the results presented in Sect. 5 show that in practice there is a very high correlation between the breeding values calculated based on a few embryonic cells compared to those from the corresponding calf, especially when imputation is utilized. Furthermore, the developmental consequences of relatively low levels of mixoploidy in blastocysts remain equivocal (King et al. 2006). Nevertheless, chromosome screening avoids the transfer of aneuploid embryos that are unlikely to result in a successful pregnancy (Scott et al. 2013).

The embryo biopsy can also be used to directly determine gene expression profiles that may be predictive of in vivo survival and so assist selection decisions on which embryos to transfer. The possibilities and challenges to identify these molecular markers have been reviewed elsewhere (Bermejo-Alvarez et al. 2011; Orozco-Lucero and Sirard 2014). Studies have identified genes that were either up- or downregulated in biopsies (representing a portion of both the inner cell mass and trophectoderm) that were retrospectively pooled, depending on the subsequent pregnancy outcome (El-Sayed et al. 2006; Ghanem et al. 2011). In the situation where only a trophectoderm biopsy is taken, utilizing informative lineage-specific transcripts prognostic of developmental fate, as well as sharing the precious sample for genomic analyses, remains a considerable challenge for reproductive biotechnologies.

With the present reliance on subjective morphological assessment of embryo quality, the relatively low chance of obtaining a live-born calf from each individual genomically selected embryo is a significant limitation. While there is a negligible decrease in embryo survival as a result of biopsy and vitrification (see Sect. 6), each IVP embryo still only has around a 45% chance of resulting in a viable calf. Future improvements in IVP systems aim to increase the developmental competence of embryos and, combined with identifying competent recipients (McMillan and Donnison 1999), increase pregnancy rates toward some biological limit or to at least result in more consistent outcomes (Vajta et al. 2010). However, even with ideal recipients, some embryos will have an inherently poorer chance of survival due to chromosomal errors or aberrant developmental programming, incompatible with a viable pregnancy. Notwithstanding these cases, options to multiply each genomically selected embryo may increase the chance of obtaining a live-born calf, or calves if required, from that desired genotype. This might include simply bisecting the remaining embryo following biopsy and cryopreservation and then transferring both demi-embryos (Oback et al. 2017). Other methods include utilizing nuclear transfer to multiply elite embryos from donor blastomeres (Misica-Turner et al. 2007) or embryonic cultures derived from either the biopsy (Ramos-Ibeas et al. 2014) or (pluripotent) cultures of the inner cell mass (Verma et al. 2013). An alternative approach may be to transplant embryos and recover the resulting fetuses at a few weeks of age, in order to establish fetal cell lines for subsequent genotyping. Somatic cell nuclear transfer can then be used with selected frozen cell stocks to potentially obtain large numbers of calves of the desired genotype from the original embryo (Kasinathan et al. 2015).

Conclusions

It can be concluded that embryo genotyping from a representative biopsy sample has clear advantages for cattle breeding programs in the following ways:

(a) Predicting the gender of the embryo
(b) Predicting the status for specific phenotypes, for example, coat color, polled, milk protein variants, etc.
(c) Predicting the status for genetic disease
(d) Predicting the genomic breeding value for economically important traits

Predicting genomic breeding values is especially relevant where there is a shortage of recipients and a selection has to be made in the embryos that can be transferred. Furthermore, in cases where embryos are transferred on a recipient farm with a fixed capacity, knowing which embryos are the best genotypes is very important to accelerate the rates of genetic gain achievable and reducing the costs associated with producing elite sires.

The combination of genomic and reproductive technologies provides the option to produce large numbers of low-cost IVP embryos from multifactorial in vitro matings, genotype them, and then only transfer the best. Thus, the current rate of genetic gain can be further accelerated by increasing the selection intensity among multiple full- and half-sib embryos produced from elite parents, compared to the single progeny born from conventional breeding with animal-based genomic selection (Ponsart et al. 2013). Another advantage of embryo genotyping is that it allows for the careful use of bulls that are in fact carriers of a genetic disease. After embryo genotyping, only those embryos free of particular disease-causing or lethal mutations are transferred, essentially recovering otherwise valuable genetics.

Finally, to increase the chance of obtaining a calf (or calves, if required) the selected embryo might be bisected or otherwise multiplied following nuclear transfer of donor blastomeres or embryonic or fetal cell cultures derived from the original embryo. However, the current efficiencies of these procedures still need to be increased for routine commercial use.

References

Bermejo-Alvarez P, Pericuesta E, Miranda A et al (2011) New challenges in the analysis of gene transcription in bovine blastocysts. Reprod Domest Anim 46(Suppl 3):2–10

Bredbacka P (1998) Recent developments in embryo sexing and its field application. Reprod Nutr Dev 38(6):605–613

Browning SR, Browning BL (2007) Rapid and accurate haplotype phasing and missing-data inference for whole-genome association studies by use of localized haplotype clustering. Am J Hum Genet 81(5):1084–1097

Calus MP, Schrooten C, Veerkamp RF (2014) Genomic prediction of breeding values using previously estimated SNP variances. Genet Sel Evol 46:52

De Roos APW, Schrooten C, Mullaart E et al (2009) Genomic selection at CRV. Interbulletin 39:47–50

Destouni A, Zamani Esteki M, Catteeuw M et al (2016) Zygotes segregate entire parental genomes in distinct blastomere lineages causing cleavage-stage chimerism and mixoploidy. Genome Res 26(5):567–578

Druet T, Georges M (2010) A hidden Markov model combining linkage and linkage disequilibrium information for haplotype reconstruction and quantitative trait locus fine mapping. Genetics 184(3):789–798

El-Sayed A, Hoelker M, Rings F et al (2006) Large-scale transcriptional analysis of bovine embryo biopsies in relation to pregnancy success after transfer to recipients. Physiol Genomics 28(1):84–96

Fisher PJ, Hyndman DL, Bixley MJ et al (2012) Potential for genomic selection of bovine embryos. Proc N Z Soc Anim Prod 72:156–158

Garcia-Herreros M, Carter TF, Villagmez DAF et al (2010) Incidence of chromosomal abnormalities in bovine blastocysts derived from unsorted and sex-sorted spermatozoa. Reprod Fertil Dev 22(8):1272–1278

Ghanem N, Salilew-Wondim D, Gad A et al (2011) Bovine blastocysts with developmental competence to term share similar expression of developmentally important genes although derived from different culture environments. Reproduction 142(4):551–564

Kasinathan P, Wei H, Xiang T et al (2015) Acceleration of genetic gain in cattle by reduction of generation interval. Sci Rep 5:8674

King WA, Coppola G, Alexander B et al (2006) The impact of chromosomal alteration on embryo development. Theriogenology 65(1):166–177

Lund MS, Roos AP, Vries AG et al (2011) A common reference population from four European Holstein populations increases reliability of genomic predictions. Genet Sel Evol 43:43

McMillan WH, Donnison MJ (1999) Understanding maternal contributions to fertility in recipient cattle: development of herds with contrasting pregnancy rates. Anim Reprod Sci 57(3–4):127–140

Misica-Turner PM, Oback F, Eichenlaub M et al (2007) Aggregating embryonic but not somatic nuclear transfer embryos increases cattle cloning efficiency. Biol Reprod 76:268–278

Mullaart E (2002) Biopsying and genotyping cattle embryos. In: Van Soom A, Boerjan M (eds) Assessment of mammalian embryo quality. Springer, Netherlands, pp 178–194

Oback FC, Wei J, Popovic L et al (2017) Blastocyst bisection to multiply biopsied and vitrified bovine embryos. Reprod Fertil Dev 29:154

Orozco-Lucero E, Sirard M-A (2014) Molecular markers of fertility in cattle oocytes and embryos: progress and challenges. Anim Reprod 11(3):183–194

Ponsart C, Le Bourhis D, Knijn H et al (2013) Reproductive technologies and genomic selection in dairy cattle. Reprod Fertil Dev 26(1):12–21

Ramos-Ibeas P, Calle A, Pericuesta E et al (2014) An efficient system to establish biopsy-derived trophoblastic cell lines from bovine embryos. Biol Reprod 91(1):15, 1–10

Robertson I, Nelson R (1998) Certification and identification of the embryo. In: Stringfellow DA, Seidel SM (eds) Manual of the International Embryo Transfer Society, 3rd edn. International Embryo Transfer Society, Illinois, pp 103–134

Scott RT, Upham KM, Forman EJ et al (2013) Blastocyst biopsy with comprehensive chromosome screening and fresh embryo transfer significantly increases in vitro fertilization implantation and delivery rates: a randomized controlled trial. Fertil Steril 100(3):697–703

Shojaei Saadi HA, Vigneault C, Sargolzaei M et al (2014) Impact of whole-genome amplification on the reliability of pre-transfer cattle embryo breeding value estimates. BMC Genomics 15(889):1–16

Treff NR, Thompson K, Rafizadeh M et al (2016) SNP array-based analyses of unbalanced embryos as a reference to distinguish between balanced translocation carrier and normal blastocysts. J Assist Reprod Genet 33(8):1115–1119

Vajta G, Rienzi L, Cobo A et al (2010) Embryo culture: can we perform better than nature. Reprod Biomed Online 20:453–469

Verma V, Huang B, Kallingappa PK et al (2013) Dual kinase inhibition promotes pluripotency in finite bovine embryonic cell lines. Stem Cells Dev 22:1728–1742

Viuff D, Palsgaard A, Rickords L et al (2002) Bovine embryos contain a higher proportion of polyploid cells in the trophectoderm than in the embryonic disc. Mol Reprod Dev 62(4):483–488

Zhang Y, Li N, Wang L et al (2016) Molecular analysis of DNA in blastocoele fluid using next-generation sequencing. J Assist Reprod Genet 33(5):637–645

Production of Transgenic Livestock: Overview of Transgenic Technologies

6

Götz Laible

Abstract

Transgenic livestock came into existence in 1985 with the production of the first transgenic pigs and sheep (Hammer et al., 1985). Since then, various technologies have been developed for the generation of transgenic livestock (Fig. 6.1). Sometimes entirely new technologies emerged, seemingly superseding an established technique, while occasionally new improvements to older methods saw them brought back to the forefront. Nevertheless, most approaches are still in use today albeit in combinations with other developments enhancing the original methods by making use of improvements. Each will offer a unique set of advantages and disadvantages. Hence, the choice of technique will be dependent on the specifics of the application and not least on the species of livestock considered for transgenesis. The following will provide an overview of available technologies for the genetic engineering of livestock species. The chapter will describe the main approaches and their advantages and disadvantages and will portray the technical advancements that today allow for the efficient and precise engineering of livestock genomes almost without limitations.

Generally, transgenic technologies can be divided into embryo-mediated and cell-mediated approaches. The first strategy introduces the genetic modification into an embryo (Fig. 6.2a), while for the latter, the genetic information is introduced into a cell which is subsequently used to generate an entire animal based on the genetics of this cell (Fig. 6.2b). The following overview will describe the various enabling technologies that have been successfully used for the generation of transgenic livestock and will discuss their relative strengths and weaknesses. The order in which the different methods are described is intended to provide an overview and does not reflect the chronological order of their development (Fig. 6.1).

G. Laible
AgResearch, Ruakura Research Centre, Hamilton, New Zealand
e-mail: goetz.laible@agresearch.co.nz

© Springer International Publishing AG, part of Springer Nature 2018
H. Niemann, C. Wrenzycki (eds.), *Animal Biotechnology 2*,
https://doi.org/10.1007/978-3-319-92348-2_6

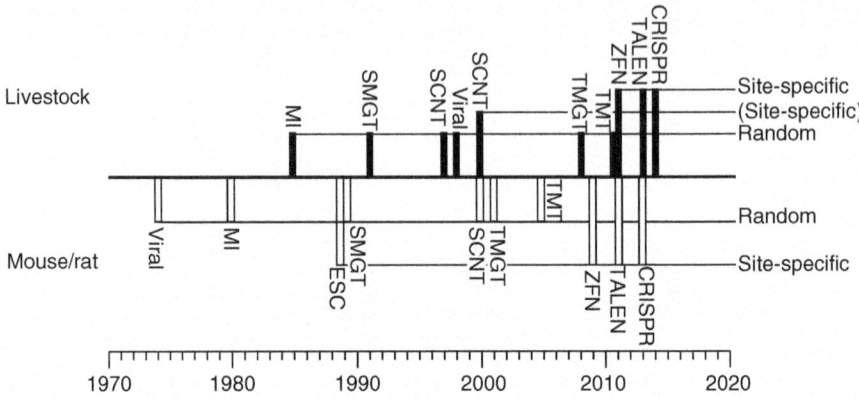

Fig. 6.1 Timeline of methods for the production of transgenic mice/rats and livestock with the ability to introduce random, low-efficient (site-specific), and efficient site-specific genome modifications. *Viral* viral vectors; *MI* microinjection; *ESCT* embryonic stem cell transgenesis; *SMGT* sperm-mediated gene transfer; *SCNT* somatic cell nuclear transfer; *TMGT* testis-mediated gene transfer; *TMT* transposon-mediated transgenesis

6.1 Embryo-Mediated Transgenesis Methods

Embryo-mediated approaches rely on the availability of reproductive technologies to access in vivo embryos or generate them in vitro and established protocols for their culture and manipulation in vitro and subsequent transfer of the manipulated embryos into recipient animals for development to term (Fig. 6.2a). Embryo-mediated methods commonly have the advantage of a relative high efficiency for the embryos to develop into live transgenic animals. This is due to the involvement of only moderate manipulations that generate embryos retaining a high level of developmental competence. The main disadvantage of this approach lies in the lack of total control of when and what modification is introduced.

6.1.1 Transgenesis with Natural Vectors for Genetic Information

Systems that have naturally evolved to deliver genetic materials into cells are obvious targets that could be exploited for transgenesis offering simplicity, efficiency, and cost-effective mechanisms for the purposeful introduction of exogenous DNA.

6.1.1.1 Viral Vectors
Viruses have evolved very efficient mechanisms to infect and deliver their own genome into eukaryotic cells. This natural ability to introduce exogenous RNA or DNA into cells and efficiently integrate the exogenous genetic information into the genome of the infected host made them an attractive tool for the stable genetic modification of mammals. The development of viral vectors for transgenesis from disease-causing viruses followed two main considerations. One was to maximize

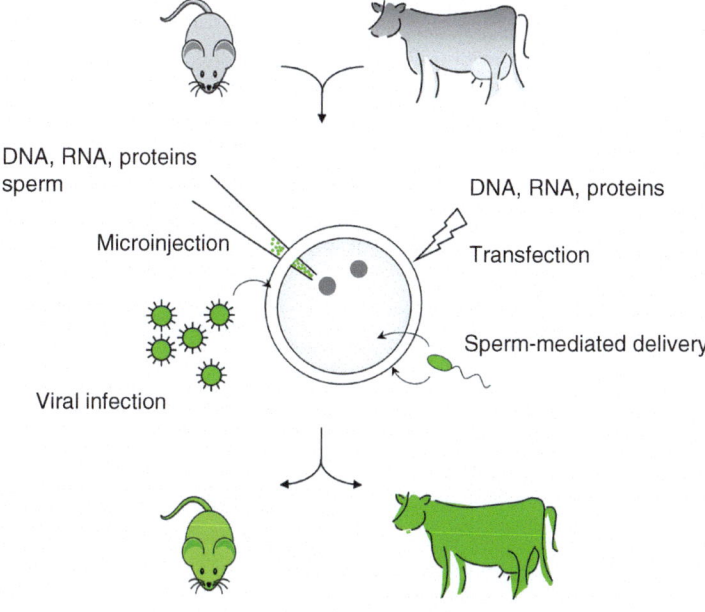

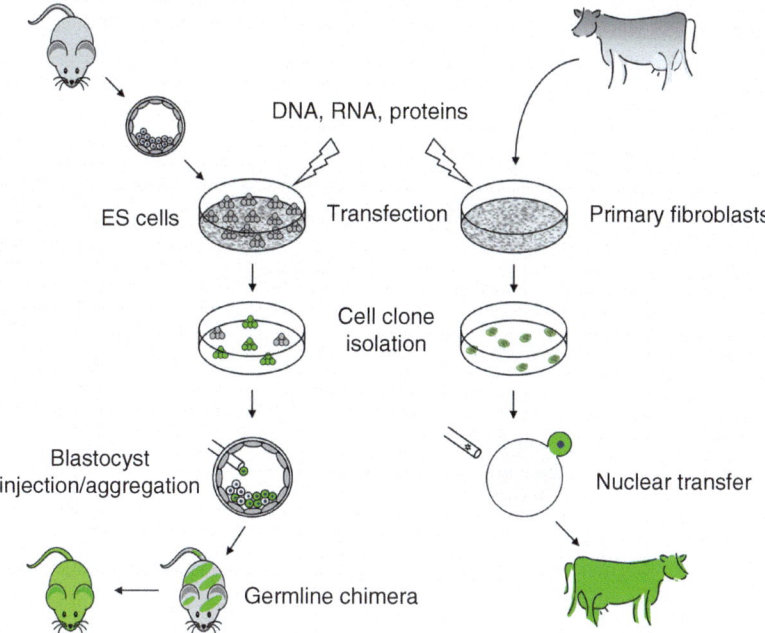

Fig. 6.2 Schematic overview of embryo-mediated (**a**) and cell-mediated (**b**) methods to generate transgenic mice and livestock. Green coloring indicates genetically engineered cells or animals. Please refer to the main text for a detailed description of these techniques

the cargo space available for an exogenous transgene and the other to minimize the biosafety risks associated with vectors developed from disease-causing pathogens. The size of the viral particle is sufficient to hold the viral genome but presents a physical limit of how much genetic information can fit into such a particle. To generate cargo space for a transgene, crucial viral genes involved in packaging and replication were typically deleted and replaced by transgene sequences. This changes a virus into a so-called viral vector which is unable to replicate itself and produce viral particles. However, these vectors still retain the ability to be packaged into infectious particles, provided the genes encoding the viral packaging machinery are supplied in trans. Based on the viral mechanism for infection, the viral vectors can then be introduced into mammalian cells with high efficiency. Because the vectors are replication- and packaging deficient, they can do so only once and are unable to produce any further infectious particles.

Thus, the development of viral vectors generated crucial cargo space for transgenes and retained the ability for efficient infection of host cells while greatly reducing the biosafety risk compared to a replication-competent, disease-causing virus (Verma and Weitzman 2005).

Viral-mediated transgenesis was the first approach that was successfully used to modify mammalian embryos by exogenous DNA (Jaenisch and Mintz 1974). Early strategies utilized the potential of retroviruses. These viruses belong to the class of RNA viruses which insert a reverse-transcribed DNA copy of their genetic material into the host genome. They have a broad host range and hence have general applicability for many different species. With this method, stable integration of the transgene appears to be relatively efficient, though the integration is into random sites and with variable timing which can be mutagenic and lead to a high degree of mosaicism in the transgenic founder animals. In particular, for larger animals with long generation times, the segregation of mosaic genotypes in subsequent generations is problematic. In addition, the residual viral sequences of these vectors are often recognized by the cellular machinery. As a consequence the exogenous sequences, including the transgene, are prone to become transcriptionally silenced through epigenetic host defense mechanisms (Wells et al. 1999; Chan et al. 1998; Gilboa et al. 1986). Partly, these limitations have been overcome with the application of replication-defective vectors based on adenovirus (Tsukui et al. 1996; Kubisch et al. 1997). Further improvements were made with the development of advanced lentiviral vector systems (Dull et al. 1998). In contrast to the retroviral vectors mentioned above, lentiviruses belong to a class of retroviruses which have the unique ability to infect not only dividing but also nondividing cells. This makes these vectors very efficient gene delivery tools reducing the risk for the generation of mosaic animals.

Their potential for mammalian transgenesis was first demonstrated in mice and rats (Lois et al. 2002). The injection of viral particles into the perivitelline space underneath the zona pellucida of one-cell embryos yielded transgenic mice with an efficiency of 80% of which 90% expressed the transgene at high levels. Similarly, incubation of denuded early-stage mouse embryos in a virus-containing solution yielded transgenic mice showing strong transgene expression and not affected by gene silencing (Pfeifer et al. 2002). The success with this method, in some instances producing up to 100% transgenic offspring, could be readily transferred from mouse to pigs (Hofmann et al. 2003; Whitelaw et al. 2004) and sheep (Ritchie et al. 2009). For cattle (Hofmann et al. 2004) and chickens (McGrew et al. 2004), the method had to be slightly modified with cattle requiring the injection of the lentiviral particles under the zona of oocytes rather than zygotes, while for the generation of transgenic chickens, the injections were done into the subgerminal cavity below the embryonic disc.

A major opportunity was seen in using lentiviral vectors in combination with RNA interference for efficient delivery of interfering RNAs that can block or destroy the viral RNA of invading viruses to generate transgenic animals with resistance to viral diseases (Clark and Whitelaw 2003). Though RNA interference is very versatile and can also be combined with other transgenesis methods, embryo mediated and cell mediated, to deliver an expression construct for a small interfering RNA, it can be applied for the purpose of precisely knocking down the expression of particular endogenous genes (Jabed et al. 2012).

Despite some attractive advantages of viral-mediated approaches such as technical simplicity and the reported high efficiencies, the method did not find wide support in the field, probably a reflection of various drawbacks holding back greater uptake by the scientific community. The viral system has a fixed limit for the size of transgenes that will not be compatible with applications requiring long transgene sequences. Furthermore, the characteristic of the lentiviral method to generate single-copy insertions into multiple loci which will segregate in the following generations complicates applications in animals with long generation intervals and few offspring per breeding cycle. Others found that a high proportion of the segregating transgenes became epigenetically silenced in the next generation (Hofmann et al. 2006) but can also affect transgene expression in the founder animals (Tian et al. 2013). However a major deterrent might have been the viral nature of the system. Producing and working with high viral titers can be difficult and require extra safety precautions. Even though lentiviral vectors are designed to ensure safety and also include a self-inactivating element to prevent transcription of viral sequences (Miyoshi et al. 1998; Zufferey et al. 1998), the general acceptance of viral systems for livestock applications by regulators and the public remains very low due to concerns for potential recombination events that may recreate replication-competent viruses. A recent study in sheep which found no evidence for the inadvertent transfer of viral vectors to other animals or generation of recombination-competent virus provides evidence that the technology is safe (Cornetta et al. 2013).

While lentiviral vector systems are continuously improved and evaluated for their safety in human gene therapy applications (Schambach et al. 2013), many more studies addressing these safety concerns from transgenic food-producing animals will be required to demonstrate the safety of the technology and satisfy regulatory demands. Despite many having experienced difficulties getting the technology to work, a few research groups have mastered the technology and report the consistent production of transgenic livestock with high efficiencies (Lillico et al. 2011). This shows that while lentiviral transgenesis may not be the technology of choice for some applications, it is an important tool to produce transgenic livestock.

6.1.1.2 Sperm-Mediated Gene Transfer

Sperm has evolved as a natural vector for the transmission of the paternal DNA as part of the sexual reproduction of animals. Therefore it appears feasible to harness the sperm's natural ability for the delivery of exogenous DNA into the germline of the zygote or developing embryo. The concept was first demonstrated in a study by Brackett and co-workers (1971). Rabbit sperm, incubated with SV40 DNA, was used for inseminating females, and from the one- and two-cell embryos that were obtained, infectious SV40 could be recovered. Some 18 years later, this sperm-mediated gene transfer (SMGT) approach was validated with the production of the first transgenic mouse using spermatozoa that were extensively washed and then incubated with DNA prior to using them for in vitro fertilization (Lavitrano et al. 1989). With a reported 30% efficiency for the production of transgenic offspring, the technique not only appeared efficient but promised absolute simplicity and low-cost production of transgenic animals with the potential for large-scale applications. In spite of the apparent simplicity of SMGT, it proved to be a highly unreliable technique with inconsistent outcomes (Brinster et al. 1989). When the SMGT methodology was later transferred to a range of other species, including rabbit (Wang et al. 2003), chicken (Nakanishi and Iritani 1993), pig (Lavitrano et al. 1997), goat (Zhao et al. 2010), and cattle (Perez et al. 1991), it was met with varying success and was frequently plagued by low efficiencies. In attempts to improve the efficiency of SMGT, the method was modified in various ways to maximize the delivery of the exogenous DNA which was comprehensively reviewed recently (Lavitrano et al. 2013). A sperm-reactive antibody with high DNA-binding capacity has been included as a linker to improve the binding of the exogenous DNA to spermatozoa (Chang et al. 2002). To aid with the stable integration of the exogenous DNA, sperm was also incubated with lentivirus (Zhang et al. 2012), or transfection of sperm was combined with restriction enzyme-mediated integration (Harel-Markowitz et al. 2009; Shemesh et al. 2000) prior to their use for artificial insemination. Despite the reports of the successful production of transgenic embryos and farm animals, this form of SMGT is associated with relatively low efficiencies and high variability of success rates. Only few laboratories have consistently reported positive results, and the cause for the inherent reliability issues remains unresolved. Many mechanistic barriers exist that may prevent the efficient uptake of exogenous DNA into sperm or the internalization of transgenes into oocytes and embryos via sperm. This has been discussed in a recent study as a major cause for the failure to produce transgenic bovine embryos following in vitro fertilization of oocytes with sperm incubated or transfected with exogenous DNA and might be at the core of the observed inconsistencies with SMGT (Eghbalsaied et al. 2013).

Any barriers to the delivery of the sperm-associated exogenous DNA into the oocyte can be readily overcome by a technique called intracytoplasmic sperm injection (ICSI). With this method, freeze-thawed or detergent-treated sperm is complexed with exogenous double-stranded DNA fragments and then directly injected into the oocyte cytoplasm for the production of a transgenic animal (Moisyadi et al. 2009). This technique, termed ICSI-mediated transgenesis (ICSI-Tr), was first demonstrated in mice where it enabled the production of a high percentage of transgene-expressing offspring (Perry et al. 1999). In contrast to viral vectors, it could also accommodate the transfer and insertion of large DNA fragments (Moreira et al. 2004). After the success with the production of transgenic mice, ICSI-Tr was applied to a variety of livestock species. However, transgenic efficiencies were quite low compared to the mouse system (Yanagimachi 2005). The inclusion of an additional chemical activation step improved the typically poor embryo development following ICSI observed with domestic animals and enabled the production of blastocysts with up to 80% confirmed for the expression of the GFP reporter gene (Bevacqua et al. 2010; Pereyra-Bonnet et al. 2008). A goat study that evaluated the impact of goat sperm on ICSI results found that immotile sperm delivered better results with ICSI-Tr than motile sperm (Shadanloo et al. 2010). Other improvement strategies were aimed at actively assisting with the integration of transgenes into the host genome by combining ICSI-Tr with the activity of a recombinase, integrase, and transposase which boosted transgenesis rates severalfold over unassisted ICSI-Tr in mice (Shinohara et al. 2007). Applied in pigs, ICSI-Tr supported by the recombinase recA was shown to increase the number of transgene-expressing embryos and achieved a 46.6% efficiency for the production of transgenic piglets (Garcia-Vazquez et al. 2010). The accumulation of more and more studies reporting on the successful production of transgenic embryos and animals with SMGT and ICSI-Tr indicates that these approaches can work. Yet, the observed high intra- and interspecies success variabilities show that it is not the most reliable and robust technique and will require further development. Though, the need for additional variations such as ICSI forfeits the promised simplicity, which had been one of the main attractions of this technology. SMGT/ICSI-Tr has also some intrinsic drawbacks and gives very limited control over the number of integrated transgene copies and time and site of integration with the potential for generating high proportions of mosaic animals.

However, other technologies supporting only random insertions described below will share these shortcomings to various degrees.

6.1.2   Transgenesis by Direct Introduction of DNA

6.1.2.1 Microinjection of Zygotes

Instead of using a natural vector, this method uses the direct delivery of exogenous DNA by manual injection with a fine glass needle into the pronucleus of a one-cell embryo. This microinjection of many copies of a linear DNA construct into one of the two pronuclei of a recently fertilized egg for the generation of transgenic mice was first demonstrated by Gordon and co-workers (1980). The technology went on to revolutionize the transgenic animal field and gave rise to a vast number of different transgenic mouse models that enabled the study of mammalian gene function and disease. Encouraged by the success in the mouse system, the technology was swiftly applied in larger animals (with the prospect for immediate improvements of livestock genetics supporting agricultural and biomedical applications) and led to the generation of transgenic rabbits, sheep, and pigs (Hammer et al. 1985). But the mouse system did not directly transfer to livestock. The high lipid content in embryos from domestic animal species obscured the pronuclei and made pronuclear microinjection more demanding compared to mice. While a centrifugation step allowed visualization of pronuclei in a large proportion of the embryos, low transgene integration efficiency and low embryo survival resulted in overall very low production efficiencies for transgenic livestock by this method with approximately only 1% of injected zygotes giving rise to transgenic offspring (Wall 2001). Pronuclear microinjection is limited to random transgene integrations, commonly associated with concatemerization and integration of multiple copies. Random insertions can cause interference with or complete loss of the function of an affected endogenous gene or might restrict the transgene's expression when it is integrated in a heterochromatic location. Moreover, lack of control over the timing of events means that integration does not necessarily occur at the one-cell stage. It can happen after the first DNA replication and cell division which will generate a mosaic animal with only some but not all cells containing the transgene. A review of the results for the production of transgenic livestock by pronuclear microinjection between 1985 and 2000 revealed the extent of the issues (Wall 2001). Only approximately 70% of transgenic founder animals were able to transmit the transgene through the germline to the next generation. For the founders that could establish a transgenic line, just over 50% expressed their transgenes at levels suitable for the respective application. This enormous inefficiency made the production of transgenic livestock very expensive and shifted the focus from agricultural to biomedical applications which generally offer greater economic benefits and thus provide better justification for the required effort to produce such animals. In the absence of better alternatives, the technology remained the dominant technique for the production of transgenic livestock until the advent of cloning from adult cells in 1997 (Wilmut et al. 1997) which provided the platform for a cell-mediated technology described in detail in Sect. 6.2.4.

Several strategies were investigated to improve the integration of transgenes following pronuclear microinjection. In combination with restriction enzyme-mediated integration, the co-injection of a transgene construct and a restriction enzyme into the pronucleus boosted the production of transgenic mouse embryos and pubs by about twofold (Seo et al. 2000). As with SMGT, active integration supported by exogenous enzymes was tested. Microinjection of DNA that was coated with the bacterial recombinase RecA resulted in better embryo survival and transgene integration in goats and pigs, increasing the overall efficiency for producing transgenic offspring (Maga et al. 2003). More recent developments with transposable delivery systems and site-specific nucleases, which offer improved control and efficiency for the integration of transgenes, have substantially augmented microinjection as a means of producing transgenic livestock and brought it back into focus as a method of choice. This warrants the description of these two improvement strategies and how they enhance conventional microinjection in a bit more detail below.

Transposon-Mediated Transgenesis

Transposons are naturally occurring mobile genetic elements which can be mobilized by a cognate transposase through an active, cut-and-paste-like process. The transposase first catalyzes the excision of the transposon and then the integration of the transposon, also called transposition, into a target site of the host genome. Initially applied for gene transfer in invertebrates, the regeneration of the Sleeping Beauty transposon system from nonfunctional elements provided the first transposase suitable for mammalian systems (Ivics et al. 1997). Since then, additional transposases have been tested for mammalian transgenesis, including PiggyBac and Tol2 (Clark et al. 2007; Wu et al. 2006), and were further developed into hyperactive variants for increased transgene integration efficiencies (Germon et al. 2009; Lampe et al. 1999; Mates et al. 2009; Yusa et al. 2011).

For transposase-mediated transgenesis, transgenes need to be flanked with short inverted terminal repeat elements that are recognized and bound by the transposase. Most commonly, the transposase is then supplied in trans with a separate expression construct to execute the cut-and-paste action and stably integrate the transgene (reviewed in Bosch et al. 2015). But also single-vector systems have been described where transposon and transposase act in cis. A design with self-inactivating properties safeguards against undesirable genetic instabilities due to remobilization of integrated transgenes should the transposase itself become unintentionally integrated at random locations (Urschitz et al. 2010). Alternatively, the transposase can be delivered as mRNA or recombinant protein to avoid possible integration and continuous presence of an active transposase.

Combining transposon systems with pronuclear injection greatly increased transgene integration rates similar to the high efficiencies achievable with viral approaches resulting in severalfold enhancement of the production efficiencies for transgenic animals over conventional pronuclear microinjection (Bosch et al. 2015). In mammalian species, it was successfully applied in mouse (Ding et al. 2005; Dupuy et al. 2002; Mates et al. 2009), rat (Jang and Behringer 2007), rabbit (Katter et al. 2013), and pig (Carlson et al. 2011). As mentioned earlier, pronuclei are obscured by lipid droplets in species such as pig and cattle, a limitation which was soon resolved by adapting the methodology to the much simpler injection into the cytoplasm (Fig. 6.3). This was made possible by the use of hyperactive transposases which had sufficiently high transposition efficiencies to support delivery by cytoplasmic injection and readily produced transgenic mice (Marh et al. 2012), pigs (Garrels et al. 2011; Li et al. 2014), and cattle (Garrels et al. 2016) with high efficiencies. More recently, this approach was also validated for sheep although the majority of lambs born had the transgene only integrated in extraembryonic tissues and the two transgenic lambs generated appeared to be highly mosaic (Bevacqua et al. 2017).

The increased integration efficiency of this transposon-mediated transgenesis strategy overcomes a major limitation of traditional pronuclear injection which is associated with very low stable chromosomal integration rates. In comparison with viral systems, it has the added advantage of a much lower biosafety risk profile and the ability to transfer even very large transgenes (Balciunas et al. 2006; Li et al. 2011; Rostovskaya et al. 2013). While enhanced integration with the aid of transposases also relies on integration into random chromosomal sites, it catalyzes the integration of monomeric transgene copies instead of multimeric, concatemerized transgenes typically observed with the more traditional approach.

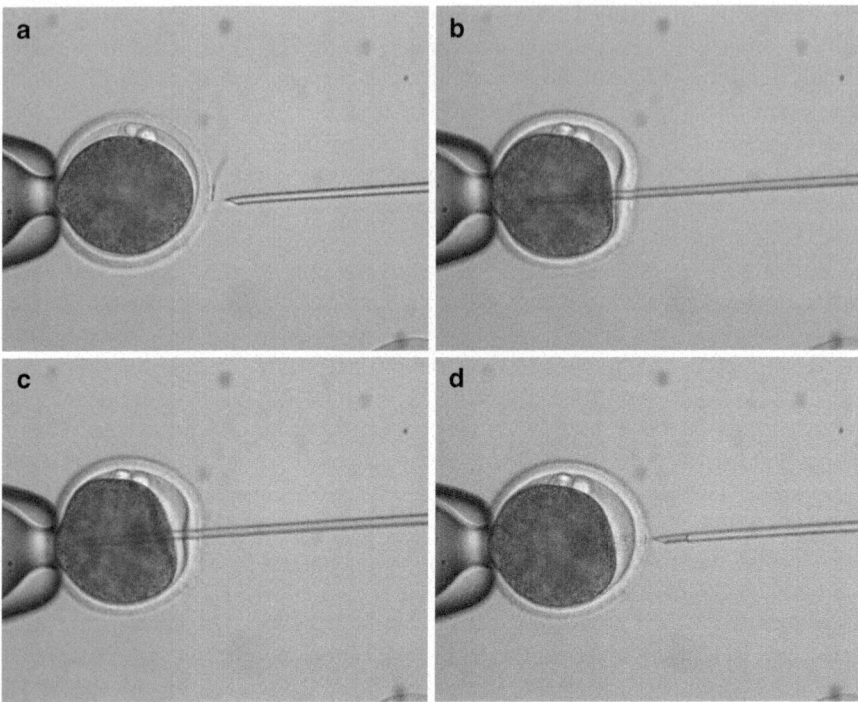

Fig. 6.3 Cytoplasmic injection of a bovine zygote. Shown are (**a**) the immobilization of the zygote with a holding pipette, (**b**) penetration of the zona pellucida and underlying plasma membrane with the injection needle, (**c**) injection of a small volume containing relevant effector molecules, and (**d**) retraction of the injection needle after completion of the delivery

However, transposases not necessarily just integrate a single copy but can integrate monomeric transgene copies into multiple different chromosomal sites. Although it is possible to segregate these through breeding, in large animals with long generation times and few offspring, this could be impractical. Thus, for many transgenic livestock applications, the aim will be single-transgene insertions which can be favored by modulating amount and ratio of transgene and transposase accordingly (Carlson et al. 2011).

Most transposases interact with short consensus sequences in the genome, such as Sleeping Beauty and PiggyBac with TA and TTAA, respectively, where also the insertion takes place. The detailed mapping of insertion sites further revealed that PiggyBac has a tendency for integrations in or near transcription units (Ding et al. 2005; Wilson et al. 2007) making it prone to insertional mutagenesis which would be an undesirable outcome for animal transgenesis. In contrast Sleeping Beauty is lacking such a preference (Liu et al. 2005; Yant et al. 2005). Based on the high proportion of transposon-produced transgenic animals expressing their transgene, it appears that chromosomal integrations are preferentially directed into locations with an open, transcriptionally permissive chromatin conformation (Garrels et al. 2012).

Thus, enhancing pronuclear injection with transposons is less prone to variegated expression and epigenetic gene silencing that are often observed with multi-copy insertions into heterochromatic regions with non-assisted methods.

Programmable, Site-Specific Nucleases

As described before, microinjection was limited by lacking precise control over where a transgene integrates into the genome. Site-specific sequence alterations in farm animals were only possible with cell-mediated approaches using primary cells (described in detail in Sect. 6.2.4). This was however a highly inefficient technique which severely dented its practicality (Laible and Alonso-Gonzalez 2009). The remarkable advent that has seen the development of site-specific nucleases, designed to target a single site in the genome, has now spectacularly changed the feasibility for targeted changes of livestock genomes by microinjection. The following provides a brief description of the main concepts and offers insights into the advanced technical capabilities associated with site-specific nucleases.

The inclusion of programmable nucleases greatly enhanced conventional microinjection to an entirely new level that makes it now possible to introduce site-specific alterations into livestock genomes via microinjection with high efficiency previously only achievable in the mouse embryonic stem cell (ESC) system. When microinjected into the cytoplasm of zygotes, these site-specific nucleases introduce a double-strand break at the desired target site. This is typically repaired by the cellular repair machinery via nonhomologous end-joining (NHEJ) repair process. NHEJ is an error-prone process due to endogenous nuclease activity at the site of the double-strand break (DSB) and as a consequence often results in the introduction of small insertions or deletions, so-called indels, at the target site. Although this still involves some aspect of randomness because the actual mutation being introduced is not predefined, it provides a very efficient and simple way to generate functional gene disruptions in livestock that were not possible before (Lillico et al. 2013; Proudfoot et al. 2015). Similarly, genomic region can be efficiently deleted by introducing two DSBs resulting in the deletion of the intervening sequences (Xiao et al. 2013; Yang et al. 2013). As a consequence of the introduction of such mutations, also the binding site for the nuclease is lost, and the mutated allele is no longer an editing target. Together with the strong cleavage activities commonly exerted by these nucleases, this approach readily enables the efficient introduction of bi-allelic modifications at the target site. But because the exact time, duration, and extent of the nuclease activity cannot be controlled, animals can be generated that contain more than two different mutated alleles and may show variable degrees of mosaicism.

The programmable nucleases are commonly delivered as circular expression plasmids which are then transcribed and translated in the embryo. To eliminate the potential risk for unwanted vector integration, they can be also injected as mRNA and recombinant protein. This technology, often referred to as genome editing, enables targeted mutagenesis without leaving any technology-associated footprint. Nuclease-mediated genome editing can be even more sophisticated when a homologous repair template is provided in trans which enables site-specific introduction of precise mutations or targeted insertions of transgenes (Tan et al. 2016). This is due to DSB-induced stimulation of homology-directed repair (HDR) mechanisms. In the presence of a homologous repair template, defined mutations can be introduced at the target site according to intended sequence changes specified by the repair template.

The first site-specific nucleases that were developed into an effective genome-editing tool were zinc finger nucleases (ZFNs) (Hauschild-Quintern et al. 2013). The toolbox was soon expanded with the development of additional editing platforms, namely, transcription activator-like effector nucleases (TALENs) which, instead of zinc fingers, use a DNA-binding domain comprised of an array of repeat units whose intrinsic polymorphic sites determine DNA-binding specificity (Mussolino and Cathomen 2012) and the clustered, regularly interspaced short palindromic repeat (CRISPR)—CRISPR-associated nuclease 9 (Cas9) system (Sander and Joung 2014). Today, scientists have an increasing range of editing enzymes at their disposal that provide a variety of options for the precise and efficient editing of livestock genomes. Because of its importance for contemporary approaches to precisely engineer livestock genomes, there is a dedicated chapter about this technology, and the interested reader is referred to Chap. 7 for a detailed description of how these editors work and can be applied to edit livestock genomes.

6.1.2.2 Electroporation of Zygotes

With the development of enhanced transgenesis options, microinjection has become again a method of choice, particularly in combination with programmable nucleases (see Chap. 7). However, microinjection is technically demanding, requires specialized micromanipulation skills, is labor intensive, and is limited to slow throughput due to the restriction for one-by-one injections of individual embryos. Electroporation potentially offers an attractive alternative as a technology which is suitable to introduce DNA and RNA into one-cell embryos (Grabarek et al. 2002). Introduction of site-specific nucleases as RNA into mouse and rat zygotes by electroporation resulted in the successful production of mice and rats with targeted gene disruptions (Kaneko et al. 2014) and precise mutations as specified by HDR templates (Qin et al. 2015). It is also possible to electroporate CRISPR/Cas9 as preassembled ribonucleoprotein complex into zygotes which achieved efficiencies of 88% for bi-allelic gene disruptions and 42% for the introduction of HDR-mediated precise mutations in mice (Chen et al. 2016). Moreover, it was shown that editing following electroporation of ribonucleoprotein complexes can occur before the first genome replication, ensuring the generation of mice no longer containing any non-edited wild-type alleles (Hashimoto et al. 2016).

So far electroporation for the delivery of genome editors has been mainly investigated in rodents and has only started to be applied in livestock species. Electroporation for the generation of CRISPR-mediated indels was successfully applied in pigs (Tanihara et al. 2016). In bovine, electroporation of zona-free zygotes (Fig. 6.4) was reported to also enable the introduction of precise mutations as a result of HDR (Wei et al. 2018). Although increasing overall editing

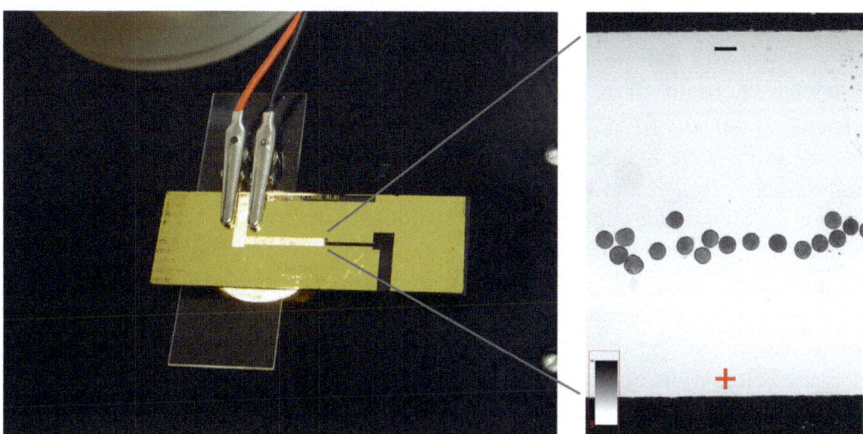

Fig. 6.4 Electroporation of zona-free bovine zygotes. The left panel shows a coplanar electrode that is positioned under a stereo microscope and connected to a square wave generator. The magnified (100×) view on the right shows the main channel of the electrode loaded with zona-free bovine zygotes prior to electroporation

efficiencies and reducing the level of mosaicism still need to be addressed, electroporation of livestock embryos holds much promise for the future. In contrast to microinjecting individual embryos, the electroporation process is very simple, and groups of tens of embryos can be simultaneously edited which provide scope for high throughput due to a much greater potential for automation of the electroporation process.

6.2 Cell-Mediated Transgenesis Methods

A prerequisite for cell-mediated transgenesis is immediate access to cells, which can be cultured and easily manipulated to introduce genetic changes in vitro. Most importantly, these cells must be suitable and provide an avenue to subsequently generate an entire animal with the same genetic blueprint from essentially a single transgenic cell (Fig. 6.2b). Compared to the above-described embryo-mediated transgenesis methods, cell-mediated approaches have several distinct advantages. The sex and genetic background of the modified animal can be simply chosen with the selection of the cells to be engineered, and because genetic engineering is done in cells cultured in vitro, it provides the opportunity to fully characterize the introduced genetic modification prior to generating a transgenic animal.

Moreover, the cell-mediated strategy avoids the issue of generating mosaic animals and enables the production of multiple founders from a characterized cell line. This contrasts the necessity to breeding from an individual founder, identified to be suitable following full genotypic characterization of multiple candidates, to establish small cohorts of animals for phenotypic analysis.

6.2.1   Embryonic Stem Cell Technology

ESCs have to be one of the most ideal types of cells for engineering the genome and generating animals from such engineered cells due to their unique characteristics. Mouse ESCs were first derived in 1981 from the inner cell mass of a blastocyst-stage embryo (Evans and Kaufman 1981; Martin 1981) and have the ability to self-renew which provides them with an unlimited life-span in culture. The quasi immortality is coupled with a high efficiency for single-cell clonal expansion greatly facilitating the isolation and characterization of transgenic cell clones. Moreover, while homologous recombination is a rare event in mammalian cells, its relative frequency in ESCs is approximately two orders of magnitude higher compared to somatic cells (Wells et al. 2003). To increase the probability for precise modifications by homologous recombination and to facilitate the identification of rare homologous recombination events, targeting vectors had to be equipped with long stretches of homologous sequences from isogenic DNA and positive- and negative-selection cassettes and were typically delivered into cells as linear fragments by electroporation. This proven strategy provided almost unlimited opportunities to precisely modify the genome of ESCs by even complex, multistep procedures and was the main driver for the success of the mouse as model organism. But even in ESCs, homologous recombinations remain relatively rare events which limited the introduction of site-specific alterations to a single allele only. Since then, gene targeting in ESCs has become much simpler with the application of programmable nucleases (Wang et al. 2013). Gene disruptions can be done without the need for complex, homologous sequences containing targeting vectors, and simple knockins can be achieved with short oligonucleotides as homology repair template to even target multiple genes all in the absence of selection. For more demanding applications such as the knockin of entire transgenes, the efficiency appears to be generally lower but is still sufficient to be achievable without the need for selection (Moehle et al. 2007).

Another equally important characteristic of ESCs is their pluripotent nature. ESCs can develop into any cell type of an adult mouse, essentially representing a one-cell equivalent of a mouse. When injected back into blastocysts or aggregated with morula-stage embryos, they can contribute to all tissues of the developing animal, including the germline, which provides an efficient route for the generation of an animal entirely derived from the injected ESCs (Robertson et al. 1986).

Initially, ESCs were only available from mouse, and it took many years until ESCs could be established from two other species, namely, rhesus monkey and human (Thomson 1998; Thomson et al. 1995). However, when compared to murine ESCs, their culture was dependent on different growth factors, and they differed in the expression of key genes and other characteristics. Today they are recognized as representing a primed state of a later embryonic state, while mouse ESCs capture the naïve or ground state of an early, preimplantation embryo (Soto and Ross 2016). The development of a new culture system, based on two inhibitors and commonly referred to as 2i, that facilitates the maintenance of a naïve pluripotent state (Ying et al. 2008) was instrumental for the successful isolation of naïve ESCs from rats

(Buehr et al. 2008; Li et al. 2008). Although this raised the expectations that ESCs from livestock species were soon to follow, this has turned out to be extremely difficult, and no livestock ESCs suitable for generating transgenic animals in this way are available to date (Ogorevc et al. 2016; Soto and Ross 2016).

6.2.2   Induced Pluripotent Stem Cells for Transgenesis

The discovery that the transient expression of a set of specific transcription factors can reprogram somatic cells into pluripotency has unlocked a new promising source of cells (Takahashi and Yamanaka 2006). The efficiency of reprogramming somatic cells back into a pluripotent state is however very low, and viral vectors were initially used to take advantage of their high transduction rate and introduce the reprogramming factors into a maximum number of cells. Since then a number of alternative non-viral methods were developed for the delivery of the reprogramming factors including direct delivery of mRNA or recombinant proteins (Kumar et al. 2015). The resulting induced pluripotent stem cells or iPSCs closely resemble the characteristics of naïve ESCs (Nichols and Smith 2009). Like ESCs, mouse and rat iPSCs were shown to be germline competent and could be used to produce iPSC-derived offspring (Hamanaka et al. 2011; Okita et al. 2007). In contrast, similar efforts with livestock iPSCs were less successful with only one published study describing germline transmission at a very low rate from a chimeric pig produced with iPSCs (West et al. 2010). In this study, no viable offspring were produced that could survive past the first 3 days (West et al. 2011). Adversely affecting the competence of livestock iPSCs could be their state of incomplete reprogramming indicated by their dependence on the continued expression of the exogenous reprogramming factors (Du et al. 2015). This is not the case with rodent and human iPSCs which become independent of the exogenous factors once fully reprogrammed. Hence, livestock cells may require different culture conditions for their derivation, continued proliferation, and maintenance of pluripotency which have not been fully identified yet and are the main reasons why there are no livestock iPSCs available for transgenesis at present (Ogorevc et al. 2016; Soto and Ross 2016). This is an area of intense research, and although neither the isolation of suitable ESCs nor iPSCs has been reported for livestock species, these cells hold considerable promise for the future. Further, more in-depth information on the current state of livestock stem cells can be found in Chap. 10.

6.2.3   Testis-Mediated Gene Transfer

The concept is based on the transfection of male spermatogonial stem cells (SSCs) that can later be transplanted into the testes of recipients which had been treated to suppress or deplete the pool of endogenous germ cells. Successful colonization of the recipient testes with the exogenous SSCs will then allow the production of transgenic sperm and thus offspring. Following the successful transplantation of exogenous mouse SSCs into a compromised recipient testis (Brinster and Avarbock 1994; Brinster and Zimmermann 1994), production of transgenic mice was demonstrated by transplantation of transfected SSCs (Nagano et al. 2001). Because it represented an attractive new route for the production of transgenic offspring, the technology was investigated for its applicability for livestock transgenesis (reviewed in Honaramooz and Yang (2011)), and first transgenic goats (Honaramooz et al. 2008) and pigs (Zeng et al. 2013) have been produced with this method. Though one should not forget that the preparation of recipients and transplantation of SSCs in large animals pose significant challenges, further improvements will be required to develop testis-mediated transgenesis into a more widely accessible and practical tool for the production of transgenic livestock.

6.2.4   Nuclear Transfer with Transfected Somatic Cells

Because of the unavailability of suitable PSCs for a long time, there was no cell-mediated route for livestock transgenesis. Readily available were differentiated primary somatic cells which could be genetically engineered by standard methods for random insertions, including viral transduction and transfection or electroporation of transgenes. A method with the potential to generate transgenic animals from such differentiated cells is presented by nuclear transfer (NT) technology (reviewed in Chap. 1). With this approach, the nucleus from a suitable transgenic donor cell is transferred into an enucleated oocyte and artificially stimulated to undergo embryonic development. The latter will be dependent on the successful reprogramming of the cell's nucleus to a state where it again is compatible with orchestrating correct development. Subsequent transfer of NT embryos into surrogate females would then allow for in vivo development to term and production of transgenic offspring (Fig. 6.5). However, early nuclear transplantation experiments in amphibians strongly indicated that it is not possible to reprogram a differentiated cell back into an embryonic state that would support the development into an adult animal from such a cell (Briggs and King 1952; Di Berardino 2001; Gurdon 1999). Essentially, this appeared to rule out any possibility to generate transgenic livestock from engineered primary somatic cells. This was then shown to at least be still possible with undifferentiated preimplantation embryonic cells as nuclear donors which could be successfully reprogrammed to produce live cloned mice (McGrath and Solter 1983) and sheep (Willadsen 1986). To be suitable for transgenesis, such undifferentiated embryonic cells would need to be available in large numbers for transfection and maintain their developmental stage during in vitro culture which is not the case and hence offered no practical solution.

The big breakthrough came with the cloning of Dolly the sheep from a mammary gland cell by NT (Wilmut et al. 1997). This demonstrated for the first time that it is possible to reverse the developmental clock and reprogram a differentiated cell back into a totipotent state with the ability to support the development of an entire animal. It broke a long-standing dogma and encouraged the investigation of alternative reprogramming strategies which later resulted in the development of iPSCs

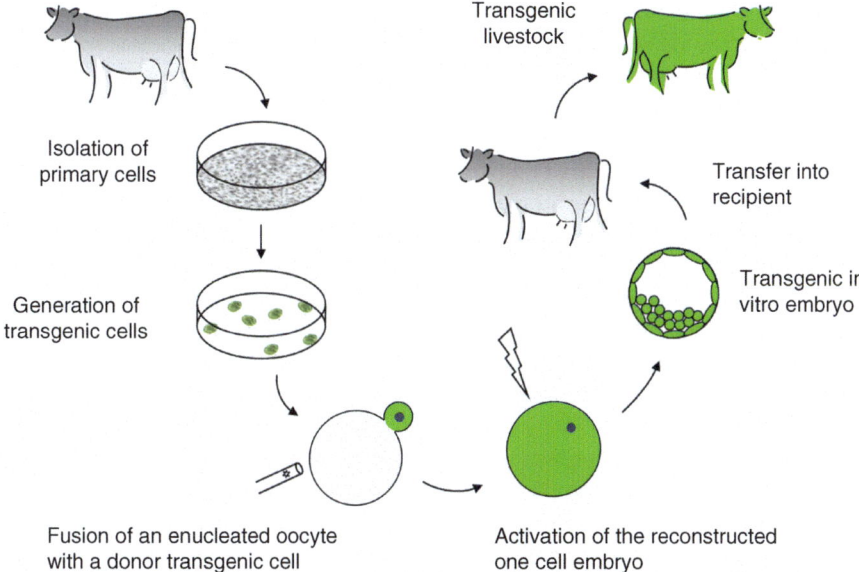

Fig. 6.5 Schematic illustration of the NT process to produce transgenic cattle from transgenic cells. The transgene is introduced into primary somatic cells, typically derived from fetal tissues. Individual transgenic cells are then used as donor cells for NT and fused with an enucleated oocyte. Following artificial activation, the reconstructed NT embryos are cultured to the blastocyst stage and transferred to recipient cows for development to term. Green coloring indicates the transgenic nature of cells or animals

(Takahashi and Yamanaka 2006). Once NT with differentiated cells had been demonstrated, it was only a small step to combine NT with transgenic cells as donors and establish a novel route for the production of transgenic sheep (Schnieke et al. 1997) which was readily applicable in other livestock species (Baguisi et al. 1999; Cibelli et al. 1998; Polejaeva et al. 2000). Random insertion of transgenes by standard transfection approaches was dependent on the use of an antibiotic selection marker to identify and isolate transgenic cell clones. Development of transposon systems (Alessio et al. 2016; Clark et al. 2007) and application of other exogenous enzymes (Bosch et al. 2015) facilitate an active integration process and can greatly enhance the production of transgenic cell clones to an extent where the use of selection could be avoided. In addition, there is no size limitation, and artificial chromosomes such as bacterial artificial chromosomes (BACs) with the capacity to carry fragments of more than one megabase can be used for integrating large transgenes. Their use for transgenesis offers insulation of functional transgene sequences from potentially interfering neighboring endogenous sequences and can protect against unwanted position effects and variegated expression (McCormick and Nielsen 1998; Peterson et al. 1997). BACs were used for the generation of cattle (Yang et al. 2008) and pigs (Ma et al. 2016; Xu et al. 2016) for the controlled expression of specific transgenes which was recently extended with the generation of multi-transgenic pigs by introducing multiple BACs into a single location to avoid possible segregation of the transgenes following breeding into the next generation (Fischer et al. 2016). The issue of insertion can be entirely avoided when mammalian artificial chromosomes are used which can be established episomally as independent chromosomes in mammalian cells (Ikeno et al. 1998). The approach was later applied to establish a 10 Mb human artificial chromosome, containing the entire repertoire for the production of human antibodies, in primary bovine cells. NT with these cells generated "transchromosomic" cattle with the ability to produce functional human antibodies (Kuroiwa et al. 2002).

A great attraction of SCNT as a cell-meditated transgenesis method for livestock was that it enabled for the first time to generate livestock with site-specific genome modifications. Gene-targeting strategies by homologous recombination, established and known from the mouse ESC system, could be applied in primary livestock cells and result in the successful disruption of an endogenous gene in sheep (McCreath et al. 2000). Although other studies followed describing the successful knockout of genes in pigs, cattle, and goats, the technology was not the expected breakthrough for site-specific engineering of livestock genomes (Laible and Alonso-Gonzalez 2009). The limited growth potential of primary cells and their intrinsically low homologous recombination frequency resulted in very poor efficiencies for the isolation of correctly targeted cell clones which made gene targeting in livestock quite challenging. To some extent, rejuvenation of a senescent cell clone's proliferative life-span is possible by re-deriving fresh cells from a cloned NT fetus (Clark et al. 2003). Although time-consuming and costly, it provides the potential for introducing complex modifications requiring multiple steps which was demonstrated with the sequential targeting of each allele of two different genes in cattle (Kuroiwa et al. 2004). The limitations due to generally very low targeting efficiencies were first addressed by switching from electroporation of cells to introduce double-stranded DNA vectors for targeting to viral delivery of single-stranded DNA-targeting constructs. Targeting of mammalian cells with vectors based on recombinant adeno-associated virus (rAAV) achieved significantly improved targeting efficiencies (Russell and Hirata 1998) which, as was demonstrated in a pig study, can be the crucial difference between failing and succeeding in the isolation of correctly targeted cell clones (Rogers et al. 2008). One factor for the observed enhancement is the ability of rAAV to transduce primary cells with high efficiency. In addition, it was speculated that the presentation by the rAAV system of linear, single-stranded molecules might provide a superior substrate for the cellular machinery that facilitates HR at the target locus (Laible and Alonso-Gonzalez 2009).

The real breakthrough in lifting targeting efficiencies in somatic cells came with the application of programmable nucleases (Whyte et al. 2011; Yu et al. 2011). The site-specific introduction of nuclease-mediated DSB made it now possible to disrupt genes in somatic cells of livestock species with ease and sufficient efficiencies for the isolation of bi-allelically modified cell clones (Carlson et al. 2012). Moreover, the flexibility of the RNA-guided CRISPR/Cas9 system even allows for the simultaneous disruption of multiple genes (Cong et al. 2013; Wang et al. 2013; Ni et al. 2014). While this is also possible with cytoplasmic injection, the cell-mediated route involves the transfection of large numbers of cells that can be selected, screened, and fully characterized which is advantageous for the less efficient intro-duction of more complex modifications (Carlson et al. 2016; Liu et al. 2013; Wu et al. 2015). However, there is a trade-off in using NT which is ultimately required to generate an animal from cells verified for possessing a specific mutation due to its reliance on somatic cells instead of ESCs (Wells et al. 2003). The reprogram-ming of a somatic cell by NT is quite a drastic process that attempts to reconfigure the gene expression profile of the specialized somatic donor cell to a state capable to support normal development. The process is however very inefficient and fre-quently hampered by faulty or incomplete epigenetic reprogramming of the trans-ferred nucleus. The resulting aberrant expression of genes is associated with a high rate of losses throughout pregnancy and NT-related phenotypes affecting the viabil-ity of these animals (Wells 2005).

Conclusion

Today, researchers have access to a great selection of different transgenic tech-nologies offering a wide range of different attributes including simplicity, cost-effectiveness, high efficiency, and precise control. With this toolbox at hand, it appears that the initial technical limitations have been overcome and livestock genomes can now be modified with ease, precision, and complexity previously only possible in the mouse system. Still, the costs of large animals, longer gen-eration times, and regulatory and acceptability aspects are important consider-ations for genetically engineered livestock which can greatly vary dependent on the intended application. The interested reader is referred to subsequent chapters which will discuss the latest technological advancements and main agricultural and biomedical applications and provide the context for legal, ethical, and public concerns and considerations.

References

Alessio AP, Fili AE, Garrels W, Forcato DO, Olmos Nicotra MF, Liaudat AC, Bevacqua RJ, Savy V, Hiriart MI, Talluri TR, Owens JB, Ivics Z, Salamone DF, Moisyadi S, Kues WA, Bosch P (2016) Establishment of cell-based transposon-mediated transgenesis in cattle. Theriogenology 85(7):1297–1311, e1292. https://doi.org/10.1016/j.theriogenology.2015.12.016

Baguisi A, Behboodi E, Melican DT, Pollock JS, Destrempes MM, Cammuso C, Williams JL, Nims SD, Porter CA, Midura P, Palacios MJ, Ayres SL, Denniston RS, Hayes ML, Ziomek CA, Meade HM, Godke RA, Gavin WG, Overstrom EW, Echelard Y (1999) Production of goats by somatic cell nuclear transfer. Nat Biotechnol 17(5):456–461

Balciunas D, Wangensteen KJ, Wilber A, Bell J, Geurts A, Sivasubbu S, Wang X, Hackett PB, Largaespada DA, McIvor RS, Ekker SC (2006) Harnessing a high cargo-capacity transposon for genetic applications in vertebrates. PLoS Genet 2(11):e169. https://doi.org/10.1371/jour-nal.pgen.0020169

Bevacqua RJ, Pereyra-Bonnet F, Fernandez-Martin R, Salamone DF (2010) High rates of bovine blastocyst development after ICSI-mediated gene transfer assisted by chemical activation. Theriogenology 74(6):922–931. https://doi.org/10.1016/j.theriogenology.2010.04.017

Bevacqua RJ, Fernandez-Martin R, Canel NG, Gibbons A, Texeira D, Lange F, Vans Landschoot G, Savy V, Briski O, Hiriart MI, Grueso E, Ivics Z, Taboga O, Kues WA, Ferraris S, Salamone DF (2017) Assessing Tn5 and Sleeping Beauty for transpositional transgenesis by cytoplasmic injection into bovine and ovine zygotes. PLoS One 12(3):e0174025. https://doi.org/10.1371/journal.pone.0174025

Bosch P, Forcato DO, Alustiza FE, Alessio AP, Fili AE, Olmos Nicotra MF, Liaudat AC, Rodriguez N, Talluri TR, Kues WA (2015) Exogenous enzymes upgrade transgenesis and genetic engineering of farm animals. Cell Mol Life Sci 72(10):1907–1929. https://doi.org/10.1007/s00018-015-1842-1

Brackett BG, Baranska W, Sawicki W, Koprowski H (1971) Uptake of heterologous genome by mammalian spermatozoa and its transfer to ova through fertilization. Proc Natl Acad Sci U S A 68(2):353–357

Briggs R, King TJ (1952) Transplantation of living nuclei from blastula cells into enucleated frogs' eggs. Proc Natl Acad Sci U S A 38:455–463

Brinster RL, Avarbock MR (1994) Germline transmission of donor haplotype following spermatogonial transplantation. Proc Natl Acad Sci U S A 91(24):11303–11307. https://doi.org/10.1073/pnas.91.24.11303

Brinster RL, Zimmermann JW (1994) Spermatogenesis following male germ-cell transplantation. Proc Natl Acad Sci U S A 91(24):11298–11302. https://doi.org/10.1073/pnas.91.24.11298

Brinster RL, Sandgren EP, Behringer RR, Palmiter RD (1989) No simple solution for making transgenic mice. Cell 59(2):239–241

Buehr M, Meek S, Blair K, Yang J, Ure J, Silva J, McLay R, Hall J, Ying QL, Smith A (2008) Capture of authentic embryonic stem cells from rat blastocysts. Cell 135(7):1287–1298. https://doi.org/10.1016/j.cell.2008.12.007

Carlson DF, Garbe JR, Tan W, Martin MJ, Dobrinsky JR, Hackett PB, Clark KJ, Fahrenkrug SC (2011) Strategies for selection marker-free swine transgenesis using the Sleeping Beauty transposon system. Transgenic Res 20(5):1125–1137. https://doi.org/10.1007/s11248-010-9481-7

Carlson DF, Tan W, Lillico SG, Stverakova D, Proudfoot C, Christian M, Voytas DF, Long CR, Whitelaw CB, Fahrenkrug SC (2012) Efficient TALEN-mediated gene knockout in livestock. Proc Natl Acad Sci U S A 109(43):17382–17387. https://doi.org/10.1073/pnas.1211446109

Carlson DF, Lancto CA, Zang B, Kim ES, Walton M, Oldeschulte D, Seabury C, Sonstegard TS, Fahrenkrug SC (2016) Production of hornless dairy cattle from genome-edited cell lines. Nat Biotechnol 34(5):479–481. https://doi.org/10.1038/nbt.3560

Chan AW, Homan EJ, Ballou LU, Burns JC, Bremel RD (1998) Transgenic cattle produced by reverse-transcribed gene transfer in oocytes. Proc Natl Acad Sci U S A 95(24):14028–14033

Chang K, Qian J, Jiang M, Liu Y-H, Wu M-C, Chen C-D, Lai C-K, Lo H-L, Hsiao C-T, Brown L et al (2002) Effective generation of transgenic pigs and mice by linker based sperm-mediated gene transfer. BMC Biotechnol 2(1):5

Chen S, Lee B, Lee AY, Modzelewski AJ, He L (2016) Highly efficient mouse genome editing by CRISPR ribonucleoprotein electroporation of zygotes. J Biol Chem 291(28):14457–14467. https://doi.org/10.1074/jbc.M116.733154

Cibelli JB, Stice SL, Golueke PJ, Kane JJ, Jerry J, Blackwell C, Ponce de Leon FA, Robl JM (1998) Cloned transgenic calves produced from nonquiescent fetal fibroblasts. Science 280(5367):1256–1258

Clark J, Whitelaw B (2003) A future for transgenic livestock. Nat Rev Genet 4(10):825–833. https://doi.org/10.1038/nrg1183

Clark AJ, Ferrier P, Aslam S, Burl S, Denning C, Wylie D, Ross A, de Sousa P, Wilmut I, Cui W (2003) Proliferative lifespan is conserved after nuclear transfer. Nat Cell Biol 5(6):535–538

Clark KJ, Carlson DF, Fahrenkrug SC (2007) Pigs taking wing with transposons and recombinases. Genome Biol 8(Suppl 1):S13. https://doi.org/10.1186/gb-2007-8-s1-s13

Cong L, Ran FA, Cox D, Lin S, Barretto R, Habib N, Hsu PD, Wu X, Jiang W, Marraffini LA, Zhang F (2013) Multiplex genome engineering using CRISPR/Cas systems. Science 339(6121):819–823. https://doi.org/10.1126/science.1231143

Cornetta K, Tessanne K, Long C, Yao J, Satterfield C, Westhusin M (2013) Transgenic sheep generated by lentiviral vectors: safety and integration analysis of surrogates and their offspring. Transgenic Res 22(4):737–745. https://doi.org/10.1007/s11248-012-9674-3

Di Berardino MA (2001) Animal cloning--the route to new genomics in agriculture and medicine. Differentiation 68(2–3):67–83

Ding S, Wu X, Li G, Han M, Zhuang Y, Xu T (2005) Efficient transposition of the piggyBac (PB) transposon in mammalian cells and mice. Cell 122(3):473–483. https://doi.org/10.1016/j.cell.2005.07.013

Du X, Feng T, Yu D, Wu Y, Zou H, Ma S, Feng C, Huang Y, Ouyang H, Hu X, Pan D, Li N, Wu S (2015) Barriers for deriving transgene-free pig iPS cells with episomal vectors. Stem Cells 33(11):3228–3238. https://doi.org/10.1002/stem.2089

Dull T, Zufferey R, Kelly M, Mandel RJ, Nguyen M, Trono D, Naldini L (1998) A third-generation lentivirus vector with a conditional packaging system. J Virol 72(11):8463–8471

Dupuy AJ, Clark K, Carlson CM, Fritz S, Davidson AE, Markley KM, Finley K, Fletcher CF, Ekker SC, Hackett PB, Horn S, Largaespada DA (2002) Mammalian germ-line transgenesis by transposition. Proc Natl Acad Sci U S A 99(7):4495–4499. https://doi.org/10.1073/pnas.062630599

Eghbalsaied S, Ghaedi K, Laible G, Hosseini SM, Forouzanfar M, Hajian M, Oback F, Nasr-Esfahani MH, Oback B (2013) Exposure to DNA is insufficient for in vitro transgenesis of live bovine sperm and embryos. Reproduction 145(1):97–108. https://doi.org/10.1530/REP-12-0340

Evans MJ, Kaufman MH (1981) Establishment in culture of pluripotential cells from mouse embryos. Nature 292(5819):154–156

Fischer K, Kraner-Scheiber S, Petersen B, Rieblinger B, Buermann A, Flisikowska T, Flisikowski K, Christan S, Edlinger M, Baars W, Kurome M, Zakhartchenko V, Kessler B, Plotzki E, Szczerbal I, Switonski M, Denner J, Wolf E, Schwinzer R, Niemann H, Kind A, Schnieke A (2016) Efficient production of multi-modified pigs for xenotransplantation by 'combineering', gene stacking and gene editing. Sci Rep 6:29081. https://doi.org/10.1038/srep29081

Garcia-Vazquez FA, Ruiz S, Matas C, Izquierdo-Rico MJ, Grullon LA, De Ondiz A, Vieira L, Aviles-Lopez K, Gutierrez-Adan A, Gadea J (2010) Production of transgenic piglets using ICSI-sperm-mediated gene transfer in combination with recombinase RecA. Reproduction 140(2):259–272. https://doi.org/10.1530/REP-10-0129

Garrels W, Mates L, Holler S, Dalda A, Taylor U, Petersen B, Niemann H, Izsvak Z, Ivics Z, Kues WA (2011) Germline transgenic pigs by Sleeping Beauty transposition in porcine zygotes and targeted integration in the pig genome. PLoS One 6(8):e23573. https://doi.org/10.1371/journal.pone.0023573

Garrels W, Ivics Z, Kues WA (2012) Precision genetic engineering in large mammals. Trends Biotechnol 30(7):386–393. https://doi.org/10.1016/j.tibtech.2012.03.008

Garrels W, Talluri TR, Apfelbaum R, Carratala YP, Bosch P, Potzsch K, Grueso E, Ivics Z, Kues WA (2016) One-step multiplex transgenesis via Sleeping Beauty transposition in cattle. Sci Rep 6:21953. https://doi.org/10.1038/srep21953

Germon S, Bouchet N, Casteret S, Carpentier G, Adet J, Bigot Y, Auge-Gouillou C (2009) Mariner Mos1 transposase optimization by rational mutagenesis. Genetica 137(3):265–276. https://doi.org/10.1007/s10709-009-9375-x

Gilboa E, Eglitis MA, Kantoff PW, Anderson WF (1986) Transfer and expression of cloned genes using retroviral vectors. Biotechniques 4:504–512

Gordon JW, Scangos GA, Plotkin DJ, Barbosa JA, Ruddle FH (1980) Genetic transformation of mouse embryos by microinjection of purified DNA. Proc Natl Acad Sci U S A 77(12):7380–7384

Grabarek JB, Plusa B, Glover DM, Zernicka-Goetz M (2002) Efficient delivery of dsRNA into zona-enclosed mouse oocytes and preimplantation embryos by electroporation. Genesis 32(4):269–276

Gurdon JB (1999) Genetic reprogramming following nuclear transplantation in Amphibia. Semin Cell Dev Biol 10(3):239–243

Hamanaka S, Yamaguchi T, Kobayashi T, Kato-Itoh M, Yamazaki S, Sato H, Umino A, Wakiyama Y, Arai M, Sanbo M, Hirabayashi M, Nakauchi H (2011) Generation of germline-competent rat induced pluripotent stem cells. PLoS One 6(7):e22008. https://doi.org/10.1371/journal.pone.0022008

Hammer RE, Pursel VG, Rexroad CE Jr, Wall RJ, Bolt DJ, Ebert KM, Palmiter RD, Brinster RL (1985) Production of transgenic rabbits, sheep and pigs by microinjection. Nature 315(6021):680–683

Harel-Markowitz E, Gurevich M, Shore LS, Katz A, Stram Y, Shemesh M (2009) Use of sperm plasmid DNA lipofection combined with REMI (restriction enzyme-mediated insertion) for production of transgenic chickens expressing eGFP (enhanced green fluorescent protein) or human follicle-stimulating hormone. Biol Reprod 80(5):1046–1052. https://doi.org/10.1095/biolreprod.108.070375

Hashimoto M, Yamashita Y, Takemoto T (2016) Electroporation of Cas9 protein/sgRNA into early pronuclear zygotes generates non-mosaic mutants in the mouse. Dev Biol 418(1):1–9. https://doi.org/10.1016/j.ydbio.2016.07.017

Hauschild-Quintern J, Petersen B, Cost GJ, Niemann H (2013) Gene knockout and knockin by zinc-finger nucleases: current status and perspectives. Cell Mol Life Sci 70(16):2969–2983. https://doi.org/10.1007/s00018-012-1204-1

Hofmann A, Kessler B, Ewerling S, Weppert M, Vogg B, Ludwig H, Stojkovic M, Boelhauve M, Brem G, Wolf E, Pfeifer A (2003) Efficient transgenesis in farm animals by lentiviral vectors. EMBO Rep 4(11):1054–1060

Hofmann A, Zakhartchenko V, Weppert M, Sebald H, Wenigerkind H, Brem G, Wolf E, Pfeifer A (2004) Generation of transgenic cattle by lentiviral gene transfer into oocytes. Biol Reprod 71(2):405–409

Hofmann A, Kessler B, Ewerling S, Kabermann A, Brem G, Wolf E, Pfeifer A (2006) Epigenetic regulation of lentiviral transgene vectors in a large animal model. Mol Ther 13(1):59–66. https://doi.org/10.1016/j.ymthe.2005.07.685

Honaramooz A, Yang Y (2011) Recent advances in application of male germ cell transplantation in farm animals. Vet Med Int. https://doi.org/10.4061/2011/657860

Honaramooz A, Megee S, Zeng W, Destrempes MM, Overton SA, Luo J, Galantino-Homer H, Modelski M, Chen F, Blash S, Melican DT, Gavin WG, Ayres S, Yang F, Wang PJ, Echelard Y, Dobrinski I (2008) Adeno-associated virus (AAV)-mediated transduction of male germ line stem cells results in transgene transmission after germ cell transplantation. FASEB J 22(2):374–382. https://doi.org/10.1096/fj.07-8935com

Ikeno M, Grimes B, Okazaki T, Nakano M, Saitoh K, Hoshino H, McGill NI, Cooke H, Masumoto H (1998) Construction of YAC-based mammalian artificial chromosomes. Nat Biotechnol 16(5):431–439

Ivics Z, Hackett PB, Plasterk RH, Izsvak Z (1997) Molecular reconstruction of Sleeping Beauty, a Tc1-like transposon from fish, and its transposition in human cells. Cell 91(4):501–510

Jabed A, Wagner S, McCracken J, Wells DN, Laible G (2012) Targeted microRNA expression in dairy cattle directs production of β-lactoglobulin-free, high-casein milk. Proc Natl Acad Sci U S A 109(42):16811–16816. https://doi.org/10.1073/pnas.1210057109

Jaenisch R, Mintz B (1974) Simian virus 40 DNA sequences in DNA of healthy adult mice derived from preimplantation blastocysts injected with viral DNA. Proc Natl Acad Sci U S A 71(4):1250–1254

Jang CW, Behringer RR (2007) Transposon-mediated transgenesis in rats. CSH Protoc 2007:pdb prot4866. https://doi.org/10.1101/pdb.prot4866

Kaneko T, Sakuma T, Yamamoto T, Mashimo T (2014) Simple knockout by electroporation of engineered endonucleases into intact rat embryos. Sci Rep 4:6382. https://doi.org/10.1038/srep06382

Katter K, Geurts AM, Hoffmann O, Mates L, Landa V, Hiripi L, Moreno C, Lazar J, Bashir S, Zidek V, Popova E, Jerchow B, Becker K, Devaraj A, Walter I, Grzybowksi M, Corbett M, Filho AR, Hodges MR, Bader M, Ivics Z, Jacob HJ, Pravenec M, Bosze Z, Rulicke T, Izsvak Z (2013) Transposon-mediated transgenesis, transgenic rescue, and tissue-specific gene expression in rodents and rabbits. FASEB J 27(3):930–941. https://doi.org/10.1096/fj.12-205526

Kubisch HM, Larson MA, Eichen PA, Wilson JM, Roberts RM (1997) Adenovirus-mediated gene transfer by perivitelline microinjection of mouse, rat, and cow embryos. Biol Reprod 56(1):119–124

Kumar D, Talluri TR, Anand T, Kues WA (2015) Induced pluripotent stem cells: mechanisms, achievements and perspectives in farm animals. World J Stem Cells 7:315–328

Kuroiwa Y, Kasinathan P, Choi YJ, Naeem R, Tomizuka K, Sullivan EJ, Knott JG, Duteau A, Goldsby RA, Osborne BA et al (2002) Cloned transchromosomic calves producing human immunoglobulin. Nat Biotechnol 20(9):889–894

Kuroiwa Y, Kasinathan P, Matsushita H, Sathiyaselan J, Sullivan EJ, Kakitani M, Tomizuka K, Ishida I, Robl JM (2004) Sequential targeting of the genes encoding immunoglobulin-mu and prion protein in cattle. Nat Genet 36(7):775–780

Laible G, Alonso-Gonzalez L (2009) Gene targeting from laboratory to livestock: current status and emerging concepts. Biotechnol J 4(9):1278–1292. https://doi.org/10.1002/biot.200900006

Lampe DJ, Akerley BJ, Rubin EJ, Mekalanos JJ, Robertson HM (1999) Hyperactive transposase mutants of the Himar1 mariner transposon. Proc Natl Acad Sci U S A 96(20):11428–11433

Lavitrano M, Camaioni A, Fazio VM, Dolci S, Farace MG, Spadafora C (1989) Sperm cells as vectors for introducing foreign DNA into eggs: genetic transformation of mice. Cell 57(5):717–723

Lavitrano M, Forni M, Varzi V, Pucci L, Bacci ML, Di Stefano C, Fioretti D, Zoraqi G, Moioli B, Rossi M, Lazzereschi D, Stoppacciaro A, Seren E, Alfani D, Cortesini R, Frati L (1997) Sperm-mediated gene transfer: production of pigs transgenic for a human regulator of complement activation. Transplant Proc 29(8):3508–3509

Lavitrano M, Giovannoni R, Cerrito MG (2013) Methods for sperm-mediated gene transfer. Methods Mol Biol 927:519–529. https://doi.org/10.1007/978-1-62703-038-0_44

Li P, Tong C, Mehrian-Shai R, Jia L, Wu N, Yan Y, Maxson RE, Schulze EN, Song H, Hsieh CL, Pera MF, Ying QL (2008) Germline competent embryonic stem cells derived from rat blastocysts. Cell 135(7):1299–1310. https://doi.org/10.1016/j.cell.2008.12.006

Li MA, Turner DJ, Ning Z, Yusa K, Liang Q, Eckert S, Rad L, Fitzgerald TW, Craig NL, Bradley A (2011) Mobilization of giant piggyBac transposons in the mouse genome. Nucleic Acids Res 39(22):e148. https://doi.org/10.1093/nar/gkr764

Li Z, Zeng F, Meng F, Xu Z, Zhang X, Huang X, Tang F, Gao W, Shi J, He X, Liu D, Wang C, Urschitz J, Moisyadi S, Wu Z (2014) Generation of transgenic pigs by cytoplasmic injection of piggyBac transposase-based pmGENIE-3 plasmids. Biol Reprod 90(5):93. https://doi.org/10.1095/biolreprod.113.116905

Lillico S, Vasey D, King T, Whitelaw B (2011) Lentiviral transgenesis in livestock. Transgenic Res 20(3):441–442. https://doi.org/10.1007/s11248-010-9448-8

Lillico SG, Proudfoot C, Carlson DF, Stverakova D, Neil C, Blain C, King TJ, Ritchie WA, Tan W, Mileham AJ, McLaren DG, Fahrenkrug SC, Whitelaw CB (2013) Live pigs produced from genome edited zygotes. Sci Rep 3:2847. https://doi.org/10.1038/srep02847

Liu G, Geurts AM, Yae K, Srinivasan AR, Fahrenkrug SC, Largaespada DA, Takeda J, Horie K, Olson WK, Hackett PB (2005) Target-site preferences of Sleeping Beauty transposons. J Mol Biol 346(1):161–173. https://doi.org/10.1016/j.jmb.2004.09.086

Liu X, Wang Y, Guo W, Chang B, Liu J, Guo Z, Quan F, Zhang Y (2013) Zinc-finger nickase-mediated insertion of the lysostaphin gene into the beta-casein locus in cloned cows. Nat Commun 4:2565. https://doi.org/10.1038/ncomms3565

Lois C, Hong EJ, Pease S, Brown EJ, Baltimore D (2002) Germline transmission and tissue-specific expression of transgenes delivered by lentiviral vectors. Science 295(5556):868–872

Ma J, Li Q, Li Y, Wen X, Li Z, Zhang Z, Zhang J, Yu Z, Li N (2016) Expression of recombinant human alpha-lactalbumin in milk of transgenic cloned pigs is sufficient to enhance intestinal growth and weight gain of suckling piglets. Gene 584(1):7–16. https://doi.org/10.1016/j.gene.2016.02.024

Maga EA, Sargent RG, Zeng H, Pati S, Zarling DA, Oppenheim SM, Collette NM, Moyer AL, Conrad-Brink JS, Rowe JD, BonDurant RH, Anderson GB, Murray JD (2003) Increased efficiency of transgenic livestock production. Transgenic Res 12(4):485–496

Marh J, Stoytcheva Z, Urschitz J, Sugawara A, Yamashiro H, Owens JB, Stoytchev I, Pelczar P, Yanagimachi R, Moisyadi S (2012) Hyperactive self-inactivating piggyBac for transposase-enhanced pronuclear microinjection transgenesis. Proc Natl Acad Sci U S A 109(47):19184–19189. https://doi.org/10.1073/pnas.1216473109

Martin GR (1981) Isolation of a pluripotent cell line from early mouse embryos cultured in medium conditioned by teratocarcinoma stem cells. Proc Natl Acad Sci U S A 78(12):7634–7638

Mates L, Chuah MK, Belay E, Jerchow B, Manoj N, Acosta-Sanchez A, Grzela DP, Schmitt A, Becker K, Matrai J, Ma L, Samara-Kuko E, Gysemans C, Pryputniewicz D, Miskey C, Fletcher B, VandenDriessche T, Ivics Z, Izsvak Z (2009) Molecular evolution of a novel hyperactive Sleeping Beauty transposase enables robust stable gene transfer in vertebrates. Nat Genet 41(6):753–761. https://doi.org/10.1038/ng.343

McCormick SP, Nielsen LB (1998) Expression of large genomic clones in transgenic mice: new insights into apolipoprotein B structure, function and regulation. Curr Opin Lipidol 9(2):103–111

McCreath KJ, Howcroft J, Campbell KH, Colman A, Schnieke AE, Kind AJ (2000) Production of gene-targeted sheep by nuclear transfer from cultured somatic cells. Nature 405(6790):1066–1069

McGrath J, Solter D (1983) Nuclear transplantation in the mouse embryo by microsurgery and cell fusion. Science 220(4603):1300–1302

McGrew MJ, Sherman A, Ellard FM, Lillico SG, Gilhooley HJ, Kingsman AJ, Mitrophanous KA, Sang H (2004) Efficient production of germline transgenic chickens using lentiviral vectors. EMBO Rep 5(7):728–733. https://doi.org/10.1038/sj.embor.7400171

Miyoshi H, Blomer U, Takahashi M, Gage FH, Verma IM (1998) Development of a self-inactivating lentivirus vector. J Virol 72(10):8150–8157

Moehle EA, Rock JM, Lee YL, Jouvenot Y, DeKelver RC, Gregory PD, Urnov FD, Holmes MC (2007) Targeted gene addition into a specified location in the human genome using designed zinc finger nucleases. Proc Natl Acad Sci U S A 104(9):3055–3060. https://doi.org/10.1073/pnas.0611478104

Moisyadi S, Kaminski JM, Yanagimachi R (2009) Use of intracytoplasmic sperm injection (ICSI) to generate transgenic animals. Comp Immunol Microbiol Infect Dis 32(2):47–60. https://doi.org/10.1016/j.cimid.2008.05.003

Moreira PN, Giraldo P, Cozar P, Pozueta J, Jimenez A, Montoliu L, Gutierrez-Adan A (2004) Efficient generation of transgenic mice with intact yeast artificial chromosomes by intracytoplasmic sperm injection. Biol Reprod 71(6):1943–1947. https://doi.org/10.1095/biolreprod.104.032904

Mussolino C, Cathomen T (2012) TALE nucleases: tailored genome engineering made easy. Curr Opin Biotechnol 23(5):644–650. https://doi.org/10.1016/j.copbio.2012.01.013

Nagano M, Brinster CJ, Orwig KE, Ryu BY, Avarbock MR, Brinster RL (2001) Transgenic mice produced by retroviral transduction of male germ-line stem cells. Proc Natl Acad Sci U S A 98(23):13090–13095. https://doi.org/10.1073/pnas.231473498

Nakanishi A, Iritani A (1993) Gene transfer in the chicken by sperm-mediated methods. Mol Reprod Dev 36(2):258–261. https://doi.org/10.1002/mrd.1080360225

Ni W, Qiao J, Hu S, Zhao X, Regouski M, Yang M, Polejaeva IA, Chen C (2014) Efficient gene knockout in goats using CRISPR/Cas9 system. PLoS One 9(9):e106718. https://doi.org/10.1371/journal.pone.0106718

Nichols J, Smith A (2009) Naive and primed pluripotent states. Cell Stem Cell 4(6):487–492. https://doi.org/10.1016/j.stem.2009.05.015

Ogorevc J, Orehek S, Dovč P (2016) Cellular reprogramming in farm animals: an overview of iPSC generation in the mammalian farm animal species. J Anim Sci Biotechnol 7(1). https://doi.org/10.1186/s40104-016-0070-3

Okita K, Ichisaka T, Yamanaka S (2007) Generation of germline-competent induced pluripotent stem cells. Nature 448(7151):313–317. https://doi.org/10.1038/nature05934

Pereyra-Bonnet F, Fernandez-Martin R, Olivera R, Jarazo J, Vichera G, Gibbons A, Salamone D (2008) A unique method to produce transgenic embryos in ovine, porcine, feline, bovine and equine species. Reprod Fertil Dev 20(7):741–749

Perez A, Solano R, Castro FO, Lleonart R, De Armas R, Martinez R, Aguilar A, Herrera L, De La Fuente J (1991) Sperm cell mediated gene transfer in cattle. Biotecnol Apl 8(1):90–94

Perry AC, Wakayama T, Kishikawa H, Kasai T, Okabe M, Toyoda Y, Yanagimachi R (1999) Mammalian transgenesis by intracytoplasmic sperm injection. Science 284(5417):1180–1183

Peterson KR, Clegg CH, Li Q, Stamatoyannopoulos G (1997) Production of transgenic mice with yeast artificial chromosomes. Trends Genet 13(2):61–66

Pfeifer A, Ikawa M, Dayn Y, Verma IM (2002) Transgenesis by lentiviral vectors: lack of gene silencing in mammalian embryonic stem cells and preimplantation embryos. Proc Natl Acad Sci U S A 99(4):2140–2145

Polejaeva IA, Chen SH, Vaught TD, Page RL, Mullins J, Ball S, Dai Y, Boone J, Walker S, Ayares DL, Colman A, Campbell KH (2000) Cloned pigs produced by nuclear transfer from adult somatic cells. Nature 407(6800):86–90

Proudfoot C, Carlson DF, Huddart R, Long CR, Pryor JH, King TJ, Lillico SG, Mileham AJ, McLaren DG, Whitelaw CB, Fahrenkrug SC (2015) Genome edited sheep and cattle. Transgenic Res 24(1):147–153. https://doi.org/10.1007/s11248-014-9832-x

Qin W, Dion SL, Kutny PM, Zhang Y, Cheng AW, Jillette NL, Malhotra A, Geurts AM, Chen YG, Wang H (2015) Efficient CRISPR/Cas9-mediated genome editing in mice by zygote electroporation of nuclease. Genetics 200(2):423–430. https://doi.org/10.1534/genetics.115.176594

Ritchie WA, King T, Neil C, Carlisle AJ, Lillico S, McLachlan G, Whitelaw CB (2009) Transgenic sheep designed for transplantation studies. Mol Reprod Dev 76(1):61–64. https://doi.org/10.1002/mrd.20930

Robertson E, Bradley A, Kuehn M, Evans M (1986) Germ-line transmission of genes introduced into cultured pluripotential cells by retroviral vector. Nature 323(6087):445–448. https://doi.org/10.1038/323445a0

Rogers CS, Hao Y, Rokhlina T, Samuel M, Stoltz DA, Li Y, Petroff E, Vermeer DW, Kabel AC, Yan Z, Spate L, Wax D, Murphy CN, Rieke A, Whitworth K, Linville ML, Korte SW, Engelhardt JF, Welsh MJ, Prather RS (2008) Production of CFTR-null and CFTR-DeltaF508 heterozygous pigs by adeno-associated virus-mediated gene targeting and somatic cell nuclear transfer. J Clin Invest 118(4):1571–1577. https://doi.org/10.1172/JCI34773

Rostovskaya M, Naumann R, Fu J, Obst M, Mueller D, Stewart AF, Anastassiadis K (2013) Transposon mediated BAC transgenesis via pronuclear injection of mouse zygotes. Genesis 51(2):135–141. https://doi.org/10.1002/dvg.22362

Russell DW, Hirata RK (1998) Human gene targeting by viral vectors. Nat Genet 18(4):325–330. https://doi.org/10.1038/ng0498-325

Sander JD, Joung JK (2014) CRISPR-Cas systems for editing, regulating and targeting genomes. Nat Biotechnol 32(4):347–350. https://doi.org/10.1038/nbt.2842

Schambach A, Zychlinski D, Ehrnstroem B, Baum C (2013) Biosafety features of lentiviral vectors. Hum Gene Ther 24(2):132–142. https://doi.org/10.1089/hum.2012.229

Schnieke AE, Kind AJ, Ritchie WA, Mycock K, Scott AR, Ritchie M, Wilmut I, Colman A, Campbell KH (1997) Human factor IX transgenic sheep produced by transfer of nuclei from transfected fetal fibroblasts. Science 278(5346):2130–2133

Seo BB, Kim CH, Yamanouchi K, Takahashi M, Sawasaki T, Tachi C, Tojo H (2000) Co-injection of restriction enzyme with foreign DNA into the pronucleus for elevating production efficiencies of transgenic animals. Anim Reprod Sci 63(1–2):113–122

Shadanloo F, Najafi MH, Hosseini SM, Hajian M, Forouzanfar M, Ghaedi K, Abedi P, Ostadhosseini S, Hosseini L, Eskandari-Nasab MP, Esfahani MH (2010) Sperm status and DNA dose play key roles in sperm/ICSI-mediated gene transfer in caprine. Mol Reprod Dev 77(10):868–875. https://doi.org/10.1002/mrd.21228

Shemesh M, Gurevich M, Harel-Markowitz E, Benvenisti L, Shore LS, Stram Y (2000) Gene integration into bovine sperm genome and its expression in transgenic offspring. Mol Reprod Dev 56(2):306–308

Shinohara ET, Kaminski JM, Segal DJ, Pelczar P, Kolhe R, Ryan T, Coates CJ, Fraser MJ, Handler AM, Yanagimachi R, Moisyadi S (2007) Active integration: new strategies for transgenesis. Transgenic Res 16(3):333–339. https://doi.org/10.1007/s11248-007-9077-z

Soto DA, Ross PJ (2016) Pluripotent stem cells and livestock genetic engineering. Transgenic Res 25(3):289–306. https://doi.org/10.1007/s11248-016-9929-5

Takahashi K, Yamanaka S (2006) Induction of pluripotent stem cells from mouse embryonic and adult fibroblast cultures by defined factors. Cell 126(4):663–676. https://doi.org/10.1016/j.cell.2006.07.024

Tan W, Proudfoot C, Lillico SG, Whitelaw CBA (2016) Gene targeting, genome editing: from Dolly to editors. Transgenic Res 25(3):273–287. https://doi.org/10.1007/s11248-016-9932-x

Tanihara F, Takemoto T, Kitagawa E, Rao S, Do LT, Onishi A, Yamashita Y, Kosugi C, Suzuki H, Sembon S, Suzuki S, Nakai M, Hashimoto M, Yasue A, Matsuhisa M, Noji S, Fujimura T, Fuchimoto D, Otoi T (2016) Somatic cell reprogramming-free generation of genetically modified pigs. Sci Adv 2(9):e1600803. https://doi.org/10.1126/sciadv.1600803

Thomson JA (1998) Embryonic stem cell lines derived from human blastocysts. Science 282(5391):1145–1147

Thomson JA, Kalishman J, Golos TG, Durning M, Harris CP, Becker RA, Hearn JP (1995) Isolation of a primate embryonic stem cell line. Proc Natl Acad Sci U S A 92(17):7844–7848. https://doi.org/10.1073/pnas.92.17.7844

Tian Y, Li W, Wang L, Liu C, Lin J, Zhang X, Zhang N, He S, Huang J, Jia B, Liu M (2013) Expression of 2A peptide mediated tri-fluorescent protein genes were regulated by epigenetics in transgenic sheep. Biochem Biophys Res Commun 434(3):681–687. https://doi.org/10.1016/j.bbrc.2013.04.009

Tsukui T, Kanegae Y, Saito I, Toyoda Y (1996) Transgenesis by adenovirus-mediated gene transfer into mouse zona-free eggs. Nat Biotechnol 14(8):982–985

Urschitz J, Kawasumi M, Owens J, Morozumi K, Yamashiro H, Stoytchev I, Marh J, Dee JA, Kawamoto K, Coates CJ, Kaminski JM, Pelczar P, Yanagimachi R, Moisyadi S (2010) Helper-independent piggyBac plasmids for gene delivery approaches: strategies for avoiding potential genotoxic effects. Proc Natl Acad Sci U S A 107(18):8117–8122. https://doi.org/10.1073/pnas.1003674107

Verma IM, Weitzman MD (2005) Gene therapy: twenty-first century medicine. Annu Rev Biochem 74:711–738. https://doi.org/10.1146/annurev.biochem.74.050304.091637

Wall RJ (2001) Pronuclear microinjection. Cloning Stem Cells 3(4):209–220. https://doi.org/10.1089/15362300152725936

Wang HJ, Lin AX, Chen YF (2003) Association of rabbit sperm cells with exogenous DNA. Anim Biotechnol 14(2):155–165. https://doi.org/10.1081/ABIO-120026485

Wang H, Yang H, Shivalila CS, Dawlaty MM, Cheng AW, Zhang F, Jaenisch R (2013) One-step generation of mice carrying mutations in multiple genes by CRISPR/Cas-mediated genome engineering. Cell 153(4):910–918. https://doi.org/10.1016/j.cell.2013.04.025

Wei J, Gaynor P, Cole S, Brophy B, Oback B, Laible G (2018) Developing the condition for bovine zygote-mediated genome editing by electroporation. Proceedings of the World Congress on Genetics Applied to Livestock Production 11.1118

Wells DN (2005) Animal cloning: problems and prospects. Rev Sci Tech 24(1):251–264

Wells K, Moore K, Wall R (1999) Transgene vectors go retro. Nat Biotechnol 17(1):25–26

Wells DN, Oback B, Laible G (2003) Cloning livestock: a return to embryonic cells. Trends Biotechnol 21(10):428–432

West FD, Terlouw SL, Kwon DJ, Mumaw JL, Dhara SK, Hasneen K, Dobrinsky JR, Stice SL (2010) Porcine induced pluripotent stem cells produce chimeric offspring. Stem Cells Dev 19(8):1211–1220. https://doi.org/10.1089/scd.2009.0458

West FD, Uhl EW, Liu Y, Stowe H, Lu Y, Yu P, Gallegos-Cardenas A, Pratt SL, Stice SL (2011) Brief report: Chimeric pigs produced from induced pluripotent stem cells demonstrate germline transmission and no evidence of tumor formation in young pigs. Stem Cells 29(10):1640–1643. https://doi.org/10.1002/stem.713

Whitelaw CB, Radcliffe PA, Ritchie WA, Carlisle A, Ellard FM, Pena RN, Rowe J, Clark AJ, King TJ, Mitrophanous KA (2004) Efficient generation of transgenic pigs using equine infectious anaemia virus (EIAV) derived vector. FEBS Lett 571(1–3):233–236. https://doi.org/10.1016/j.febslet.2004.06.076

Whyte JJ, Zhao J, Wells KD, Samuel MS, Whitworth KM, Walters EM, Laughlin MH, Prather RS (2011) Gene targeting with zinc finger nucleases to produce cloned eGFP knockout pigs. Mol Reprod Dev 78(1):2. https://doi.org/10.1002/mrd.21271

Willadsen SM (1986) Nuclear transplantation in sheep embryos. Nature 320(6057):63–65

Wilmut I, Schnieke AE, McWhir J, Kind AJ, Campbell KH (1997) Viable offspring derived from fetal and adult mammalian cells. Nature 385(6619):810–813

Wilson MH, Coates CJ, George AL Jr (2007) PiggyBac transposon-mediated gene transfer in human cells. Mol Ther 15(1):139–145. https://doi.org/10.1038/sj.mt.6300028

Wu SC, Meir YJ, Coates CJ, Handler AM, Pelczar P, Moisyadi S, Kaminski JM (2006) piggyBac is a flexible and highly active transposon as compared to sleeping beauty, Tol2, and Mos1 in mammalian cells. Proc Natl Acad Sci U S A 103(41):15008–15013. https://doi.org/10.1073/pnas.0606979103

Wu H, Wang Y, Zhang Y, Yang M, Lv J, Liu J (2015) TALE nickase-mediated SP110 knockin endows cattle with increased resistance to tuberculosis. Proc Natl Acad Sci U S A 112(13):E1530–E1539. https://doi.org/10.1073/pnas.1421587112

Xiao A, Wang Z, Hu Y, Wu Y, Luo Z, Yang Z, Zu Y, Li W, Huang P, Tong X, Zhu Z, Lin S, Zhang B (2013) Chromosomal deletions and inversions mediated by TALENs and CRISPR/Cas in zebrafish. Nucleic Acids Res 41(14):e141. https://doi.org/10.1093/nar/gkt464

Xu P, Li Q, Jiang K, Yang Q, Bi M, Jiang C, Wang X, Wang C, Li L, Qiao C, Gong H, Xing Y, Ren J (2016) BAC mediated transgenic Large White boars with FSHalpha/beta genes from Chinese Erhualian pigs. Transgenic Res 25(5):693–709. https://doi.org/10.1007/s11248-016-9963-3

Yanagimachi R (2005) Intracytoplasmic injection of spermatozoa and spermatogenic cells: its biology and applications in humans and animals. Reprod Biomed Online 10(2):247–288

Yang P, Wang J, Gong G, Sun X, Zhang R, Du Z, Liu Y, Li R, Ding F, Tang B, Dai Y, Li N (2008) Cattle mammary bioreactor generated by a novel procedure of transgenic cloning for large-scale production of functional human lactoferrin. PLoS One 3(10):e3453. https://doi.org/10.1371/journal.pone.0003453

Yang H, Wang H, Shivalila CS, Cheng AW, Shi L, Jaenisch R (2013) One-step generation of mice carrying reporter and conditional alleles by CRISPR/Cas-mediated genome engineering. Cell 154(6):1370–1379. https://doi.org/10.1016/j.cell.2013.08.022

Yant SR, Wu X, Huang Y, Garrison B, Burgess SM, Kay MA (2005) High-resolution genome-wide mapping of transposon integration in mammals. Mol Cell Biol 25(6):2085–2094. https://doi.org/10.1128/MCB.25.6.2085-2094.2005

Ying QL, Wray J, Nichols J, Batlle-Morera L, Doble B, Woodgett J, Cohen P, Smith A (2008) The ground state of embryonic stem cell self-renewal. Nature 453(7194):519–523. https://doi.org/10.1038/nature06968

Yu S, Luo J, Song Z, Ding F, Dai Y, Li N (2011) Highly efficient modification of beta-lactoglobulin (BLG) gene via zinc-finger nucleases in cattle. Cell Res 21(11):1638–1640. https://doi.org/10.1038/cr.2011.153

Yusa K, Zhou L, Li MA, Bradley A, Craig NL (2011) A hyperactive piggyBac transposase for mammalian applications. Proc Natl Acad Sci U S A 108(4):1531–1536. https://doi.org/10.1073/pnas.1008322108

Zeng W, Tang L, Bondareva A, Honaramooz A, Tanco V, Dores C, Megee S, Modelski M, Rodriguez-Sosa JR, Paczkowski M, Silva E, Wheeler M, Krisher RL, Dobrinski I (2013) Viral transduction of male germline stem cells results in transgene transmission after germ cell transplantation in pigs. Biol Reprod 88(1). https://doi.org/10.1095/biolreprod.112.104422

Zhang Y, Xi Q, Ding J, Cai W, Meng F, Zhou J, Li H, Jiang Q, Shu G, Wang S, Zhu X, Gao P, Wu Z (2012) Production of transgenic pigs mediated by pseudotyped lentivirus and sperm. PLoS One 7(4):e35335. https://doi.org/10.1371/journal.pone.0035335

Zhao Y, Wei H, Wang Y, Wang L, Yu M, Fan J, Zheng S, Zhao C (2010) Production of transgenic goats by sperm-mediated exogenous DNA transfer method. Asian-Aust J Anim Sci 23(1):33–40. https://doi.org/10.5713/ajas.2010.90216

Zufferey R, Dull T, Mandel RJ, Bukovsky A, Quiroz D, Naldini L, Trono D (1998) Self-inactivating lentivirus vector for safe and efficient in vivo gene delivery. J Virol 72(12):9873–9880

DNA Nucleases and their Use in Livestock Production

7

Abstract

DNA nucleases, including zinc-finger nucleases (ZFN), transcription activator-like endonucleases (TALENS), and meganucleases, possess long recognition sites and cutting domains and are thus capable of cutting DNA in a very specific manner. These molecular scissors mediate targeted genetic alterations by enhancing the DNA mutation rate via induction of double-strand breaks at a predetermined genomic site. Compared to conventional homologous recombination-based gene targeting, DNA nucleases can increase the targeting rate up to 10,000-fold, and gene disruption via mutagenic DNA repair is stimulated at a similar frequency. The successful application of different DNA nucleases has been demonstrated in a multitude of organisms, including insects, amphibians, plants, nematodes, and mammals, including livestock animals. Recently, another novel class of molecular scissors was described that uses short RNA sequences to target a specific genomic site (Fig. 7.1). The CRISPR/CAS9 originates from a bacterial defense mechanism and can be programmed to target almost any site within a genome. The ease and low costs to create very specific genetic alterations by DNA nucleases have revolutionized the production of genetically modified livestock. Current results indicate that DNA nucleases can be successfully employed in a broad range of organisms which renders them useful for improving the understanding of complex physiological systems, producing genetically modified animals, including creating large animal models for human diseases and creating specific cell lines. Genetic modifications could also increase animal welfare by making dehorning and sexing obsolete or by making farm animals resistant/resilient against specific pathogens. Livestock with a desired phenotype or trait can now be produced with previously unknown precision and ease and within a very short time frame

B. Petersen
Institute of Farm Animal Genetics, Friedrich-Loeffler-Institut (FLI),
Neustadt, a.Rbge, Mariensee, Germany
e-mail: bjoern.petersen@fli.de

© Springer International Publishing AG, part of Springer Nature 2018
H. Niemann, C. Wrenzycki (eds.), *Animal Biotechnology 2*,
https://doi.org/10.1007/978-3-319-92348-2_7

considered to be impossible before their advent. This chapter provides an update on DNA nucleases and their underlying mechanism and focuses on their use in livestock production. It has to be kept in mind that, at the time of writing this chapter, none of the genetically modified livestock has entered the food chain or had been used for the production of livestock-derived products.

7.1 Introduction

7.1.1 The Importance of Genetically Modified Livestock Animals

Genetically modified livestock animals play an important role not only in basic research but also in highly diverse areas such as food improvement, disease resistance, human disease models, and recombinant therapeutics production. They contribute to human health by serving as models for the treatment of diseases and disorders as well as a source of biomaterials used for rebuilding or replacing tissues and organs and by producing recombinant therapeutics in their body fluids (Kues and Niemann 2004). Additionally, the enormous challenges arising from the growing global human population and the ecological consequences thereof demand new solutions. Over the next 50 years, the world's farmers and ranchers will be called upon to produce more food than has been produced in the past 10,000 years combined and to do so in an ecological sensitive manner (Data from FAO, www.fao.org).

Mice have historically been the prime medical model owing to their ease to be kept and bred. The production of numerous inbred lines improved the reproducibility of experimental results. But the main advantage of using mice was the availability of murine embryonic stem cells which display an unlimited replication capacity and are compatible with the sophisticated selection methods needed to select for the precise genetic alterations that occur at very low frequencies, i.e., 10^{-5}–10^{-8} (Mansour et al. 1988; Denning et al. 2001). This led to the establishment of thousands of knockout mouse strains for basic research and models for human diseases. However, the complete phenotype of human diseases cannot always be reflected in genetically modified mice, which might be due to the high phylogenetic distance between human and mice. A prominent example is mouse models for cystic fibrosis that do not show the identical pathological phenotype that humans encounter when carrying the same mutant gene (Welsh et al. 2009; Rogers et al. 2008). In contrast, pigs carrying the same genetic mutation for cystic fibrosis exhibited the important features of the illness in a humanlike manner (Rogers et al. 2008).

The domestic pig provides valuable resources for biomedical research as pigs share many similarities with humans with regard to genetics, body size, anatomy, physiology, diet, and also in pathophysiological responses. Compared to mice, pigs have a relatively long life expectancy, which allows longitudinal studies under conditions that mimic human patients. However, the introduction of precise genetic modifications in livestock genomes was hindered by the lack of true pluripotent stem cells from these species. Thus, techniques successfully used to generate the thousands of KO strains in mice were not applicable to livestock animals or only at very low success rates. The appearance of "Dolly," the first mammal cloned from a somatic cell,

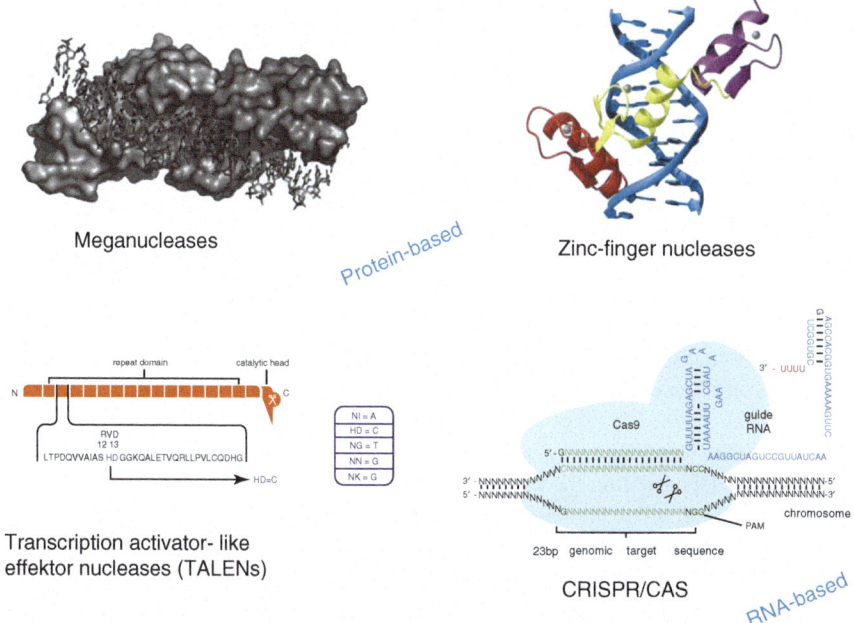

Fig. 7.1 Common DNA targeting platforms for gene editing. While meganucleases, zinc-finger nucleases, and TALENs generate a double-strand DNA break by protein-DNA pairing, the CRISPR/Cas system relies on RNA-DNA pairing. Unlike the three protein-based nucleases, CRISPR/Cas does not require the engineering of novel proteins for each DNA target site

changed the field of genome editing in livestock animals dramatically (Wilmut et al. 1997). Moreover, highly specific synthetic endonucleases such as zinc-finger nucleases (ZFNs), transcription activator-like effector endonucleases (TALENs), and CRISPR/Cas (Fig. 7.1) and new expression vectors such as transposon-based constructs together with the improved genomic sequence data lifted the field to a completely new level with previously unconceivable precision and efficiency. This chapter will provide basic information regarding DNA nucleases, their principle mechanism, and the application of DNA nucleases to modify the genome of livestock.

7.2   The Use of DNA Nucleases to Edit the Livestock Genome

Genetic modification starts with the creation of a double-strand break (DSB) of the DNA. The efficiency of a targeted genetic modification can be significantly enhanced by creating a site-specific DSB (Rouet et al. 1994). Genome editing tools normally consist of a cleavage domain and a DNA-binding domain, which can be designed to bind to nearly any DNA sequence. By selecting for different outcomes of DNA repair, either gene knockout or targeted transgene insertion can be obtained.

Homologous recombination is a rare cellular event that has numerous applications, including studies of basic mechanisms in mammalian development and physiology and the production of transgenic farm animals for xenotransplantation, as human disease models, for gene pharming or simply to increase breeding performance and/or specific agriculturally important traits. In embryonic stem cells, homologous recombination (HR) can be achieved using a positive-negative selection approach based on the presence of an antibiotic selection cassette within the homologous region, which will confer resistance against an antibiotic drug. By using a promoterless approach, the resistance cassette has to be driven by an endogenous active promoter. This approach significantly reduces the amount of false-positive cell clones. Combining the promoterless approach with a selection cassette localized outside of the homologous region further reduces the amount of false-positive selected cell clones. The overall efficiency obtained by using such an approach normally does not exceed 10^{-6} HR events. In contrast, several studies have reported 1–18% homologous recombination events per mammalian cell, when the targeted double-strand break was introduced by natural or artificial endonucleases (Choulika et al. 1995; Donoho et al. 1998; Epinat et al. 2003; Vasquez et al. 2001; Szczepek et al. 2007). Meganucleases were the first genome editing tools that were discovered and used to cut a specific DNA within the host genome. Following the discovery that induction of a double-strand break increases the frequency of homology-directed repair (HDR) by several orders of magnitude, targeted nucleases have emerged as the method of choice for improving the efficiency of HDR-mediated genetic alterations. By co-delivering a site-specific nuclease with a donor plasmid bearing locus-specific homology arms, single or multiple transgenes or mutations can be efficiently integrated into any endogenous locus. In the past few years, new genome editing tools were discovered that cut DNA in a very precise way, with unprecedented efficiency, and in a straightforward manner. These new programmable endonucleases include zinc-finger nucleases, transcription activator-like effector endonucleases (TALEN), and the most recent addition RNA-programmed genome editors (CRISPR/CAS9) (for comparison see Table 7.1).

7.2.1   Zinc-Finger Nucleases

7.2.1.1 Structure of Zinc-Finger Nucleases

The first zinc-finger (ZF) motif which had specific binding affinity to DNA was discovered as part of the transcription factor IIIa in *Xenopus* oocytes (Miller et al. 1985). A typical zinc finger (Cys_2His_2) consists of 30 amino acids which form two antiparallel β-sheets opposing an α-helix (Pabo et al. 2001). The domain is stabilized by two cysteine and two histidine residues binding a zinc ion, thus forming a compact globular domain. The zinc-finger motif uses residues in the alpha helix to bind to approximately three specific base pairs in the major groove of the DNA (Pavletich and Pabo 1991). ZFs can be designed to bind almost any base triplet (Pabo et al. 2001). Multiple ZFs can be combined to form a larger DNA-recognition domain which in turn increases specificity and efficiency of genetic modification. Specific binding of individual zinc fingers is largely independent, with some contacts between adjacent fingers altering base pair recognition. While the zinc-finger motif was discovered in the 1980s (Miller et al. 1985), ZFNs have a shorter history.

Table 7.1 Comparison of three classes of molecular scissors

	ZFN	TALEN	CRISPR
Targeting domain	Zinc-finger proteins	Transcription activator-like effector	CRISPR RNA or single-chain guide RNA
Nuclease	FokI	FokI	Cas9/FokI
Biallelic knockout achieved	Yes	Yes	Yes
Average mutation rate	++	+++	+++
Length of recognition domain	18–36 bp	30–40 bp	20 bp
Restriction in target site	G-rich	Start with T, and end with A	Protospacer adjacent motif (NGG or NAG) at end of target sequence
Complexity to design vector	–	+	+++
Off-target events	Variable	Low	Variable, to be determined
Cytotoxicity	Variable to high	Low	low
Number of plasmids necessary	2	2	1 (2 in case of a CRISPR/FokI construct)
Costs	+++	++–	+––

The first specific zinc-finger nuclease was reported ~15 years after the discovery of the zinc-finger domain (Kim et al. 1996). A ZFN consists of a site-specific zinc-finger DNA-binding domain fused to the nonspecific cleavage domain of the *FokI* endonuclease. At least two ZFN molecules are required for genetic modification, since the *FokI* nuclease must dimerize to cut the DNA. The need of two ZFN molecules doubles the number of specifically targeted base pairs (Smith et al. 2000). The two ZFN molecules bind to the targeted DNA in a tail-to-tail orientation separated by 5–7 bp, with double-stranded DNA cleavage occurring in the spacer region.

7.2.1.2 Genetical Modification by Zinc-Finger Nucleases

To employ a specific ZFN for genetic engineering, the plasmid DNA or mRNA encoding a specific ZFN is introduced into cells or embryos via microinjection or transfection (Hauschild-Quintern et al. 2013a). After translation, the ZFN pair binds to its specific target, the *FokI* nucleases dimerize, and the DNA is cleaved. ZFN activity can be enhanced by incubating transfected cells at 30 °C for a few days (Doyon et al. 2010). Cultivation at sub-physiological temperatures slows down the cell cycle, giving the ZFNs more time to bind and cut at the targeted locus (personal communication Greg Cost, Sangamo Biosciences, CA, USA). A ZFN pair induces a site-specific DSB only at the genomic site for which the molecule had been designed.

After ZFN-mediated DNA cleavage in eukaryotic cells, double-strand break repair is initiated, either by nonhomologous end joining (NHEJ) or homology-directed repair (HDR). NHEJ is error-prone and often creates short insertions or deletions

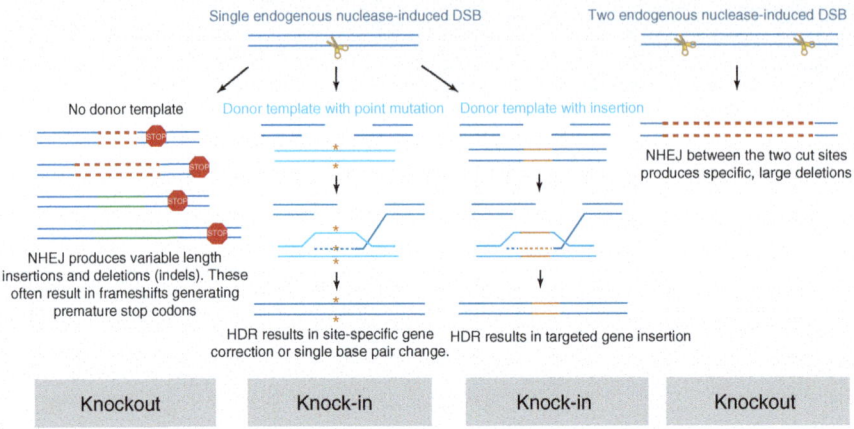

Fig. 7.2 Mechanisms of double-strand break repair. Modified from Maeder and Gersbach, 2016

(indels) of a few base pairs (10–20 bp) at the sealed break (Bibikova et al. 2002). Such mutations can cause frameshift or disruption of a gene, which in turn leads to the genetic knockout. Since the frequency of genetic modification is generally >1%, isolation of knockout cells is readily achieved by interrogation of cell clones generated by limiting dilution. Fluorescence-activated cell sorting (FACS) or magnetic bead selection has been successfully employed to enrich the targeted non-immortalized and other poorly clonable cells lines (Yu et al. 2011; Whyte et al. 2011; Hauschild et al. 2011; Li et al. 2013).

Mitotic cells often repair a DSB using homology-based DNA repair. In such a case, the cell normally uses the sister chromosome as a template to repair the DSB, which faithfully restores the original sequence (Fig. 7.2). The molecule can be used as template when a donor DNA molecule containing homologous arms to both sides of the DSB is co-transfected with the ZFNs. The exogenous DNA sequence placed between the two regions of homology will be copied into the chromosome during the DNA repair process (Moehle et al. 2007). In the absence of a site-specific break, the donor DNA must contain a large region (6–7 kb) homologous to the targeted region for capturing one of the rare spontaneous breaks (Deng and Capecchi 1992). In contrast, ZFN-based targeting strategy is compatible with a significantly shorter stretch of homologous DNA. Typically, 500–1500 bp is used. Even linear <50 bp homologous donor sequences (Orlando et al. 2010) and single-stranded DNA oligonucleotides can be used to induce mutations, deletions, or insertions at the target site (Chen et al. 2011).

Integration of the ZFN has to be avoided as it would result in permanent transcription of the ZFN and thereby would likely lead to permanent nonspecific DNA cleavage. Usually, ZFN plasmids are rapidly diluted and disappear from the treated cells when a transient transfection protocol is applied. Besides the high efficiency, a major advantage of ZFN-mediated targeting is the lack of random integration, which prevents negative side effects such as insertional mutagenesis. Nuclease-mediated targeted integration normalizes for positional effects that typically confound many types of genetic analysis and enables study of structure-function relations in the complex and native chromosomal environment. ZFNs have been broadly applied in basic research, biotechnology, and medicine, but genome engineering with ZFNs is limited by the random generation of unwanted indels at homology sites (Liu et al. 2013). One potential strategy to overcome this limitation is the targeted introduction into DNA containing a single-strand break (SSB) or nick. A nick can be equivalent to a double-strand break (DSB) and stimulate the HDR pathway (Meselson and Radding 1975; Radding 1982). In contrast to a DSB, a nick is not a bona fide substrate for repair by the NHEJ pathway. Thus, a targeted nick has the potential to restrict repair to the HDR pathway (Wang et al. 2012).

7.2.1.3 Genetical Modifications of Farm Animals Using Zinc-Finger Nucleases

Genetically modified farm animals, specifically the domestic pig, increasingly serve as a model in human medicine, including xenotransplantation, basic research, and human disease models. The latter is an important complementation to the laboratory mouse where it has frequently been shown that the typical disease manifestation does not fully mimic the human disease symptoms. Pigs share many genetic, anatomical, and physiological features with humans and have rapidly emerged as a suitable model for specific diseases, including cystic fibrosis, diabetes, cancer, and several neurological disorders (Flisikowska et al. 2014). Pigs are also considered as suitable organ donors for xenotransplantation to reduce or even eliminate the shortage of suitable human organs (Cooper and Ayares 2011; Petersen et al. 2009). This requires genetic modification of the donor pigs to overcome the severe immunological rejection responses occurring after pig-to-primate xenotransplantation. Conventional targeting is extremely inefficient and usually does not lead to a biallelic KO. Moreover, true germ line competent pluripotent cells are not yet available from pigs and other domestic animals, which prevents selection for rare HDR events as it is feasible in laboratory mice (Nowak-Imialek and Niemann 2012). The production of transgenic farm animals is significantly facilitated by effective somatic cell nuclear transfer (SCNT) protocols (Petersen et al. 2008). This cell-mediated transgenesis is compatible with screening for genetic modifications and analysis of the transgenic genotype in vitro rather than in animals "on the farm." These cells are then used to produce animals with the desired phenotype. While cell-mediated transgenesis is more labor intensive than direct transgenesis, in vitro genetic manipulation of cells followed by detailed genome analysis bears significant advantages. First, it reduces the total number of animals required to generate a useful transgenic offspring. Second, it increases dramatically the number of independent transgene integration events that can be screened and investigated. Third, it facilitates the engineering of precisely controlled genetic alterations (gene targeting) by allowing selection and isolation of rare integration events resulting from homologous recombination. Fourth, the use of a selected cell clone as cell donor for SCNT leads to a syngenic clone cohort, which facilitates detailed analysis of the phenotype. Finally, biallelic knockout via ZFN provides a significant time advantage compared with traditional knockout via homologous recombination, which significantly stream-lines the production of relevant large animal models. Genome editing technology has been successfully applied to zebrafish (Bedell et al. 2012), rabbits (Flisikowska et al. 2011), and rodents (Geurts et al. 2009) by direct injection of mRNA or DNA of genome editors into embryos. Injection of ZFN mRNA or DNA into zygotes can also be used to generate null phenotype offspring in large animals with high effi-ciencies (Lillico et al. 2013). This high versatility of genome editing tools allowed many laboratories worldwide the use of this technology.

Pig

The first experiments with ZFNs to modify the pig genome were conducted to target the transgenic *eGFP* (pCX-*eGFP*) locus in the domestic pig, with ~10 genomic integration sites. After targeting, the rate of non-fluorescent cells increased from 6% (control) to 21% (ZFN-targeted cells), showing that in ~15% of the cells nearly all copies of the *eGFP* gene had been disrupted. Sequencing of several non-fluorescent cell clones revealed that wild-type DNA (non-mutated *eGFP*) variants remained, implying that at least one intact *eGFP* copy was silenced (Watanabe et al. 2010).

The first live ZFN-mediated KO pig carried a hemizygous transgenic *eGFP allele*. Porcine fibroblasts were co-transfected with a pair of ZFN plasmids and a red fluorescent CAG-tomato plasmid (transient selectable fluorophore). Two percent of the cells showed red fluorescence and could be sorted by FACS. A second round of selection for green cells by FACS led to 5% *eGFP*-negative cells. Selected cells used in SCNT led to the delivery of six out of seven piglets in which eGFP fluores-cence was knocked out. Sequencing revealed several deletions and insertions at the targeted locus. A third litter with six piglets was entirely *eGFP* negative. One piglet carried an unusual large deletion of 700 base pairs deleting nearly the entire *eGFP* coding sequence (Whyte et al. 2011). Yang et al. showed that ZFNs can be used to target endogenous genes and that these cells are capable to generate live offspring (Yang et al. 2011a). They targeted the endogenous *peroxisome proliferator-activated receptor-γ* (*PPAR-γ*) locus by using ZFNs. *PPAR-γ*$^{-/-}$ animals could be a useful model for studies on cardiovascular diseases. Male fibroblasts were co-transfected with a *PPAR-γ*-specific ZFN pair and a neomycin resistance gene. After selection with G418, 4% of screened cell clones carried a mutated *PPAR-γ* gene and served as donors in SCNT. Two live-born piglets carried a mutation in one of the *PPAR-γ* alleles. Western blotting analysis confirmed the successful production of heterozy-gous *PPAR-γ* KO animals (Yang et al. 2011b).

The first pigs with a biallelic KO of an endogenous gene via ZFN targeting were produced by our laboratory (Hauschild et al. 2011). Transfection of fetal fibroblasts with a pair of ZFN plasmids directed against exon 9 (catalytic domain) of the *α1,3-galactosyltransferase* (*GGTA1*, Gal) gene induced biallelic mutations in 1% of the cells. The *α1,3-galactosyltransferase* synthesizes galactose epitopes on the surface of porcine cells, which are the major antigen in a xenotransplantation setting and are recognized by 1% of all antibodies circulating in human blood flow. Magnetic beads were used to counter-select for Gal-negative cells, reaching a purity of >99% Gal-negative cells. The Gal-negative cells served as donor cells in somatic cell nuclear transfer (SCNT) and led to the birth of nine live *GGTA1*-knockout piglets. Sequencing revealed five different haplotypes with two homozygous and three heterozygous (individual mutations on each allele) mutations. The *GGTA1* gene showed deletions from one to seven base pairs in size and one unusual large deletion of 96 bp. The *GGTA1*-KO fibroblasts derived from the ZFN approach were protected against lysis in a complement in vitro assay. Disruption of both alleles by conventional HR generally involves production of mono-allelic knockout clones followed by breeding with other heterozygous knockouts to obtain a homozygous knockout in 25% of the offspring (Whyte and Prather 2012). Compared to conventional gene targeting, the use of ZFNs to generate a functional gene knockout led to a 10,000-fold efficiency increase.

In a follow-up study, we showed that the efficiency of the ZFNs is not influenced by the gender of the cells (Hauschild-Quintern et al. 2013b), thus allowing production of knockout pigs of both sexes with similar efficiency. Our results to disrupt the porcine *GGTA1*-locus by using ZFNs have been confirmed by other groups (Li et al. 2013; Bao et al. 2014), showing the robustness and reproducibility of this technology.

After knockout of *GGTA1*, the Hanganutziu-Deicher antigen remains a major antigen that is implicated in subsequent xenograft rejection (Ezzelarab et al. 2005). The responsible porcine gene for the generation of the HD antigen on porcine cells is the CMP-N-acetylneuraminic acid hydroxylase (*CMAH*). ZFNs designed to target exon 8 of the *CMAH* locus led to 9.1% targeted alleles when donor DNA coding for a neomycin resistance cassette was not added to the transfection mix. A dramatic increase of targeting frequency (41.7%) was observed when donor DNA with a 789 bp homologous 5'arm and a 763 bp homologous 3'arm was added. Biallelic knockouts were in all cases associated with integration of the exogenous DNA (Kwon et al. 2013). A possible explanation for this difference is the difficulty of separating non-transfected from total cells used for the transfection without a selection marker. This study demonstrated, for the first time, gene targeting using ZFN-assisted HR of donor DNA in porcine somatic cells (Kwon et al. 2013).

In these studies, the ZFN-encoding DNA was introduced into nuclear donor cells for SCNT to produce genetically modified pigs. However, plasmid DNA can also be integrated into the genome of cells, which may result in disruption of endogenous genes and constitutive expression of ZFNs. This drawback of plasmid DNA can be eliminated by the use of ZFN-encoding mRNA, which cannot be inserted into the host genome. Watanabe et al. (2013) applied ZFN-encoding mRNA to knock out the interleukin-2 receptor gamma (*IL2RG*) gene on the X chromosome of male porcine fibroblasts; these cells supported development to live offspring after SCNT (Watanabe et al. 2013). The *IL2RG*-KO pigs obtained in this study lacked T and NK cells but showed normal B cell populations which mimic adequately the phenotype of human XSCID patients. Due to the limited capacity of their immune system, *IL2RG*-KO pigs are susceptible to any kind of pathogen. To keep such pigs for longer studies, expensive gnotobiotic housing conditions are necessary. Nevertheless, IL2RG-KO pigs represent the first step toward developing a porcine SCID model and can contribute not only to cancer and stem cell research but also to preclinical evaluation of the transplantation of pluripotent stem cells such as iPS cells (Watanabe et al. 2013).

Lillico et al. (2013) tested a pair of ZFNs with a targeted location of 1330–1338 bp (NM_001114281) relative to the translational start site in the porcine RELA cDNA sequence (Lillico et al. 2013). The *RELA* locus might play an important role in rendering pigs resistant against African swine fever. In order to investigate whether cytoplasmic microinjection into zygotes could also result in HDR if combined with a DNA template, they co-injected porcine zygotes with mRNA encoding the pair of RELA-specific ZFN and a single-stranded oligodeoxynucleotide (ssODN) or plasmid DNA bearing the warthog SNPs. Analysis of the offspring revealed three live piglets, which bore HDR-generated alleles of RELA. All of them resulted from the cohort of animals injected with ZFN-encoding mRNA and a plasmid repair template. They did not observe any HDR integration in piglets injected with a repair ssODN. Only one piglet was homozygous for the five intended base changes. As a negative side effect, the use of plasmid DNA as repair template resulted in the randomly genomic integration of the template in addition to HDR (Lillico et al. 2016). Unexpectedly, RELA gene-edited pigs did not show any resilience against infection with African swine fever virus (personal communication).

Cattle

In cattle, ZFN-mediated gene targeting was conducted to produce *beta-lactoglobulin (BLG)*-KO animals. *BLG* is the major whey protein in bovine milk and is the critical milk allergen. Bovine fetal fibroblasts were transfected with mRNA coding for ZFNs designed against the *BLG* gene. Sequencing revealed that ~15% of the cells carried a mutated variant and 3% of the single cell colonies showed a biallelic *BLG* gene knockout. Homozygous KO cells were used in SCNT, and eight cloned animals were born; one survived the postnatal period. The mutated *BLG* gene was shorter (nine and 15 base pairs deletion, no frameshift) than the wild-type version. Off-target site mutations induced by the ZFN pair were also analyzed for *BLG*. While one base pair mismatch with the targeting sequence led to 7% gene targeting (single-nucleotide polymorphism in cattle), three and seven base pair mismatches did not result in a mutated phenotype in sheep and pigs. Results suggest that ZFN-mediated targeting is promising for specific gene editing in large domestic animals with little risk of off-target site cleavage (Yu et al. 2011).

Mastitis costs the dairy industry billions of dollars annually and is the most consequential disease of dairy cattle. Therefore, the use of genome editing tools to integrate mastitis resistance via transgenes such as human lysozyme (Liu et al. 2014) or lysostaphin (Liu et al. 2013) in the β-casein locus may open a unique avenue for the creation of transgenic cows with enhanced mastitis resistance and improved health and welfare of livestock. ZFNickase-stimulated gene addition at the endogenous bovine locus is feasible and compatible with the production of cloned bacterial lysostaphin-transgenic cows (Liu et al. 2013). In this particular study, the *FokI* catalytic domain was mutated at amino acid D450 in one of the two ZFNs necessary for dimerization and subsequent DNA cleavage, leading to nickase activity of the ZFN pair. A lysostaphin-coding vector was transfected into bovine fetal fibroblasts along with expression plasmids encoding ZFN/ZFNickase to introduce a nick in intron 2 of the *CSN2* locus. Finally, 69 cell clones with a correct integration of the lysostaphin vector at the *CSN2* locus were obtained and using these cells as donors in somatic cell nuclear transfer resulted in 16 transgenic calves. When calves were induced to lactate, the milk contained lysostaphin-mediated antibacterial activity.

The relatively high percentage of integration of a long DNA fragment into a predetermined locus in the bovine genome demonstrates that ZFNickases are active in bovine cells and can be used to further minimize the risk of potential off-target events. Thereby, the use of ZFNickases ensures that only a single copy of the transgene is integrated into the host genome. This can further facilitate a range of new transgenic technologies beneficial for both agriculture and biomedicine.

7.2.2 Transcription Activator-Like Effector Endonucleases (TALEN)

7.2.2.1 Structure of TALENs

TALEs (transcription activator-like effectors) are naturally produced by plant pathogens such as *Xanthomonas*, gram-negative bacteria, that can infect a wide variety of plant species including pepper, rice, citrus, cotton, tomato, and soybeans (Boch et al. 2009; Boch and Bonas 2010). TALEs bind to their host DNA, act as transcription factors, and activate the expression of plant genes that aid bacterial infection. Plants have developed a defense mechanism against type III effectors that includes resistance genes triggered by these effectors. Some of these genes appear to have evolved to contain TAL-effector-binding sites similar to sites in the intended target genes. This competition between pathogenic bacteria and the host plant has been hypothesized to account for the malleability of the TAL effector DNA-binding domain (Voytas and Joung 2009). TALEs consist of repeats, each consisting of 33–35 amino acids with two polymorphisms at positions 12 and 13 within the module, which are called the repeat variable diresidue (RVD). One RVD binds specifically to one nucleotide of genomic DNA (Moscou and Bogdanove 2009; Boch et al. 2009), hence establishing a 1:1 code for protein-DNA interaction (Fig. 7.2).

Individual TALE repeats can be used to engineer DNA-binding domains capable of recognizing endogenous sequences in mammalian cells. By linking the binding domain with the nonspecific cleavage domain from the type II restriction endonuclease *FokI*, TALENS can be used as a tool for stimulating NHEJ and HR (Cermak et al. 2011; Christian et al. 2010; Li et al. a, b; Mahfouz et al. 2011; Miller et al. 2011). Given the modular nature of this DNA-binding domain, RVDs with different specificities can be assembled into arrays in order to target user-defined DNA sequences.

TALENs can be successfully used to target endogenous genes and efficiently cleave DNA leading to NHEJ (Hockemeyer et al. 2011). A comparative study with human ES cells and induced pluripotent stem cells and three different target genes *AAVS1*, *OCT4*, and *PITX3* revealed that TALENs and ZFNs had a similar targeting efficiency (Hockemeyer et al. 2011). TALENs have been used to knock out genes in rats and zebrafish (Tesson et al. 2011; Sander et al. 2011; Huang et al. 2011) and in cattle, sheep, and pigs, thus demonstrating that TALENs are effective in inducing genetic modifications in a broad range of different species (Carlson et al. 2012; Proudfoot et al. 2015).

ZFNs and TALENs differ in three main aspects: (1) TALE repeats are three to four times larger than ZFNs, when recognized per base pair of the targeted DNA. This may interfere with viral delivery methods, particularly adeno-associated virus. (2) The spacer length (the gap between two binding sites) is variable and not restricted to a specific length, which complicates TALEN design and could lead to greater off-target activity relative to an identical nuclease with a fixed spacer length. (3) ZFNs' assembly requires an archive of high-quality, well-characterized modules to achieve specific gene targeting because cross talk between the individual fingers can lead to imperfect DNA recognition (Defrancesco 2011). Context-dependent effects between the repeat units, as reported for ZFNs (Cathomen and Joung 2008), have not been reported so far for TALENs.

Various assembly methods have indicated that TALE repeats can be combined to recognize potentially any target sequence, the only limitation is that TALE binding sites must start with thymidine (Boch and Bonas 2010). This needs to be considered when screening a locus for potential target sites. TALENs appear to be superior to ZFNs in terms of simplicity and straightforwardness in design and assembly strategies. Manufacture of effective TALENs is cheaper and faster compared to ZFNs. The relative simple TALE assembly is displayed in a recent study reporting the construction of a library of TALENs targeting 18,740 different human protein-coding genes (Kim et al. 2013). Active, custom-designed TALENs have been reported to induce indel frequencies between 2% and 55% of targeted chromosomes (Carlson et al. 2012). TALENs can be easily designed and assembled using molecular biology techniques available in most laboratories around the world.

7.2.2.2 Modification of Livestock Genomes Using TALENs

Pig

The LDL receptor gene was targeted with the aid of TALEN in pigs to create a model for familial hypercholesterolemia (Carlson et al. 2012), and the porcine *DMD* gene was targeted to create a porcine model for Duchenne muscular dystrophy. The most active TALEN pair targeting the *DMD* gene had a cleavage efficiency of 38%. The *DMD* gene was successfully targeted in 41% of analyzed cell clones with ~30% of these carrying a biallelic mutation. The combined transfection of TALEN pairs targeting exons 6 and 7 of the *DMD* locus resulted in deletion of 6.5 kb DNA in 10.3% of selected colonies. Mono- and biallelic *LDLR* gene-modified cell clones were pooled and used as donors for somatic cell nuclear transfer. Pregnancies were established in seven of nine transfers. Six pregnancies were maintained to term and yielded 18 live-born piglets of which eight contained mono-allelic mutations and ten contained biallelic mutations of the *LDLR* gene. To enhance disease resistance in pigs, 20 ng/μL-specific TALEN mRNA were microinjected into porcine zygotes to target the porcine *RELA* gene (*p65*) which is critically involved in tolerance against African swine fever virus infection (Palgrave et al. 2011). Sixteen out of 56 successfully injected embryos revealed indels detected by Surveyor assay and/or sequence analysis. One-third of the mutants were either homozygous or heterozygous mutants. *RELA*-mutated porcine embryos were not transferred to assess in vivo development.

These results demonstrate the robustness and reproducibility of TALEN to nearly any genomic locus for which the genomic sequence is available. These results clearly show that TALENs are not only compatible with the deletion at a defined genomic locus but also allow precise allelic introgression and large chromosomal deletions/inversions, rendering TALEN a valuable tool for genetic modification of farm animals.

Cattle

Carlson et al. (2012) designed TALENs to target the bovine *ACAN* and *GDF8* genes in fibroblasts. *ACAN*, also known as *Aggrecan*, is thought to play an important role in the formation of congenital achondroplasia, while *GDF8* (*growth differentiation factor 8, myostatin*) is a regulator of muscle growth. A nonfunctional *myostatin* gene is known to cause muscular hypertrophy in Belgian Blue and Piedmontese cattle. *GDF8*-targeted bovine fibroblasts showed a modification of the gene in seven out of 24 cell clones (29%). None of the cell clones carried a biallelic modification. The *ACAN* gene was targeted in 27 out of 35 cell clones (77%). Two cell clones showed a biallelic modification. Modified cells could be used as donor cells for somatic cell nuclear transfer to produce live offspring carrying the desired genetic modification (Carlson et al. 2012).

Physical dehorning of dairy cattle is practiced to protect animals and the farmers. Genetic analyses have identified variants that are associated with hornlessness (polled) in cattle, a trait that is common in beef but rare in dairy breeds. Identification of the genetic cause of hornlessness in cattle has been the subject of intensive genetic and genomic research. The candidate allele in beef breeds is a simple allele of Celtic origin (Pc) corresponding to a duplication of 212 bp in place of a 10 bp deletion on chromosome 1 of the bovine genome (Carlson et al. 2016). Only about 5% of all Holstein Friesians have a polled phenotype. A research group in the USA introgressed a candidate POLLED allele into dairy cattle by genome editing via TALENs and reproductive cloning, providing both evidence for genetic causation and a means to introduce POLLED into livestock (Carlson et al. 2016).

Sheep

Proudfoot et al. (2015) recently described the generation of gene-edited sheep. As the bovine and ovine genomic sequences of the *MSTN* locus have high similarity, they used the same TALENs that successfully targeted the bovine *MSTN* locus. Transient transfection of TALEN mRNA into ovine cells and subsequent analysis by the Surveyor nuclease assay showed similar levels of activity in both species. To generate living offspring, they microinjected TALEN mRNA in in vitro-produced ovine zygotes and transferred blastocysts to synchronized recipient ewes. In total,

26 blastocysts were transferred to nine recipients resulting in eight pregnancies and 12 live births. One of the offspring carried a heterozygous mutation in the targeted gene. This study further exemplifies the utility and ease with which TALENs can be used to engineer the genome of livestock (Proudfoot et al. 2015).

7.2.3 RNA-Guided Genomic Engineering (CRISPR/Cas9)

7.2.3.1 Structure of CRISPR/CAS9

The CRISPR/Cas9 system has recently emerged as potentially facile and efficient alternative to ZFNs, TALENs, and other meganucleases for inducing targeted genetic alterations and has revolutionized the field for targeted genomic engineering in the short time since its appearance. In bacteria and archaea, CRISPR (clustered regularly interspaced short palindromic repeats)/Cas (CRISPR-associated) loci encode RNA-guided adaptive immune systems that can destroy foreign DNA (Bhaya et al. 2011; Terns and Terns 2011; Wiedenheft et al. 2012). The *Streptococcus pyogenes* SF370 type II CRISPR locus consists of four genes, including the Cas9 nuclease and two noncoding RNAs. TracrRNA and a pre-crRNA array containing nuclease-guided sequences interspaced by identical direct repeats (Cong et al. 2013). In vitro reconstitution of the *S. pyogenes* CRISPR system demonstrated that crRNA fused to a normally trans-coded tracrRNA is sufficient to direct Cas9 protein to highly specific cleavage of target DNA sequences matching the crRNA (Mali et al. 2013). This redesign as a single transcript (single-guide RNA or guide RNA (gRNA)) encompasses the features required for both Cas9 binding and DNA target site recognition. Using sgRNA, Cas 9 can be programmed to cleave double-stranded DNA at any genomic site defined by the guide RNA sequence and a protospacer adjacent motif (PAM). The PAM is an essential targeting component that also serves as a self-versus-non-self-recognition system to prevent the CRISPR locus itself from being targeted. Many type II systems have different PAM requirements, which may affect their usefulness and targeting efficiency. The most commonly engineered system, from *Streptococcus pyogenes*, requires a NGG protospacer adjacent motif (PAM), where N can be any nucleotide. In bacterial systems CRISPR/Cas can be used as it is, while in humans it involves expression of a human-codon-optimized Cas9 protein with an appropriate nuclear localization signal. Moreover, the crRNA and tracrRNA must be expressed either individually or as a single chimera via a RNA polymerase III promoter (Cong et al. 2013; Mali et al. 2013; Jinek et al. 2013). The typical features of CRISPR/Cas9 suggest that is a simple and versatile system for generating double-stranded breaks that facilitate site-specific genome editing. Moreover, CRISPR/Cas can target multiple loci by the sgRNA, potentially allowing simultaneous targeting of multiple genomic loci. CRISPR/Cas9 vectors are commercially available and can be used after introducing the specific gRNA sequence, which is a nucleotide of 20–30 bp (Fig. 7.3).

CRISPR/Cas vectors showed high activity in embryos making them a perfect tool for genome editing by simple cytoplasmic microinjection of CRISPR into mammalian zygotes. With CRISPR/Cas, the generation of biallelic knockout

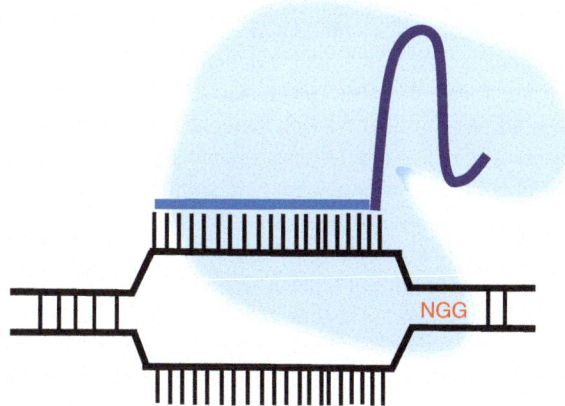

Fig. 7.3 CRISPR/Cas9 vectors are commercially available (addgene.org) and can be used after introducing the specific guide RNA sequence (blue), which is a nucleotide of 20–30 bp. The Cas9 nuclease from the microbial CRISPR adaptive immune system is localized to specific DNA sequences via the guide sequence on its guide RNA (blue), directly base pairing with the DNA target. Binding of a protospacer adjacent motif (PAM, red) downstream of the target locus helps to direct Cas9-mediated double-strand DNA breaks

animals does not rely on somatic cell nuclear transfer anymore. The number of reports, since the first description of the successful use of CRISPR/Cas for targeting a specific genomic locus, has dramatically increased. Current data suggest that CRISPRs have similar specificity and efficiency as ZFNs and TALENS. In addition, CRISPRs have the advantage of being simple to generate, easy to handle, efficient, and cost-effective. Open questions regarding their specificity have to be addressed in future experiments. CRISPR vectors with nickase activity, to avoid off-target events (Shen et al. 2014), or vectors that have an inactivated version of the Cas-motif connected to the *FokI* endonuclease, which has to dimerize before cutting and thereby increases the specificity (Tsai et al. 2014; Guilinger et al. 2014), are already available. The Cas9 nuclease can be converted to nickase (Cas9n) or nuclease-deficient mutant (dCas9). The dCas9 variant has broad applications such as single-base editing systems without introducing DSBs, mediated by cytidine deaminase combined with dCas9, also known as "base editors" (Komor et al. 2016). The specificity can be further enhanced by the use of a truncated gRNA (Fu et al. 2014) or by using high-fidelity variants of Cas harboring alterations designed to reduce nonspecific DNA contacts (Kleinstiver et al. 2016a).

7.2.3.2 Evolution of CRISPR/Cas

Although multiplex gene editing is possible with Cas9 nuclease, it requires relatively large constructs or simultaneous delivery of multiple plasmids, both of which are problematic for multiplex screens or in vivo applications. Recently, a Cas protein named Cpf1, a type V CRISPR/Cas system, has been identified that can also be programmed to cleave target DNA sequences (Zetsche et al. 2015). Unlike Cas9, Cpf1 requires only a single 42-nt crRNA, not coupled with a trcrRNA, which has

23nt at its 3'end that are complementary to the protospacer of the target DNA sequence. Whereas Cas9 recognizes an NGG PAM sequence that is 3' of the protospacer, *Acidaminococcus* (As)Cpf1 and *Lachnospiraceae* (Lb)Cpf1 recognize TTTN PAMs that are found 5' of the protospacer. This feature of Cpf1 leads to higher affinity to bind and cut also AT-rich sequences which are hard to target by Cas9. The sensitivity of Cpf1 to single-base mismatches in certain positions of the protospacer might mean that these nucleases are suitable for allele-specific editing of heterozygous alleles. Analyses suggest that the specificities of Cpf1 nucleases may approach that of the described high-fidelity Cas9 variants (Kleinstiver et al. 2016b). Another important feature of Cpf1 is that Cpf1 nuclease produces cohesive ends with 4–5-nt overhangs, while Cas9 produces blunt ends. In this regard, NHEJ-mediated knock-in might be facilitated using Cpf1. Unlike Cas9, Cpf1 contains not only the DSB-inducing activity but also an RNase III activity involved in pre-crRNA processing. This activity can be utilized for the efficient multiplex genome editing via a tandemly arrayed pre-crRNA expressing construct, producing multiple mature crRNAs by Cpf1.

Recently, 53 new class 2 CRISPR/Cas candidates were discovered and categorized into three groups defined by the context characteristics: C2c1, C2c2, and C2c3 (Shmakov et al. 2015). C2c2 and C2c3 were later grouped in Type V, and C2c2 was grouped in Type VI. C2c2 nucleases have a unique feature. Their potential target is not double-stranded DNA but single-stranded RNA; thus they can be applied for gene knockdown applications or potential knockout applications at the mRNA level, leaving the DNA sequence unmodified (Abudayyeh et al. 2016). C2c2 possesses a unique RNase activity responsible for CRISPR RNA maturation that is distinct from its RNA-activated single-stranded RNA degradation activity (East-Seletsky et al. 2016). These dual RNase functions are chemically and mechanistically different from each other and form the crRNA-processing behavior of the evolutionary unrelated CRISPR enzyme Cpf1 (Fonfara et al. 2016). The two RNase activities of C2c2 enable multiplexed processing and loading of guide RNAs that in turn allow sensitive detection of cellular transcripts. Cpf1 and C2c2 are only two examples of the further increasing toolbox for genome editing. Current genome editing technologies introduce double-stranded DNA breaks at a target locus as the first step to gene correction. Besides that, new approaches were developed such as "base editing," which enables direct, irreversible conversion of one target DNA base into another in a programmable manner, without requiring dsDNA backbone cleavage or a donor template (Komor et al. 2016). These CRISPR constructs were designed by fusion CRISPR/Cas and a cytidine deaminase enzyme that retain the ability to be programmed with a guide RNA and mediate the direct conversion of cytidine to uridine, thereby effecting a C to T (or G to A) substitution (Fig. 7.4a). The most common base editors are third-generation designs (BE3) comprising a catalytically impaired CRISPR/Cas mutant that cannot make DSBs, a single-strand-specific cytidine deaminase that converts C to uracil (U) within a five-nucleotide window, a uracil glycosylase inhibitor (UGI) that impedes uracil excision and nickase activity to nick the non-edited DNA strand, directing cellular DNA repair processes to replace the G-containing DNA strand. Base-editing capabilities have

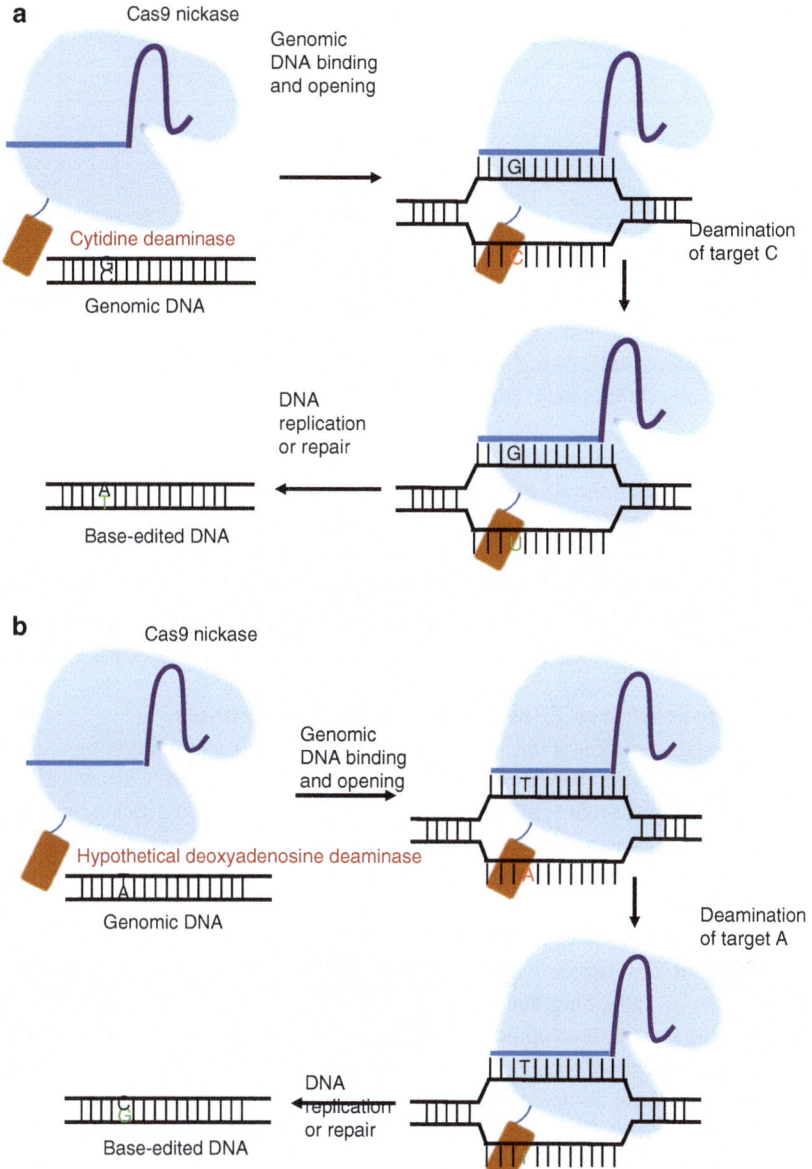

Fig. 7.4 (**a**) DNA with a target C (red) at a locus specified by a guide RNA (blue) is bound by Cas9 nickase, which mediates local DNA strand separation. Cytidine deamination by a tethered APOBEC1 enzyme (cytidine deaminase, brown) converts the single-stranded target C to U (uracil). The resulting G/U heteroduplex can be permanently converted to an A/T base pair following DNA replication or DNA repair. Modified from Komor et al. (2016). (**b**) ABEs contain a hypothetical deoxyadenosine deaminase, which does not exist in nature, and a catalytically impaired Cas9. They bind target DNA in a guide RNA-programmed manner, exposing a small bubble of single-stranded DNA. The following deamination of adenosine (A) forms inosine (I), which is read as guanosine (G) by polymerase enzymes. Following DNA repair or replication, the original A/T base pair is replaced with a G/C base pair at the target site. Modified from Gaudelli et al. (2017)

Table 7.2 Base pair changes required to correct pathogenic human SNPs in the ClinVar database (https://www.ncbi.nlm.nih.gov/clinvar/)

Correction	Percentage of pathogenic human SNPs (%)
A/T -> G/C	48
A/T -> C/G	15
C/G -> T/A	14
C/G -> G/C	11
A/T -> T/A	7
C/G -> A/T	6

expanded through the development of base editors with different protospacer adjacent motifs, narrowed editing windows, and enhanced DNA specificity. Fourth-generation base editors (BE4) further improve editing efficiency and product purity (Gaudelli et al. 2017). Recently, new base editors were designed called adenine base editors (ABEs) that convert A to G (or T to C) in bacteria and human cells (Fig. 7.4b). ABEs greatly expand the scope of base editing and, together with BE4, enable programmable installation of all four transitions (C to T, A to G, T to C, and G to A) in genomic DNA without any DNA cleavage (Gaudelli et al. 2017). Base editors hold great promise to repair single-nucleotide polymorphisms (SNP) in the human genome correlated with a pathogenic phenotype. About 48% of the described 32,044 pathogenic human SNPs could be corrected by ABEs alone (Table 7.2).

7.2.3.3 Application of CRISPR/Cas in Domestic Animals

Since the first description of the successful use of CRISPR/Cas for targeting a specific genomic locus, the number of reports has dramatically increased. Current, still preliminary data suggest that CRISPRs have similar specificity and efficiency as ZFNs and TALENs. The CRISPR/Cas technology was soon adapted to modify the genome of livestock animals, such as pigs, cattle, goat, and sheep. For a good overview, see Tan et al. (2016). Hai et al. (2014) targeted exon 5 of the *vWF* (von Willebrand factor) gene by microinjecting Cas9 mRNA and sgRNA into the cytoplasm of porcine zygotes. In vitro development of the injected embryos did not differ from that of the control embryos, indicating that the Cas9 mRNA/sgRNA injection did not interfere with early embryonic development. A total of 76 injected embryos were transferred into five surrogate sows; three pregnancies were established, 16 piglets were delivered, and ten of these contained indels in the targeted site. Six piglets carried a biallelic knockout of the *vWF* gene. Expression of vWF at the protein level was not detectable in biallelic *vWF*-KO pigs, and the vWF-KO pigs displayed a phenotype comparable to the human von Willebrand disease. Off-target cleavage events were not found. CRISPR/Cas9 is very efficient in modifying the genome of very early embryos at the zygote stage, prior to the first division of the embryo. Cytoplasmic microinjection of the CRISPR/Cas9 vector is sufficient to lead to a majority (2/3) of offspring born with a biallelic modification of the targeted locus (Petersen et al. 2016).

CRISPR/Cas9 was also successfully employed to target the porcine *p65* locus in fetal fibroblasts (Tan et al. 2013). Despite efficient production of double-strand breaks in the target site, the frequency of CRISPR-/Cas9-mediated homologous recombination (HR) was lower than with TALENs. Targeting the porcine *APC* gene, CRISPR/Cas9 worked more efficiently but still did not reach the level of HR induced by TALENs at the same site (30% vs. 60%). A highlight application of CRISPR-/Cas9-mediated gene editing was the generation of pigs resistant against infection with the porcine reproductive and respiratory syndrome virus (PRRSV). PRRS manifests differently in pigs of all stages but primarily causes late-term abortions and stillbirths in sows and respiratory disease in piglets. PRRSV only infects a specific subset of cells of the innate immune system of the macrophage/monocyte lineage. The virus uses a specific receptor, CD163, in order to make its own membrane fuse with the host cell membrane in an uptake vesicle to release the viral genetic information into the cytosol and achieve a successful infection. The knockout of the whole receptor (Whitworth et al. 2016) or only of the CD163 subdomain 5 (Burkard et al. 2017) rendered pigs resistant against PRRSV infections. These first reports on CRISPR/Cas suggest locus-specific differences with regard to HR efficiency which may be adjusted by modifying the conditions under which CRISPR/Cas works best. There is still a controversial debate within the scientific community with regard to efficiency and precision of cleavage of ZFNs, TALENs, and CRISPR/Cas. This may mainly be grounded on the different expertises of the labs with one or the other programmable endonuclease. More studies are required in order to fully exploit the potential of CRISPR/Cas9.

7.3 Concluding Remarks and Future Directions

Genome editing tools such as ZFNs, TALENs, and RNA-guided DNA endonucleases (CRISPR/Cas) have emerged as valuable molecular tools that have already been shown to revolutionize biological research with great benefits for personalized medicine. These emerging technologies significantly expand the ability to create and study model organisms, including large animals, and will be instrumental for correcting many genetic diseases in livestock species and humans (Table 7.3 gives a selection of important achievements). With the aid of these tools, researchers are able to develop biomedical models in species that are physiologically closer related to humans than mice. The domestic pig is particularly promising in this regard. The growing number of human disease models in pigs supports this assumption (Flisikowska et al. 2014).

Due to the high degree of physiological similarity with humans, porcine organs are considered as promising solution to satisfy the growing demand of human organs for allotransplantation. To achieve this goal and to avoid immune rejection responses, the porcine genome has to be modified to ensure long-term survival of porcine organs in patients after xenografting. ZFNs, TALENs, and CRISPR/Cas can now be used to elegantly knock out candidate pig genes or to precisely knock in transgenes at specific genomic sites in the porcine genome to produce pigs specifically tailored as organ donors.

However, to exploit the full potential of these new technologies, important questions and challenges must be addressed. A high degree of specificity is a main

Table 7.3 A selective list of gene-edited livestock

Gene	Genome editor	Method	Reference
NHEJ			
Pig			
PPARγ	ZFN	SCNT	Yang et al. (2011)
α1,3GT	ZFN	SCNT	Hauschild et al. (2011)
eGFP	ZFN	SCNT	Whyte et al. (2011), Watanabe et al. (2010)
LDLR	TALEN	SCNT	Carlson et al. (2012)
vWF	CRISPR/Cas9	CMI	Hai et al. (2014)
α1,3GT	CRISPR/Cas9	CMI	Petersen et al. (2016)
SLA-1,2,3	CRISPR/Cas9	SCNT	Reyes et al. (2014)
B2M	TALEN	CMI	Wang et al. (2016)
CD163	CRISPR/Cas9	CMI	Whitworth et al. (2016), Burkard et al. (2017)
α1,3GT,CMAH,B4GalNT2	CRISPR/Cas9	SCNT	Butler et al. (2016)
Cattle			
BLG	ZFN	SCNT	Yu et al. (2011)
GDF8	ZFN	SCNT	Luo et al. (2014)
GDF8	TALEN	CMI	Proudfoot et al. (2015)
PRPN	CRISPR/Cas9	CMI	Bevacqua et al. (2016)
Sheep			
GDF8	TALEN	CMI	Proudfoot et al. (2015)
Goat			
GDF8	CRISPR/Cas9	SCNT	Ni et al. (2014)
FGF, GDF8	CRISPR/Cas9	CMI	Wang et al. (2015)
HDR			
Pig			
RelA	ZFN	CMI	Lillico et al. (2013, 2016)
Cattle			
Polled locus	TALEN	SCNT	Carlson et al. (2016)
NRAMP1	CRISPR/Cas9	SCNT	Gao et al. (2017)
SP110	TALEN	SCNT	Wu et al. (2015)

challenge and would be a critical prerequisite for employing these technologies in human patients or for the generation of livestock animals. Comprehensive profiling of off-target cleavage sites will provide insight into the stringency of target recognition in each system, which in turn will help to increasing the specificity of the systems and to develop algorithms that calculate the most promising sequences to be targeted within a specific locus. The feasibility to use ZFNickases for genetic alterations of farm animals is a great step forward to lower the risk of off-target events, making the technology more predictable.

Although CRISPR/Cas seems to show the greatest promise and flexibility for genetic engineering, sequence requirements within the PAM sequence may constrain some applications. Therefore, evolution of the Cas9 protein should pave the way toward PAM independence and may also provide means to generate an even more efficient Cas9 endonuclease. Additional studies will also be required to evaluate the specificity and toxicity of RNA-guided DNA endonucleases in vitro and in vivo. Recent developments, in which an inactivated Cas element was conjugated to the FokI endonuclease that requires dimer formation, are promising as thereby a higher specificity can be achieved (Tsai et al. 2014; Guilinger et al. 2014). Biophysical and biochemical studies on CRISPRs could help to improve the design of next-generation genome editing tools.

The different genome editing tools have their individual advantages and disadvantages, and the selection of a specific system seems more to depend on the expertise of the individual researcher rather than on the weaknesses of one of these technologies. In summary, genome editors are valuable tools, which scientists 10 years ago could only dream of. These technologies expand and revolutionize our ability to explore and alter any genome and constitute a new and promising paradigm to understand and treat diseases.

References

Abudayyeh OO, Gootenberg JS, Konermann S, Joung J, Slaymaker IM, Cox DB, Shmakov S, Makarova KS, Semenova E, Minakhin L, Severinov K, Regev A, Lander ES, Koonin EV, Zhang F (2016) C2c2 is a single-component programmable RNA-guided RNA-targeting CRISPR effector. Science 353(6299):aaf5573. https://doi.org/10.1126/science.aaf5573

Bao L, Chen H, Jong U, Rim C, Li W, Lin X, Zhang D, Luo Q, Cui C, Huang H, Zhang Y, Xiao L, Fu Z (2014) Generation of GGTA1 biallelic knockout pigs via zinc-finger nucleases and somatic cell nuclear transfer. Sci China Life Sci 57(2):263–268. https://doi.org/10.1007/s11427-013-4601-2

Bedell VM, Wang Y, Campbell JM, Poshusta TL, Starker CG, Krug RG II, Tan W, Penheiter SG, Ma AC, Leung AY, Fahrenkrug SC, Carlson DF, Voytas DF, Clark KJ, Essner JJ, Ekker SC (2012) In vivo genome editing using a high-efficiency TALEN system. Nature 491(7422):114–118. https://doi.org/10.1038/nature11537

Bhaya D, Davison M, Barrangou R (2011) CRISPR-Cas systems in bacteria and archaea: versatile small RNAs for adaptive defense and regulation. Annu Rev Genet 45:273–297. https://doi.org/10.1146/annurev-genet-110410-132430

Bibikova M, Golic M, Golic KG, Carroll D (2002) Targeted chromosomal cleavage and mutagenesis in Drosophila using zinc-finger nucleases. Genetics 161(3):1169–1175

Boch J, Bonas U (2010) Xanthomonas AvrBs3 family-type III effectors: discovery and function. Annu Rev Phytopathol 48:419–436. https://doi.org/10.1146/annurev-phyto-080508-081936

Boch J, Scholze H, Schornack S, Landgraf A, Hahn S, Kay S, Lahaye T, Nickstadt A, Bonas U (2009) Breaking the code of DNA binding specificity of TAL-Type III effectors. Science 326(5959):1509–1512. https://doi.org/10.1126/science.1178811

Burkard C, Lillico SG, Reid E, Jackson B, Mileham AJ, Ait-Ali T, Whitelaw CB, Archibald AL (2017) Precision engineering for PRRSV resistance in pigs: macrophages from genome edited pigs lacking CD163 SRCR5 domain are fully resistant to both PRRSV genotypes while maintaining biological function. PLoS Pathog 13(2):e1006206. https://doi.org/10.1371/journal.ppat.1006206

Carlson DF, Tan W, Lillico SG, Stverakova D, Proudfoot C, Christian M, Voytas DF, Long CR, Whitelaw CB, Fahrenkrug SC (2012) Efficient TALEN-mediated gene knockout in livestock. Proc Natl Acad Sci U S A 109(43):17382–17387. https://doi.org/10.1073/pnas.1211446109

Carlson DF, Lancto CA, Zang B, Kim ES, Walton M, Oldeschulte D, Seabury C, Sonstegard TS, Fahrenkrug SC (2016) Production of hornless dairy cattle from genome-edited cell lines. Nat Biotechnol 34(5):479–481. https://doi.org/10.1038/nbt.3560

Cathomen T, Joung JK (2008) Zinc-finger nucleases: the next generation emerges. Mol Ther 16(7):1200–1207. https://doi.org/10.1038/mt.2008.114

Cermak T, Doyle EL, Christian M, Wang L, Zhang Y, Schmidt C, Baller JA, Somia NV, Bogdanove AJ, Voytas DF (2011) Efficient design and assembly of custom TALEN and other TAL effector-based constructs for DNA targeting. Nucleic Acids Res 39(12):e82. https://doi.org/10.1093/nar/gkr218

Chen F, Pruett-Miller SM, Huang Y, Gjoka M, Duda K, Taunton J, Collingwood TN, Frodin M, Davis GD (2011) High-frequency genome editing using ssDNA oligonucleotides with zinc-finger nucleases. Nat Methods 8(9):753–755. https://doi.org/10.1038/nmeth.1653

Choulika A, Perrin A, Dujon B, Nicolas JF (1995) Induction of homologous recombination in mammalian chromosomes by using the I-SceI system of Saccharomyces cerevisiae. Mol Cell Biol 15(4):1968–1973

Christian M, Cermak T, Doyle E, Schmidt C, Zhang F, Hummel A, Bogdanove A, Voytas D (2010) Targeting DNA double-strand breaks with TAL effector nucleases. Genetics 186(2):757–761

Cong L, Ran FA, Cox D, Lin S, Barretto R, Habib N, Hsu PD, Wu X, Jiang W, Marraffini LA, Zhang F (2013) Multiplex genome engineering using CRISPR/Cas systems. Science 339(6121):819–823. https://doi.org/10.1126/science.1231143

Cooper DK, Ayares D (2011) The immense potential of xenotransplantation in surgery. Int J Surg 9(2):122–129. https://doi.org/10.1016/j.ijsu.2010.11.002

Defrancesco L (2011) Move over ZFNs. Nat Biotechnol 29(8):681–684

Deng C, Capecchi MR (1992) Reexamination of gene targeting frequency as a function of the extent of homology between the targeting vector and the target locus. Mol Cell Biol 12(8):3365–3371

Denning C, Dickinson P, Burl S, Wylie D, Fletcher J, Clark AJ (2001) Gene targeting in primary fetal fibroblasts from sheep and pig. Cloning Stem Cells 3(4):221–231

Donoho G, Jasin M, Berg P (1998) Analysis of gene targeting and intrachromosomal homologous recombination stimulated by genomic double-strand breaks in mouse embryonic stem cells. Mol Cell Biol 18(7):4070–4078

Doyon Y, Choi VM, Xia DF, Vo TD, Gregory PD, Holmes MC (2010) Transient cold shock enhances zinc-finger nuclease-mediated gene disruption. Nat Methods 7(6):459–460. https://doi.org/10.1038/nmeth.1456

East-Seletsky A, O'Connell MR, Knight SC, Burstein D, Cate JH, Tjian R, Doudna JA (2016) Two distinct RNase activities of CRISPR-C2c2 enable guide-RNA processing and RNA detection. Nature 538(7624):270–273. https://doi.org/10.1038/nature19802

Epinat JC, Arnould S, Chames P, Rochaix P, Desfontaines D, Puzin C, Patin A, Zanghellini A, Paques F, Lacroix E (2003) A novel engineered meganuclease induces homologous recombination in yeast and mammalian cells. Nucleic Acids Res 31(11):2952–2962

Ezzelarab M, Ayares D, Cooper DK (2005) Carbohydrates in xenotransplantation. Immunol Cell Biol 83(4):396–404. https://doi.org/10.1111/j.1440-1711.2005.01344.x

Flisikowska T, Thorey IS, Offner S, Ros F, Lifke V, Zeitler B, Rottmann O, Vincent A, Zhang L, Jenkins S, Niersbach H, Kind AJ, Gregory PD, Schnieke AE, Platzer J (2011) Efficient immunoglobulin gene disruption and targeted replacement in rabbit using zinc finger nucleases. PLoS One 6(6):e21045. https://doi.org/10.1371/journal.pone.0021045

Flisikowska T, Kind A, Schnieke A (2014) Genetically modified pigs to model human diseases. J Appl Genet 55(1):53–64. https://doi.org/10.1007/s13353-013-0182-9

Fonfara I, Richter H, Bratovic M, Le Rhun A, Charpentier E (2016) The CRISPR-associated DNA-cleaving enzyme Cpf1 also processes precursor CRISPR RNA. Nature 532(7600):517–521. https://doi.org/10.1038/nature17945

Fu Y, Sander JD, Reyon D, Cascio VM, Joung JK (2014) Improving CRISPR-Cas nuclease specificity using truncated guide RNAs. Nat Biotechnol 32(3):279–284. https://doi.org/10.1038/nbt.2808

Gaudelli NM, Komor AC, Rees HA, Packer MS, Badran AH, Bryson DI, Liu DR (2017) Programmable base editing of A*T to G*C in genomic DNA without DNA cleavage. Nature 551(7681):464–471. https://doi.org/10.1038/nature24644

Geurts AM, Cost GJ, Freyvert Y, Zeitler B, Miller JC, Choi VM, Jenkins SS, Wood A, Cui X, Meng X, Vincent A, Lam S, Michalkiewicz M, Schilling R, Foeckler J, Kalloway S, Weiler H, Menoret S, Anegon I, Davis GD, Zhang L, Rebar EJ, Gregory PD, Urnov FD, Jacob HJ, Buelow R (2009) Knockout rats via embryo microinjection of zinc-finger nucleases. Science 325(5939):433. https://doi.org/10.1126/science.1172447

Guilinger JP, Thompson DB, Liu DR (2014) Fusion of catalytically inactive Cas9 to FokI nuclease improves the specificity of genome modification. Nat Biotechnol 32:577. https://doi.org/10.1038/nbt.2909

Hai T, Teng F, Guo R, Li W, Zhou Q (2014) One-step generation of knockout pigs by zygote injection of CRISPR/Cas system. Cell Res 24(3):372–375

Hauschild J, Petersen B, Santiago Y, Queisser AL, Carnwath JW, Lucas-Hahn A, Zhang L, Meng X, Gregory PD, Schwinzer R, Cost GJ, Niemann H (2011) Efficient generation of a biallelic knockout in pigs using zinc-finger nucleases. Proc Natl Acad Sci U S A 108(29):12013–12017. https://doi.org/10.1073/pnas.1106422108

Hauschild-Quintern J, Petersen B, Cost GJ, Niemann H (2013a) Gene knockout and knockin by zinc-finger nucleases: current status and perspectives. Cell Mol Life Sci 70(16):2969–2983. https://doi.org/10.1007/s00018-012-1204-1

Hauschild-Quintern J, Petersen B, Queisser AL, Lucas-Hahn A, Schwinzer R, Niemann H (2013b) Gender non-specific efficacy of ZFN mediated gene targeting in pigs. Transgenic Res 22(1):1–3. https://doi.org/10.1007/s11248-012-9647-6

Hockemeyer D, Wang H, Kiani S, Lai CS, Gao Q, Cassady JP, Cost GJ, Zhang L, Santiago Y, Miller JC, Zeitler B, Cherone JM, Meng X, Hinkley SJ, Rebar EJ, Gregory PD, Urnov FD, Jaenisch R (2011) Genetic engineering of human pluripotent cells using TALE nucleases. Nat Biotechnol 29(8):731–734. https://doi.org/10.1038/nbt.1927

Huang P, Xiao A, Zhou M, Zhu Z, Lin S, Zhang B (2011) Heri gene targeting in zebrafish using customized TALENs. Nat Biotechnol 29(8):699–700. https://doi.org/10.1038/nbt.1939

Jinek M, East A, Cheng A, Lin S, Ma E, Doudna J (2013) RNA-programmed genome editing in human cells. eLife 2:e00471. https://doi.org/10.7554/eLife.00471

Kim YG, Cha J, Chandrasegaran S (1996) Hybrid restriction enzymes: zinc finger fusions to Fok I cleavage domain. Proc Natl Acad Sci U S A 93(3):1156–1160

Kim Y, Kweon J, Kim A, Chon JK, Yoo JY, Kim HJ, Kim S, Lee C, Jeong E, Chung E, Kim D, Lee MS, Go EM, Song HJ, Kim H, Cho N, Bang D, Kim S, Kim JS (2013) A library of TAL effector nucleases spanning the human genome. Nat Biotechnol 31(3):251–258. https://doi.org/10.1038/nbt.2517

Kleinstiver BP, Pattanayak V, Prew MS, Tsai SQ, Nguyen NT, Zheng Z (2016a) High-fidelity CRISPR-Cas9 nucleases with no detectable genome-wide off-target effects. Nature 529:490. https://doi.org/10.1038/nature16526

Kleinstiver BP, Tsai SQ, Prew MS, Nguyen NT, Welch MM, Lopez JM, McCaw ZR, Aryee MJ, Joung JK (2016b) Genome-wide specificities of CRISPR-Cas Cpf1 nucleases in human cells. Nat Biotechnol 34(8):869–874. https://doi.org/10.1038/nbt.3620

Komor AC, Kim YB, Packer MS, Zuris JA, Liu DR (2016) Programmable editing of a target base in genomic DNA without double-stranded DNA cleavage. Nature 533(7603):420–424. https://doi.org/10.1038/nature17946

Kues WA, Niemann H (2004) The contribution of farm animals to human health. Trends Biotechnol 22(6):286–294. https://doi.org/10.1016/j.tibtech.2004.04.003

Kwon DN, Lee K, Kang MJ, Choi YJ, Park C, Whyte JJ, Brown AN, Kim JH, Samuel M, Mao J, Park KW, Murphy CN, Prather RS, Kim JH (2013) Production of biallelic CMP-Neu5Ac hydroxylase knock-out pigs. Sci Rep 3:1981. https://doi.org/10.1038/srep01981

Li T, Huang S, Jiang WZ, Wright D, Spalding MH, Weeks DP, Yang B (2011a) TAL nucleases (TALNs): hybrid proteins composed of TAL effectors and FokI DNA-cleavage domain. Nucleic Acids Res 39(1):359–372. https://doi.org/10.1093/nar/gkq704

Li T, Huang S, Zhao X, Wright DA, Carpenter S, Spalding MH, Weeks DP, Yang B (2011b) Modularly assembled designer TAL effector nucleases for targeted gene knockout and gene replacement in eukaryotes. Nucleic Acids Res 39(14):6315–6325. https://doi.org/10.1093/nar/gkr188 gkr188 [pii]

Li P, Estrada JL, Burlak C, Tector AJ (2013) Biallelic knockout of the alpha-1,3 galactosyltransferase gene in porcine liver-derived cells using zinc finger nucleases. J Surg Res 181:e39. https://doi.org/10.1016/j.jss.2012.06.035

Lillico SG, Proudfoot C, Carlson DF, Stverakova D, Neil C, Blain C, King TJ, Ritchie WA, Tan W, Mileham AJ, McLaren DG, Fahrenkrug SC, Whitelaw CB (2013) Live pigs produced from genome edited zygotes. Sci Rep 3:2847. https://doi.org/10.1038/srep02847

Lillico SG, Proudfoot C, King TJ, Tan W, Zhang L, Mardjuki R, Paschon DE, Rebar EJ, Urnov FD, Mileham AJ, McLaren DG, Whitelaw CB (2016) Mammalian interspecies substitution of immune modulatory alleles by genome editing. Sci Rep 6:21645. https://doi.org/10.1038/srep21645

Liu X, Wang Y, Guo W, Chang B, Liu J, Guo Z, Quan F, Zhang Y (2013) Zinc-finger nickase-mediated insertion of the lysostaphin gene into the beta-casein locus in cloned cows. Nat Commun 4:2565. https://doi.org/10.1038/ncomms3565

Liu X, Wang Y, Tian Y, Yu Y, Gao M, Hu G, Su F, Pan S, Luo Y, Guo Z, Quan F, Zhang Y (2014) Generation of mastitis resistance in cows by targeting human lysozyme gene to beta-casein locus using zinc-finger nucleases. Proc Biol Sci 281(1780):20133368. https://doi.org/10.1098/rspb.2013.3368

Mahfouz MM, Li L, Shamimuzzaman M, Wibowo A, Fang X, Zhu JK (2011) De novo-engineered transcription activator-like effector (TALE) hybrid nuclease with novel DNA binding specificity creates double-strand breaks. Proc Natl Acad Sci U S A 108(6):2623–2628. https://doi.org/10.1073/pnas.1019533108 1019533108 [pii]

Mali P, Yang L, Esvelt KM, Aach J, Guell M, DiCarlo JE, Norville JE, Church GM (2013) RNA-guided human genome engineering via Cas9. Science 339(6121):823–826. https://doi.org/10.1126/science.1232033

Mansour SL, Thomas KR, Capecchi MR (1988) Disruption of the proto-oncogene int-2 in mouse embryo-derived stem cells: a general strategy for targeting mutations to non-selectable genes. Nature 336(6197):348–352. https://doi.org/10.1038/336348a0

Meselson MS, Radding CM (1975) A general model for genetic recombination. Proc Natl Acad Sci U S A 72(1):358–361

Miller J, McLachlan AD, Klug A (1985) Repetitive zinc-binding domains in the protein transcription factor IIIA from Xenopus oocytes. EMBO J 4(6):1609–1614

Miller JC, Tan S, Qiao G, Barlow KA, Wang J, Xia DF, Meng X, Paschon DE, Leung E, Hinkley SJ, Dulay GP, Hua KL, Ankoudinova I, Cost GJ, Urnov FD, Zhang HS, Holmes MC, Zhang L, Gregory PD, Rebar EJ (2011) A TALE nuclease architecture for efficient genome editing. Nat Biotechnol 29(2):143–148. https://doi.org/10.1038/nbt.1755

Moehle EA, Rock JM, Lee YL, Jouvenot Y, DeKelver RC, Gregory PD, Urnov FD, Holmes MC (2007) Targeted gene addition into a specified location in the human genome using designed zinc finger nucleases. Proc Natl Acad Sci U S A 104(9):3055–3060. https://doi.org/10.1073/pnas.0611478104

Moscou MJ, Bogdanove AJ (2009) A simple cipher governs DNA recognition by TAL effectors. Science 326(5959):1501–1501. https://doi.org/10.1126/science.1178817

Nowak-Imialek M, Niemann H (2012) Pluripotent cells in farm animals: state of the art and future perspectives. Reprod Fertil Dev 25(1):103–128. https://doi.org/10.1071/RD12265

Orlando SJ, Santiago Y, DeKelver RC, Freyvert Y, Boydston EA, Moehle EA, Choi VM, Gopalan SM, Lou JF, Li J, Miller JC, Holmes MC, Gregory PD, Urnov FD, Cost GJ (2010) Zinc-finger nuclease-driven targeted integration into mammalian genomes using donors with limited chromosomal homology. Nucleic Acids Res 38(15):e152. https://doi.org/10.1093/nar/gkq512 gkq512 [pii]

Pabo CO, Peisach E, Grant RA (2001) Design and selection of novel Cys2His2 zinc finger proteins. Annu Rev Biochem 70:313–340. https://doi.org/10.1146/annurev.biochem.70.1.313

Palgrave CJ, Gilmour L, Lowden CS, Lillico SG, Mellencamp MA, Whitelaw CB (2011) Species-specific variation in RELA underlies differences in NF-kappaB activity: a potential role in African swine fever pathogenesis. J Virol 85(12):6008–6014. https://doi.org/10.1128/JVI.00331-11

Pavletich NP, Pabo CO (1991) Zinc finger-DNA recognition: crystal structure of a Zif268-DNA complex at 2.1 A. Science 252(5007):809–817

Petersen B, Lucas-Hahn A, Oropeza M, Hornen N, Lemme E, Hassel P, Queisser AL, Niemann H (2008) Development and validation of a highly efficient protocol of porcine somatic cloning using preovulatory embryo transfer in peripubertal gilts. Cloning Stem Cells 10(3):355–362. https://doi.org/10.1089/clo.2008.0026

Petersen B, Carnwath JW, Niemann H (2009) The perspectives for porcine-to-human xenografts. Comp Immunol Microbiol Infect Dis 32(2):91–105. https://doi.org/10.1016/j.cimid.2007.11.014

Petersen B, Frenzel A, Lucas-Hahn A, Herrmann D, Hassel P, Klein S, Ziegler M, Hadeler K-G, Niemann H (2016) Efficient production of biallelic GGTA1 knockout pigs by cytoplasmic microinjection of CRISPR/Cas9 into zygotes. Xenotransplantation 23(5):338–346. https://doi.org/10.1111/xen.12258

Proudfoot C, Carlson DF, Huddart R, Long CR, Pryor JH, King TJ, Lillico SG, Mileham AJ, McLaren DG, Whitelaw CB, Fahrenkrug SC (2015) Genome edited sheep and cattle. Transgenic Res 24(1):147–153. https://doi.org/10.1007/s11248-014-9832-x

Radding CM (1982) Homologous pairing and strand exchange in genetic recombination. Annu Rev Genet 16:405–437. https://doi.org/10.1146/annurev.ge.16.120182.002201

Rogers CS, Stoltz DA, Meyerholz DK, Ostedgaard LS, Rokhlina T, Taft PJ, Rogan MP, Pezzulo AA, Karp PH, Itani OA, Kabel AC, Wohlford-Lenane CL, Davis GJ, Hanfland RA, Smith TL, Samuel M, Wax D, Murphy CN, Rieke A, Whitworth K, Uc A, Starner TD, Brogden KA, Shilyansky J, PB MC Jr, Zabner J, Prather RS, Welsh MJ (2008) Disruption of the CFTR gene produces a model of cystic fibrosis in newborn pigs. Science 321(5897):1837–1841. https://doi.org/10.1126/science.1163600

Rouet P, Smih F, Jasin M (1994) Expression of a site-specific endonuclease stimulates homologous recombination in mammalian cells. Proc Natl Acad Sci U S A 91(13):6064–6068

Sander JD, Cade L, Khayter C, Reyon D, Peterson RT, Joung JK, Yeh JR (2011) Targeted gene disruption in somatic zebrafish cells using engineered TALENs. Nat Biotechnol 29(8):697–698. https://doi.org/10.1038/nbt.1934

Shen B, Zhang W, Zhang J, Zhou J, Wang J, Chen L, Wang L, Hodgkins A, Iyer V, Huang X, Skarnes WC (2014) Efficient genome modification by CRISPR-Cas9 nickase with minimal off-target effects. Nat Methods 11(4):399–402. https://doi.org/10.1038/nmeth.2857

Shmakov S, Abudayyeh OO, Makarova KS, Wolf YI, Gootenberg JS, Semenova E, Minakhin L, Joung J, Konermann S, Severinov K, Zhang F, Koonin EV (2015) Discovery and functional characterization of diverse class 2 CRISPR-Cas systems. Mol Cell 60(3):385–397. https://doi.org/10.1016/j.molcel.2015.10.008

Smith J, Bibikova M, Whitby FG, Reddy AR, Chandrasegaran S, Carroll D (2000) Requirements for double-strand cleavage by chimeric restriction enzymes with zinc finger DNA-recognition domains. Nucleic Acids Res 28(17):3361–3369

Szczepek M, Brondani V, Buchel J, Serrano L, Segal DJ, Cathomen T (2007) Structure-based redesign of the dimerization interface reduces the toxicity of zinc-finger nucleases. Nat Biotechnol 25(7):786–793. https://doi.org/10.1038/nbt1317

Tan W, Carlson DF, Lancto CA, Garbe JR, Webster DA, Hackett PB, Fahrenkrug SC (2013) Efficient nonmeiotic allele introgression in livestock using custom endonucleases. Proc Natl Acad Sci U S A 110(41):16526–16531. https://doi.org/10.1073/pnas.1310478110

Tan W, Proudfoot C, Lillico SG, Whitelaw CB (2016) Gene targeting, genome editing: from Dolly to editors. Transgenic Res 25(3):273–287. https://doi.org/10.1007/s11248-016-9932-x

Terns MP, Terns RM (2011) CRISPR-based adaptive immune systems. Curr Opin Microbiol 14(3):321–327. https://doi.org/10.1016/j.mib.2011.03.005

Tesson L, Usal C, Menoret S, Leung E, Niles BJ, Remy S, Santiago Y, Vincent AI, Meng X, Zhang L, Gregory PD, Anegon I, Cost GJ (2011) Knockout rats generated by embryo microinjection of TALENs. Nat Biotechnol 29(8):695–696. https://doi.org/10.1038/nbt.1940

Tsai SQ, Wyvekens N, Khayter C, Foden JA, Thapar V, Reyon D, Goodwin MJ, Aryee MJ, Joung JK (2014) Dimeric CRISPR RNA-guided FokI nucleases for highly specific genome editing. Nat Biotechnol 32(6):569–576. https://doi.org/10.1038/nbt.2908

Vasquez KM, Marburger K, Intody Z, Wilson JH (2001) Manipulating the mammalian genome by homologous recombination. Proc Natl Acad Sci U S A 98(15):8403–8410. https://doi.org/10.1073/pnas.111009698

Voytas DF, Joung JK (2009) Plant science. DNA binding made easy. Science 326(5959):1491–1492. https://doi.org/10.1126/science.1183604

Wang J, Friedman G, Doyon Y, Wang NS, Li CJ, Miller JC, Hua KL, Yan JJ, Babiarz JE, Gregory PD, Holmes MC (2012) Targeted gene addition to a predetermined site in the human genome using a ZFN-based nicking enzyme. Genome Res 22(7):1316–1326. https://doi.org/10.1101/gr.122879.111

Watanabe M, Umeyama K, Matsunari H, Takayanagi S, Haruyama E, Nakano K, Fujiwara T, Ikezawa Y, Nakauchi H, Nagashima H (2010) Knockout of exogenous EGFP gene in porcine somatic cells using zinc-finger nucleases. Biochem Biophys Res Commun 402(1):14–18. https://doi.org/10.1016/j.bbrc.2010.09.092

Watanabe M, Nakano K, Matsunari H, Matsuda T, Maehara M, Kanai T, Kobayashi M, Matsumura Y, Sakai R, Kuramoto M, Hayashida G, Asano Y, Takayanagi S, Arai Y, Umeyama K, Nagaya M, Hanazono Y, Nagashima H (2013) Generation of interleukin-2 receptor gamma gene knockout pigs from somatic cells genetically modified by zinc finger nuclease-encoding mRNA. PLoS One 8(10):e76478. https://doi.org/10.1371/journal.pone.0076478

Welsh MJ, Rogers CS, Stoltz DA, Meyerholz DK, Prather RS (2009) Development of a porcine model of cystic fibrosis. Trans Am Clin Climatol Assoc 120:149–162

Whitworth KM, Rowland RR, Ewen CL, Trible BR, Kerrigan MA, Cino-Ozuna AG, Samuel MS, Lightner JE, McLaren DG, Mileham AJ, Wells KD, Prather RS (2016) Gene-edited pigs are protected from porcine reproductive and respiratory syndrome virus. Nat Biotechnol 34(1):20–22. https://doi.org/10.1038/nbt.3434

Whyte JJ, Prather RS (2012) Cell biology symposium: zinc finger nucleases to create custom-designed modifications in the swine (Sus scrofa) genome. J Anim Sci 90(4):1111–U1159. https://doi.org/10.2527/jas.2011-4546

Whyte JJ, Zhao J, Wells KD, Samuel MS, Whitworth KM, Walters EM, Laughlin MH, Prather RS (2011) Gene targeting with zinc finger nucleases to produce cloned eGFP knockout pigs. Mol Reprod Dev 78(1):2. https://doi.org/10.1002/mrd.21271

Wiedenheft B, Sternberg SH, Doudna JA (2012) RNA-guided genetic silencing systems in bacteria and archaea. Nature 482(7385):331–338. https://doi.org/10.1038/nature10886

Wilmut I, Schnieke AE, McWhir J, Kind AJ, Campbell KH (1997) Viable offspring derived from fetal and adult mammalian cells. Nature 385(6619):810–813

Yang D, Yang H, Li W, Zhao B, Ouyang Z, Liu Z, Zhao Y, Fan N, Song J, Tian J, Li F, Zhang J, Chang L, Pei D, Chen YE, Lai L (2011) Generation of PPARgamma mono-allelic knockout pigs via zinc-finger nucleases and nuclear transfer cloning. Cell Res 21(6):979–982. https://doi.org/10.1038/cr.2011.70

Yu S, Luo J, Song Z, Ding F, Dai Y, Li N (2011) Highly efficient modification of beta-lactoglobulin (BLG) gene via zinc-finger nucleases in cattle. Cell Res 21(11):1638–1640. https://doi.org/10.1038/cr.2011.153

Zetsche B, Gootenberg JS, Abudayyeh OO, Slaymaker IM, Makarova KS, Essletzbichler P, Volz SE, Joung J, van der Oost J, Regev A, Koonin EV, Zhang F (2015) Cpf1 is a single RNA-guided endonuclease of a class 2 CRISPR-Cas system. Cell 163:759. https://doi.org/10.1016/j.cell.2015.09.038

Regulatory Dysfunction inhibits the Development and Application of Transgenic Livestock for Use in Agriculture

8

James D. Murray and Elizabeth A. Maga

Abstract

Since the production of the first transgenic livestock, the technology for producing the animals and controlling transgene expression has matured. Initially, the lack of knowledge about promoter, enhancer, and coding regions of genes of interest greatly hampered efforts to create transgenes that would express appropriately in livestock and be useful to industry. There have been many developments in the technology to create transgenic animals, including somatic cell nuclear transfer-based cloning and gene editing. In the 31 years since the first report of transgenic livestock, a number of potentially useful animals, including cattle, goats, pigs, and sheep, have been made. However, there still are no genetically engineered animal-based food products on the market. There has been a failure of the regulatory processes to effectively move forward across the world, with many countries adopting process-based regulations, rather than product-based, and some countries having no regulatory framework at all. Additionally, there is a perception among some consumers that transgenic technology is potentially harmful in spite of a large, and growing, body of evidence to the contrary. Estimates suggest the world will need to approximately double our current food production by 2050, including animal-based foods; that is, we will have to produce an amount of food each year equal to that consumed by mankind over the past 500 years. The practical benefits of transgenic animals in agriculture have not yet reached consumers, and in the absence of predictable, science-based regulatory programs, it is unlikely that the benefits will be realized in the short to medium term.

J. D. Murray (✉)
Department of Animal Science, University of California, Davis, CA, USA

Department of Population Health and Reproduction, University of California, Davis, CA, USA
e-mail: jdmurray@ucdavis.edu

E. A. Maga
Department of Animal Science, University of California, Davis, CA, USA
e-mail: eamaga@ucdavis.edu

© Springer International Publishing AG, part of Springer Nature 2018
H. Niemann, C. Wrenzycki (eds.), *Animal Biotechnology 2*,
https://doi.org/10.1007/978-3-319-92348-2_8

8.1   Introduction

While the development of transgenic technology started with the landmark paper of Gordon et al. (1980), it was the dramatic demonstration of the power of the technology by Palmiter et al. (1982) that truly caught the attention of scientists. In their 1982 paper on the dramatic growth of mice injected with a metallothionein-growth hormone construct, Palmiter et al. recognized the potential of genetic engineering for use in agriculture saying in the discussion:

> The implicit possibility is to use this technology to stimulate rapid growth of commercially valuable animals. Benefit would presumably accrue from a shorter production time and possibly from increased efficiency of food utilization. --- Having a regulatable promoter may be particularly advantageous for timely expression of GH. Applying these techniques to large animals will be more difficult. Nevertheless, when genes for desired traits can be isolated, this approach should provide a valuable adjunct to traditional breeding methods.

The first transgenic livestock, swine, sheep, and rabbits, were reported 3 years later by the same group (Hammer et al. 1985). Many excellent reviews have been published, both foreshadowing potential applications of transgenic animals in production agriculture (e.g., Pursel et al. 1989; Ward et al. 1990; Pursel and Rexroad 1993; Maga and Murray 1995; Pinkert and Murray 1999; Murray and Maga 1999) and, more recently, summarizing what has been accomplished over the past three decades (e.g., Laible 2009; Kues and Niemann 2011; Tan et al. 2012; Cooper et al. 2015; Garas et al. 2015; Murray and Maga 2016a). The excellent nature of these reviews over the years allows us the luxury to focus this article more on the development of this field, maturation to potential applications, and the current need for this technology in agricultural applications.

8.2   The Developmental Years

Following the report that expression of a human growth hormone transgene in mice resulted in a 50% increase in body size and weight in mice (Palmiter et al. 1982), the effort to produce transgenic livestock for use in agriculture initially focused on growth-promoting transgenes in swine and sheep (Table 8.1). The work of Hammer et al. (1985) was followed over the next decade by numerous reports from research groups in the USA (Pursel et al. 1987, 1989, 1999, 2004; Pinkert et al. 1987; Rexroad et al. 1989, 1991; Ebert et al. 1990; Wieghart et al. 1990), Australia (Vize et al. 1988; Murray et al. 1989; Nottle et al. 1999), Germany (Brem et al. 1985), and the UK (Polge et al. 1989) using pronuclear microinjection to produce transgenic swine and sheep expressing various growth factor-based transgenes (Table 8.1), with the goal of producing lines of animals with superior growth and lean carcass traits. A variety of promoter elements were used to express transgenes incorporating human, rat, bovine, ovine, and porcine growth hormone (GH) genes, growth hormone-releasing factor (GRF), and insulin-like growth factor-1 (IGF-I). While transgenic animals routinely could be produced, the results for the most part were

Table 8.1 Transgenic pigs and sheep for enhanced growth

	Transgene[a]	Reference
Pigs	mMT/hGH	Brem et al. (1985), Hammer et al. (1985), Pursel et al. (1987)
	mMT/hGRF	Pinkert et al. (1987), Pursel et al. (1989)
	mMT/bGH	Pursel et al. (1987)
	hMT/pGH	Vize et al. (1988), Nottle et al. (1999)
	MLV/rGH	Ebert et al. (1988)
	bPRL/bGH	Polge et al. (1989)
	hALB/hGRF	Pursel et al. (1989)
	mMT/hIGF-1	Miller et al. (1989), Pursel et al. (1989)
	rPEPCK/bGH	Wieghart et al. (1990)
	CMV/pGH	Ebert et al. (1990)
	MLV/pGH	Ebert et al. (1990)
	oMT/oGH	Pursel et al. (1997)
	cASK/hIGF-1	Pursel et al. (1999, 2004)
Sheep	mMT/hGH	Hammer et al. (1985), Pursel et al. (1987)
	mMT/bGH	Pursel et al. (1987), Rexroad et al. (1989)
	oMT/oGH	Murray et al. (1989)
	mMT/hGRF	Rexroad et al. (1989)
	mTF/bGH, mAlb/ hGRF	Rexroad et al. (1991)

[a]Transgene shows promoter/coding region:
Promoters: mMT, mouse metallothionein; hMT, human metallothionein; MLV, mouse leukemia virus; bPRL, bovine prolactin; hALB, human albumin; rPEPCK, rat phosphoenolpyruvate carboxykinase; CMV, cytomegalovirus; oMTla, ovine metallothionein la; cASK, chicken α-skeletal actin; mTF, mouse transferrin; mALB, mouse albumin. Transgene: hGH, human growth hormone; hGRF, human growth hormone-releasing factor; bGH, bovine growth hormone; pGH, porcine growth hormone; rGH, rat growth hormone; hIGF-1, human insulin-like growth factor-1; oGH, ovine growth hormone

unacceptable as the animals performed poorly and usually displayed a range of detrimental phenotypes due to the overexpression of GH (Pursel et al. 1989; Nancarrow et al. 1991).

Ultimately the health problems were overcome in swine by two very different approaches. Nottle et al. (1999) took the approach of producing a very large number of lines of transgenic pigs and then selected two lines that expressed very low levels of GH. Alternatively, Pursel et al. (1999) produced a line of pigs that expressed IGF-I under the control of a muscle specific promoter, thus limiting the effect of the transgene to the muscle. In both cases there were the predicted positive benefits of increased growth and muscle mass, with no documented deleterious effects on the health of the animals.

In addition to growth-promoting traits, other initial attempts at producing genetically engineered animals for use in agriculture focused on altering wool growth (Damak et al. 1996a), properties of wool (Bawden et al. 1998), or enhancing viral resistance (Clements et al. 1994). While none of these applications have been advanced, collectively these efforts lead to a greater understanding of how to configure gene constructs for better control of expression and improvements in producing transgenic livestock.

Initially all livestock were produced using pronuclear microinjection, a relatively straightforward though inefficient technique, with only 1–3% of transferred, microinjected embryos born as transgenic animals (see Wall et al. 1992). At first, while skilled operators could regularly produce genetically engineered animals, the principal limitation was the result of inefficient protocols for the production of pronuclear embryos and the ability to visualize the pronuclei in the fertilized ova of some species. During this initial developmental phase of the technology, there were technical advances in synchronization of ovulation (Nancarrow et al. 1984), centrifugation of embryos to allow visualization of the pronuclei (Wall et al. 1985), and in the development of improved protocols for in vitro production of zygotes (Lu et al. 1987).

During this period technical advances were also made in two other important areas. First, increased understanding of the control of gene expression lead to better transgene constructs. Promoter and enhancer elements were identified that gave controlled tissue-specific patterns of expression (e.g., Archibald et al. 1990; Krimpenfort et al. 1991; Ebert et al. 1994; Damak et al. 1996b; Pursel et al. 1999; Bleck et al. 1998), and insulator sequences were identified and tested (McKnight et al. 1992) that isolated the transgene from surrounding, endogenous genes. The result was improvements in the control of transgene expression in animals, such that ectopic transgene expression as a potential problem was largely eliminated.

By the mid-1990s all of our common livestock species, including cows, sheep, goats, pigs, and rabbits, could be genetically engineered (Pinkert and Murray 1999). While most transgenic livestock were produced by pronuclear microinjection, retroviral vectors (Haskell and Bowen 1995), lentiviral-based vectors (Naldini et al. 1996), and sperm-mediated gene transfer (Lavitrano et al. 1997) were developed as alternative methods. Finally transposons such as *Sleeping Beauty* (Ivics et al. 1997) and *PiggyBac* (Ding et al. 2005) were developed for use in mammalian cells. While collectively these developments improved the reliability of producing genetically engineered animals with targeted gene expression, the principal limitations to the use of transgenic animals in agriculture still applied, namely, the lack of knowledge of the genetic basis of factors limiting production traits and the need for tissue and developmentally appropriate promoters (Ward et al. 1986). The need to improve the efficiency of production of transgenic livestock was partially overcome by the use of lentiviral vectors and transposon systems, but both of these approaches are limited in terms of the amount of DNA that can be carried and the potential for multiple insertion sites.

8.3 Breakthroughs and Applications

In the late 1990s, three events, two resulting in major paradigm shifts and the other the inevitable consequence of the expanding genomics revolution, altered the transgenic livestock field. The first was the discovery by Campbell and Wilmut at the Roslin Institute that sheep could be cloned by the transfer of the nucleus from a somatic cell into an enucleated ovum (Campbell et al. 1996; Wilmut et al. 1997). Somatic cell nuclear transfer (SCNT)-based cloning was not only a new method to produce transgenic animals (Schnieke et al. 1997; Cibelli et al. 1998), it also allowed for homologous recombination, which in turn is necessary for gene targeting and rapidly resulted in gene knockouts (KO) in sheep (Denning et al. 2001) and pigs (Dai et al. 2002). The ability to target gene integration into a preselected site in cells in tissue culture followed by SCNT-based cloning to regenerate a viable, fertile offspring from those cells opened up the possibility to enact any genomic change of interest. Not only could a transgene be inserted, but it could also be integrated into a preselected spot to eliminate potential insertional mutations and to facilitate appropriate expression. Endogenous genes could be knocked out or a specific mutation introduced and all of this in any species in which SCNT cloning was available. The use of the SCNT cloning approach ensured that all offspring obtained were genetically engineered, but the inefficiencies of the cloning process itself still resulted in the overall process of obtaining a transgenic animal being inefficient.

The second revolutionary development occurring in the mid-1990s was the production of programmable endonucleases which in turn enabled gene editing. The first such endonucleases developed were zinc finger nucleases (Shi and Berg 1995) and with them came the potential to directly target an endogenous genome sequence for mutation, KO, or transgene insertion (Durai et al. 2005). The first livestock produced using ZFNs were rabbits (Flisikowska et al. 2011) and pigs (Whyte et al. 2011) in 2011.

Prior to the development of SCNT-based cloning, a number of potentially useful transgenic animals were produced by pronuclear microinjection and have subsequently been characterized for use in agriculture (Table 8.2). Notable among these

Table 8.2 Transgenic animals produced and characterized for use in agriculture

			References
Species	Transgene[a]	Production	Characterization
Pig	ba-LA/ ba-LA	Bleck et al. (1998)	Wheeler et al. (2001), Noble et al. (2002); Marshall et al. (2006)
	mPSP/ APPA	Golovan et al. (2001)	Forsberg et al. (2003, 2013, 2014a, b), Meidinger et al. (2013)
	ba-LA/ hIGF-I	Monaco et al. (2005)	Hartke et al. (2005)
Cattle	bCsn/hLF	van Berkel et al. (2002)	Thomassen et al. (2005), Simojoki et al. (2010), Cooper et al. (2012, 2014a), Garas et al. (2016), Parc et al. (2017)
Goat	baS$_1$Csn/ hLz	Maga et al. (2003)	Maga et al. (2006a, b, c, 2012), Scharfen et al. (2007), Brundige et al. (2008, 2010), Jackson et al. (2010), Cooper et al. (2011, 2013, 2014a), Carvalho et al. (2012), Clark et al. (2014), McInnis et al. (2015), Garas et al. (2016), Garas et al. (2017), Carneiro et al. (2018)

[a]Transgenes: ba-LA, bovine α-lactalbumin; mPSP/APPA, mouse parotid secretory protein/*E. coli* phytase; ba-LA/hIGF-1, bovine α-lactalbumin/human insulin-like growth factor-1; bCsn/hLF, bovine casein/human lactoferrin; bCsn/hLz, bovine αS$_1$casein/human lysozyme

were pigs expressing high levels of bovine α-lactoglobulin (Bleck et al. 1998) and IGF-I (Monaco et al. 2005) in their mammary glands, pigs expressing *E. coli* phytase in their salivary glands (Golovan et al. 2001), dairy cows expressing human lactoferrin (hLF) in their mammary glands (van Berkel et al. 2002), and goats expressing human lysozyme (hLZ) (Fig. 8.1) in their mammary glands (Maga et al. 2003). The cumulative work on the lactoferrin and lysozyme animals has been recently reviewed (Cooper et al. 2015). While other lines of animals were also produced, what makes these five lines particularly noteworthy is the amount of characterization of the health of the line and the efficacy of the resulting phenotype.

A number of general conclusions can be reached when the work on these five lines of animals is collectively appraised. First, each line has been maintained and studied over multiple generations and shown to stably transmit and consistently and

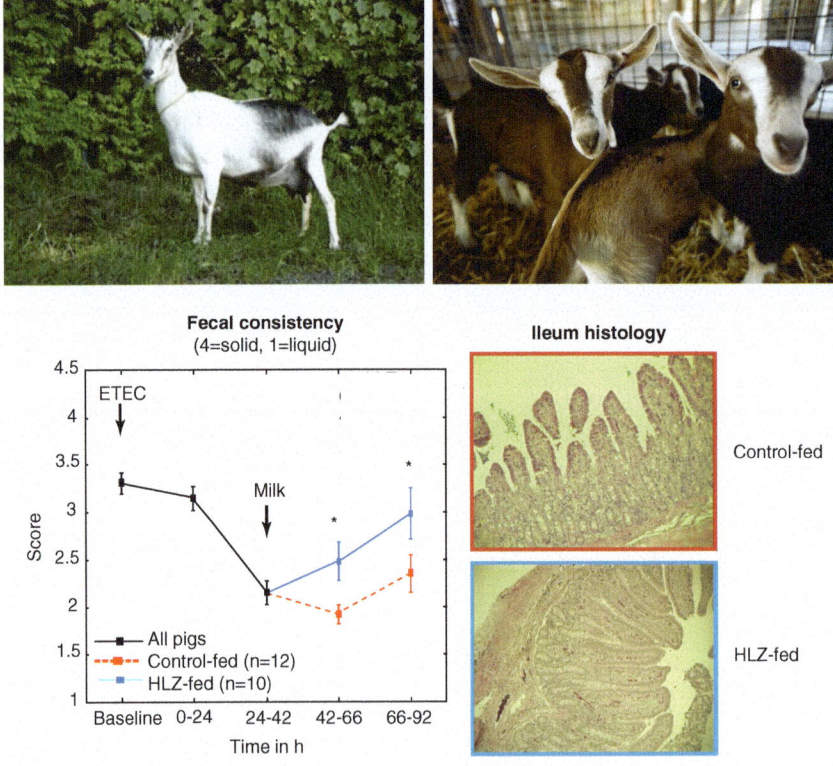

Fig. 8.1 Top left: Artemis—founder animal of lysozyme transgenic goat line born in 1999; Top right: Descendants of Artemis born in 2017. The Artemis line of transgenic goats has been propagated for eight generations with no deleterious effects on animal health, reproduction, or milk production (Maga et al. 2006a; Jackson et al. 2010; Clark et al. 2014; McInnis et al. 2015), with consistent levels of human lysozyme expression in the milk (Maga et al. 2006a). Studies have confirmed that the presence of human lysozyme in goat's milk is an effective antimicrobial in vivo, promoting significantly faster recovery from enterotoxigenic *E. coli* infection at the clinical (Bottom left) and histological level (bottom right) within 50 h of feeding (from Cooper et al. 2013)

predictably express the transgene from generation to generation. Second, the observation that each line does indeed transmit the transgene in a Mendelian pattern indicates that there are no health or reproductive impairments attributable to the mere fact that an animal carries a transgene. Third, each of these lines has expressed the incorporated transgene in a tissue-specific manner as directed by the promoter and enhancer elements used, thus illustrating that we have learned how to specifically direct transgene expression. Fourth, each line has demonstrated the predicted phenotype over several generations, and thus they represent new lines of livestock with a novel trait specifically addressing a need.

The pigs expressing high levels of bovine α-lactoglobulin (Bleck et al. 1998) and IGF-I (Monaco et al. 2005) in their mammary glands were produced to address the welfare and production problems associated with large litters and a limited milk supply from the sow following parturition. Pigs expressing *E. coli* phytase in their salivary glands (Golovan et al. 2001) were produced to decrease phosphorus excretion in feces by allowing the pig to digest phytic acid, the principal storage form of phosphorus in plant tissues and seeds. Finally, the dairy cows expressing human hLF (van Berkel et al. 2002) and the dairy goats expressing hLZ (Maga et al. 2003) in their milk were both designed to increase the availability of these naturally occurring human antimicrobial proteins for human use. In both cases, the proteins are reliably expressed in the milk and function when consumed by young pigs to enhance gastrointestinal health (for review, see Cooper et al. 2015).

8.4 Current Status

As just noted, lines of transgenic livestock (cattle, goats, pigs) have been produced and characterized that could potentially be useful in agriculture. In addition to the five lines discussed above, a number of other, less well-characterized lines are also available (Table 8.3). From Table 8.3 a number of conclusions can be drawn concerning the development of transgenic livestock for use in agriculture.

First, although a fairly large number of transgenic lines have been produced, there has been little long-term characterization of the health of the animals and the robustness and efficacy of the phenotypes. This is partly due to the long-term nature of working with large animals, the length of their life cycles, and, except for pigs, the limited number of offspring per breeding. However, this is also a reflection of the limited level of funding available and lack of industry investment in most parts of the world.

Second, applications are not just concerned with growth promotion and the properties of milk, although some new work in the later area such as reducing the level of expression of ß-lactoglobulin (Jabed et al. 2012), a major milk allergen, has been initiated. Other applications though include developing BSE prion-free cattle (Richt et al. 2007) and enhancing disease resistance/resilience in pigs (Li et al. 2014) and cattle (Wang et al. 2015).

Third, the predominant amount of work to create transgenic animals for agriculture has shifted from the developed world to Asia, with China being the site of

Table 8.3 Additional transgenic livestock produced for use in agriculture

Species	Transgene[a]	Reference
Pig	maP2/FAD2	Saeki et al. (2004)
	CAG/hfat-1	Lai et al. (2006), Liu et al. (2016)
	bCsn/hLz	Tong et al. (2011)
	U6-RNA/PRRSV[siRNA]	Li et al. (2014)
Cattle		
	bCsn/hLF	Krimpenfort et al. (1991)
	MSV/cc-ski	Bowen et al. (1994)
	bbCsn/bbCsn and bk-Csn	Brophy et al. (2003)
	KO PrP[BSE]	Richt et al. (2007)
	hLA/hLA	Wang et al. (2008)
	hLF/hLF	Yang et al. (2008)
	bCsn/hLz	Yang et al. (2011)
	?/fat-1	Guo et al. (2011)
	EF1a/anti-GDF8 shRNA	Tessanne et al. (2012)
	mMCKE-cbA/mfat-1	Wu et al. (2012)
	mWAP-BLG-miRNA	Jabed et al. (2012)
	bMSR1-Ipr1	Wang et al. (2015)
Goat	bLG/SCD	Reh et al. (2004)
	oCsn/hGH	Lee et al. (2006)
	oCsn/hLF	Zhang et al. (2008)
Sheep	RSV/CE, CK	Rogers (1990)
	oMT/CE, CK	Ward and Nancarrow (1991)
	mKER/oIGF-I	Damak et al. (1996a)
	U6-RNA/MSTN[shRNA]	Hu et al. (2013)

[a]Transgene shows promoter/coding region:
Promoters: MSV, mouse sarcoma virus LTR; maP2, mouse aP2, adipocyte lipid-binding protein P2; cASK, chicken α-skeletal actin; CAG (also called CAGG/CAGGS), human CMV early enhancer fused to b-actin promoter; bCsn, bovine casein; bbCas, bovine ß-casein; bK, bovine Κ-casein; EF1a, elongation factor 1a; mKER, mouse keratin; hLA, human α-lactalbumin; hLF, human lactoferrin; oCas, ovine casein; oMTla, ovine metallothionein la; RSV, Rous sarcoma virus LTR; mMCKE-cbA, mouse muscle creatine kinase enhancer, cytomegalovirus enhancer with a chicken β-actin promoter; mWAP, mouse whey acidic protein; bMSR1, macrophage scavenger receptor 1 (MSR1); U6-RNA, U6-RNA gene promoter; bLG, bovine β-lactoglobulin; ? unknown Transgenes: FAD2, spinach Delta-12 fatty acid desaturase; hfat-1, humanized (codon optimized) fat-1; hLz, human lysozyme; hER, human estrogen receptor; cc-ski, chicken c-ski DNA-binding protein; bbCas, bovine ß-casein; bK, bovine Κ-casein; KO PrP[BSE], knockout of BSE-causing prion; hLA, human α-lactalbumin; hLF, human lactoferrin; anti-GDF8 shRNA, anti-myostatin short hairpin RNA; rSCD, rat stearoyl-CoA desaturase; hGH, human growth hormone; CE, *E. coli* cysE; CK, *E. coli* cysK; oIGF-I, ovine insulin-like growth factor-1; Prp[BSE], anti-major bovine prion protein or CD230 short hairpin RNA; BLG-miRNA, miRNA specific for bovine ß-lactoglobulin; Ipr1, intracellular pathogen resistance; MSTN[shRNA], shRNAs targeting sheep MSTN; PRRSV[siRNA], siRNAs to target the open reading frame (ORF) 1b and six regions of PRRSV

development of the majority of new transgenic lines of animals for agriculture since 2006. This again reflects the availability of government funding for research and development, which is largely not available in the EU and has been significantly reduced in the USA and other Western countries initially involved in the production of transgenic livestock. Over the past 20 years, the number of laboratories engaged in producing transgenic animals for use in agriculture has decreased or remained stagnant in the developed, Western countries.

8.5   Regulatory Dysfunction

In order to feed the world's population in the coming decades, it will be necessary to significantly increase the production of animal-based food in an economically and environmentally sustainable manner; that is, we will need to produce more food with less land and water and in a manner that does not degrade the environment for future generations. While this is a well-understood and recognized problem, other than the AquAdvantage salmon in Canada, no other transgenic animal has been approved for use as food anywhere in the world. The data is clear that transgenic animals can be and, indeed, have been produced already that address significant issues facing animal agriculture, including animal welfare, increasing growth, and feed efficiency to increase the sustainability of animal production and to increase disease resistance/resilience in our livestock species (Maga and Murray 2010; Fahrenkrug et al. 2010; Garas et al. 2015). However, to date the social and regulatory debate concerning the use of transgenic animals for food production has focused solely on the perception that there may be some unknown risk associated with the use of the technology in general, in spite of the mounting evidence to the contrary.

We have pointed out before that a question that needs to be debated by politicians and regulators is: What benefits toward improving animal production, animal welfare, and national and worldwide food security are we prepared to forego against the possibility that genetic engineering in animals as a technology will result in some, as yet unknown, negative impact (Murray and Maga 2010)? To date, worldwide governments have for the most part chosen to regulate the process of making a GE animal instead of the resulting product, i.e., the unique phenotype of the specific animal resulting from the expression, or lack of expression, of a specific transgene, in spite of the mounting evidence that there is nothing inherently hazardous stemming from the act of making an animal transgenic. There are real benefits from using the technology as documented in the work characterizing the five lines of animals listed in Table 8.2, and for each of these lines, the animals have remained healthy and reproductively sound across multiple generations. The benefits these animals bring to agriculture and human health need to be weighed against potential adverse impacts keeping in mind three things, as we have noted before (Murray and Maga 2016a). First, the meta-data on the consumption of GE plant-based feeds by livestock has not found any evidence of any animals having adverse reactions to such feed (Van Eenennaam and Young 2014). Second, over the past 20+ years, humans have consumed billions of meals containing GE plant-based foods with no documented evidence of any adverse reactions. Third, the transgenic pig, cattle, and goat lines listed in Table 8.2 are healthy animals, with multiple generations having been studied, with no evidence of any adverse impact of the technology. Other than in China, the last 10 years has been predominantly about advances in the technology and stagnation in the production of new animals specifically intended for use in agriculture due to regulatory dysfunction and the lack of industry investment.

There have been advances in the technology, including the development of somatic cell nuclear transfer-based cloning (Campbell et al. 1996; Wilmut et al. 1997), the development of gene editing techniques based on TALENs (Boch et al. 2009; Moscou and Bogdanove 2009) and CRISPR/Cas9 (Jinek et al. 2012; Wiedenheft et al. 2012) systems, and the widespread availability of genomic sequence information and annotations for a large range of both domestic and wild animals, which has allowed the identification of alleles in a wide array of loci affecting production and disease traits, as well as loci contributing to disease resistance/resilience. For the first 15–20 years following the production of the first genetically engineered livestock, the limitations to the production of transgenic animals for use in agriculture were the lack of knowledge of what alleles/genes to transfer; the lack of suitable promoters and enhancers to ensure proper, controlled expression of inserted transgenes; and the inefficiency of the process of making genetically engineered animals (Ward et al. 1986). These advances have largely overcome these limitations, as we have the ability to integrate and express genes where and when desired, with the limitations today being a lack of investment and regulatory dysfunction (Murray and Maga 2016a).

8.6   Conclusions

The bottleneck limiting the adoption of transgenic, and potentially gene-edited, animals into agriculture is the lack of timely and efficient product-based regulatory systems worldwide, which feeds the perception that the technology is dangerous, in spite of all of the scientific evidence to the contrary. The technology and science behind the applications are not the issue as biotechnology has evolved and largely removed the technical limitations. We have produced transgenic livestock suitable for use as food, for example, some of the lines listed in Table 8.2. The applications and benefits of transgenic technologies to animal agriculture, and thus human food security and animal welfare, are real, and it will be necessary in the coming decades to employ this technology.

Transgenic livestock pose no more threat to the environment and biodiversity than our current livestock breeds. Moreover, the cumulative experience with GE plants and animals indicates that they pose little to no food safety risk to people, with GE crops being widely consumed by both animals and humans. The limitations to the use of transgenic animals in agriculture are largely political, with a diverse set of regulatory paradigms across countries, from a complete ban in the EU to any novel food triggering review in Canada. In the USA the regulatory system is essentially process-based, as it is the injection of the transgene DNA under FDA Guidance for Industry 187 (FDA 2009) that initiates regulatory action, rather than the product. In fact, with the current US regulatory system, transgenic animals made as biomedical models, or made to produce a pharmaceutical, or one made for food all enter the same regulatory pathway even though the products are very different. In those countries with applications (USA, China, Canada, and perhaps New Zealand), they are dysfunctional in that regulatory decisions are not being made in a timely manner. One consequence of long time frames for regulatory approvals is the inhibition of investment and the stifling of innovation for agricultural applications. The regulatory environment is compounded (and perhaps fueled, e.g., in the EU or New Zealand) by the opposition of groups opposed to any use of genetic engineering in agriculture in the absence of any scientific or clinical data demonstrating an adverse consequence to consuming the products of genetic engineering.

8.6.1 Recommendations

It is clear that in those countries with long-standing developers of transgenic animals for use as food that a fully functioning regulatory framework is lacking, either because developers have not submitted due to the costs or because of regulatory dysfunction within the system. The widespread consumption of food containing GE plant-derived components has demonstrated that the technology per se does not pose a food safety issue, but rather it is the product of the transgene that should be assessed for potential problems, which also holds true for GE animal-derived food products. Unless the transgene product is a known allergen, the level of the transgene product in a given tissue should not be the issue but rather the amount consumed as part of a meal. If the objective is to regulate agriculture to ensure food safety based on scientific grounds, then where are the data to suggest that any of the transgenes discussed above may pose a food safety risk? Short of such evidence, we are regulating the process based on the perception that there may be an unknown, and likely, unknowable food safety risk. Based on our experiences organizing the 11 Transgenic Animal Research Conferences, in many discussions with regulators and developers, and as developers ourselves, we have offered the following recommendations for changing the regulatory framework in the USA (Murray and Maga 2016a) and feel that they should be relevant to any country interested in developing or revising a science-based framework for regulating the use of transgenic animals in agriculture.

> First, although the fact that an animal is transgenic may be the trigger for regulatory review, the initial review should be based solely on whether or not the transgene product is expressed in the food parts of the animal. If not, then, because the consumption of DNA is generally considered to be safe (or GRAS), the product should not be subject to premarket review. Post-market monitoring should be sufficient. Expression of the transgene should be determined by the presence of detectable mRNA for the transgene product based on regular PCR. Trace amounts of transgene product, especially if the product is normally found in food, should be acceptable rather than a requirement for no or zero tolerance of the product; that is, the product should be deemed substantially equivalent to the nontransgenic product.

Second, if the transgene product is found in food products derived from the transgenic animal, the regulatory review should ask if the product is normally found in the equivalent non-transgenic food product and whether the level in the transgene-derived product significantly exceeds the normal level, that is, falls outside the upper limit normally observed in the non-transgenic product. If the level contributed by the transgene to food is within the normal range of that product in that food or in equivalent food products in general, then the application should be moved to discretionary enforcement, not regulated, or only subjected to post-market monitoring.

Third, where the product of the transgene is either found at levels significantly greater than normally found in the equivalent food or if the transgene product is an orally active compound, then the regulatory process should require further review.

Fourth, mandatory time limits should be established for each phase of the review.

Fifth, for gene editing applications of fish, poultry, or livestock for use in agriculture, i.e., targeted mutagenesis or homology-dependent repair-based allele conversion, the animal products of ZFN, TALEN, CRISPR, or similar-based gene editing technologies should not be subject to premarket review. Post-market monitoring is sufficient.

At present, a segment of the population worldwide has a biased perception of what genetic engineering technology does and what potential risks it may cause within the context of food safety. Ideally a government's regulatory policy should assure these individuals, and the general population, that their food is safe and wholesome. At the same time, the regulatory process should make scientifically defensible decisions in a timely manner and conform to international treaty obligations concerning trade. Right now, we have neither. We put forward our suggestions in the hope that it will help lead to a robust regulatory process, based on a clear and defensible understanding of the potential risks and benefits. Transgenic animal technology can contribute to the future of animal agriculture for improving food security and animal health and welfare and the improvement of the nutritional benefits of various foods for human consumption. This can only be realized through appropriate investment, and this will only occur when developers have the confidence of an appropriate regulatory process for those animals and food products. Many nations are poised to take steps forward with the application of transgenic animals in agriculture, and many benefits are clearly in sight, but for this to be realized in the short to medium term, we need to have a functioning, product-based regulatory system, both within and across national borders.

References

Archibald AL, McClenaghan M, Hornsey V, Simons JP, Clark AJ (1990) High-level expression of biologically active human alpha 1-antitrypsin in the milk of transgenic mice. Proc Natl Acad Sci U S A 87:5178–5182

Bawden CS, Powell BC, Walker SK, Rogers GE (1998) Expression of a wool intermediate filament keratin transgene in sheep fibre alters structure. Transgenic Res 7:273–287

van Berkel PH, Welling MM, Geerts M, van Veen HA, Ravensbergen B, Salaheddine M, Pauwels EK, Pieper F, Nuijens JH, Nibbering PH (2002) Large scale production of recombinant human lactoferrin in the milk of transgenic cows. Nat Biotechnol 20:484–487

Bleck GT, White BR, Miller DJ, Wheeler MB (1998) Production of bovine α-lactalbumin in the milk of transgenic pigs. J Anim Sci 76:3072–3078

Boch J, Scholze H, Schornack S, Landgraf A, Hahn S, Kay S, Lahaye T, Nickstadt A, Bonas U (2009) Breaking the code of DNA binding specificity of TAL-type III effectors. Science 326:1509–1512

Bowen RA, Reed ML, Schnieke A, Seidel GE Jr, Stacey A, Thomas WK, Kajikawa O (1994) Transgenic cattle resulting from biopsied embryos: expression of c-ski in a transgenic calf. Biol Reprod 50:664–668

Brem G, Brenig B, Goodman HM, Selden RC, Graf F, Kruff B et al (1985) Production of transgenic mice, rabbits and pigs by microinjection into pronuclei. Zuchthygiene 20:251–252

Brophy B, Smolenski G, Wheeler T, Wells D, L'Huillier P, Laible G (2003) Cloned transgenic cattle produce milk with higher levels of beta-casein and kappacasein. Nat Biotechnol 21:157–162

Brundige DR, Maga EA, Klasing KC, Murray JD (2008) Lysozyme transgenic goats' milk influences gastrointestinal morphology in young pigs. J Nutr 138:921–926

Brundige DR, Maga EA, Klasing KC, Murray JD (2010) Consumption of pasteurized human lysozyme transgenic goats' milk alters serum metabolite profile in young pigs. Transgenic Res 19:563–574

Campbell KH, McWhir J, Ritchie WA, Wilmut I (1996) Sheep cloned by nuclear transfer from a cultured cell line. Nature 380:64–68

Carneiro IS, Menezes JNR, Maia JA, Miranda AM, Oliveira VBS, Murray JD, Maga EA, Bertolini M, Bertolini LR (2018) Milk from transgenic goat expressing human lysozyme for recovery and treatment of gastrointestinal pathogens. Eur J Pharm Sci 112:79–86

Carvalho EB, Maga EA, Quetz JS, Lima IFN, Magalhaes HYF, Rodrigues FAR, Silva AVA, Prata MMG, Cavalcante PA, Havt A, Bertolini M, Bertolini LR, Lima AAM (2012) Goat milk with and without increased concentrations of lysozyme improves repair of intestinal cell damage induced by enteroaggregative Escherichia coli. BMC Gastroenterol 12:106

Cibelli JB, Stice SL, Golueke PJ, Kane JJ, Jerry J, Blackwell C, Ponce de León FA, Robl JM (1998) Cloned transgenic calves produced from nonquiescent fetal fibroblasts. Science 280:1256–1258

Clark M, Murray JD, Maga EA (2014) Assessing unintended effects of a mammary-specific transgene at the whole animal level in host and non-target animals. Transgenic Res 23:245–256

Clements JE, Wall RJ, Narayan O, Hauer D, Schoborg R, Sheffer D et al (1994) Development of transgenic sheep that express the visna virus envelope gene. Virology 200:370–380

Cooper CA, Brundige DR, Reh WA, Maga EA, Murray JD (2011) Lysozyme transgenic goats' milk positively impacts intestinal cytokine expression and morphology. Transgenic Res 20:1235–1243

Cooper CA, Nelson KM, Maga EA, Murray JD (2012) Consumption of transgenic cows' milk containing human lactoferrin results in beneficial changes in the gastrointestinal tract and systemic health of young pigs. Transgenic Res 22:571–578

Cooper CA, Garas Klobas L, Maga EA, Murray JD (2013) Consuming transgenic goats' milk containing the antimicrobial protein lysozyme helps resolve diarrhea in young pigs. PLoS One 8:e58409

Cooper CA, Maga EA, Murray JD (2014a) Consumption of transgenic milk containing the antimicrobials lactoferrin and lysozyme separately and in conjunction by 6 week old pigs improves intestinal and systemic health. J Dairy Res 81:30–37

Cooper CA, Nonnecke E, Lonnerdal B, Murray JD (2014b) The lactoferrin receptor may mediate the reduction of eosinophils in the duodenum of pigs consuming milk containing recombinant human lactoferrin. Biometals 27:1031–1038. https://doi.org/10.1007/s10534-014-9778-8

Cooper CA, Maga EA, Murray JD (2015) Production of human lactoferrin and lysozyme in the milk of transgenic dairy animals: past, present and future. Transgenic Res 24:605–614. https://doi.org/10.1007/s11248-015-9885-5

Dai Y, Vaught TD, Boone J, Chen S-H Phelps CJ, Ball S, Monahan JA, Jobst PM, McCreath KJ, Lamborn AE, Cowell-Lucero JL, Wells KD, Colman A, Polejaeva IA, Ayares DL (2002) Targeted disruption of the α1,3-galactosyltransferase gene in cloned pigs. *Nat Biotechnol* **20**:251–255

Damak S, Jay NP, Barrell GK, Bullock DW (1996a) Targeting gene expression to the wool follicle in transgenic sheep. Biotechnology 14:181–184

Damak S, Su H, Jay NP, Bullock DW (1996b) Improved wool production in transgenic sheep expressing insulin-like growth factor 1. Biotechnology 14:185–188

Denning C, Burl S, Ainslie A, Bracken J, Dinnyes A, Fletcher J, King T, Ritchie M, Ritchie WA, Rollo M, de Sousa P, Travers A, Wilmut I, Clark AJ (2001) Deletion of the l[alpha]l(1,3)galactosyl transferase (GGTA1) gene and the prion protein (PrP) gene in sheep. *Nat Biotechnol* **19**:559–562

Ding S, Wu X, Li G, Han M, Zhuang Y, Xu T (2005) Efficient transposition of the piggyBac (PB) transposon in mammalian cells and mice. Cell 122:473–483

Durai S, Mani M, Kandavelou K, Wu J, Porteus MH, Chandrasegaran S (2005) Zinc finger nucleases: custom-designed molecular scissors for genome engineering of plant and mammalian cells. Nucleic Acids Res 33:5978–5990

Ebert KM, Low MJ, Overstrom EW, Buonomo FC, Baile CA, Roberts TM et al (1988) A Moloney MLV-rat somatotropin fusion gene produces biologically active somatotropin in a transgenic pig. Mol Endocrinol 2:277–283

Ebert KM, Smith TE, Buonoma FC, Overstrom EW, Low EJ (1990) Porcine growth hormone gene expression from viral promoters in transgenic swine. Anim Biotechnol 1:145–159

Ebert KM, DiTullio P, Barry CA, Schindler JE, Ayres SL, Smith TE, Pellerin LJ, Meade HM, Denman J, Roberts B (1994) Induction of human tissue plasminogen activator in the mammary gland of transgenic goats. Bio/Technology 12:699–702

Fahrenkrug SC, Blake A, Carlson DF, Doran T, Van Eenennaam A, Faber D, Galli C, Hackett PB, Li N, Maga EA, Murray JD, Stotish R, Sullivan E, Taylor JF, Walton M, Wheeler M, Whitelaw B, Glenn BP (2010) Precision genetics for complex objectives in animal agriculture. J Anim Sci 88:2530–2539

FDA (2009) Guidance 187: regulation of genetically engineered animals containing heritable recombinant DNA constructs. www.fda.gov/RegulatoryInformation/Guidances/default.htm

Flisikowska T, Thorey IS, Offner S, Ros F, Lifke V, Zeitler B, Rottmann O, Vincent A, Zhang L, Jenkins S, Niersbach H, Kind AJ, Gregory PD, Schnieke AE, Platzer J (2011) Efficient immunoglobulin gene disruption and targeted replacement in rabbit using zinc finger nucleases. PLoS One 6:e21045. https://doi.org/10.1371/journal.pone.0021045

Forsberg CW (2001) Pigs expressing salivary phytase produce low-phosphorus manure. Nat Biotechnol 19:741–745

Forsberg CW, Phillips JP, Golovan SP, Fan MZ, Meidinger RG, Ajakaiye A, Hilborn D, Hacker RR (2003) The Enviropig physiology, performance, and contribution to nutrient management advances in a regulated environment: the leading edge of change in the pork industry12. J Anim Sci 81:E68–E77. https://doi.org/10.2527/2003.8114_suppl_2E68x

Forsberg CW, Meidinger RG, Liu M, Cottrill M, Golovan S, Phillips JP (2013) Integration, stability and expression of the E. coli phytase transgene in the Cassie line of Yorkshire Enviropig™. Transgenic Res 22:379–389

Forsberg CW, Meidinger RG, Ajakaiye A, Murray D, Fan MZ, Mandell IB, Phillips JP (2014a) Comparative carcass and tissue nutrient composition of transgenic Yorkshire pigs expressing phytase in the saliva and conventional Yorkshire pigs. J Anim Sci 92:4417–4439

Forsberg CW, Meidinger RG, Murray D, Keirstead ND, Hayes MA, Fan MZ, Ganeshapillai J, Monteiro MA, Golovan SP, Phillips JP (2014b) Phytase properties and locations in tissues of transgenic pigs secreting phytase in the saliva. J Anim Sci 92:3375–3387

Garas L, Murray JD, Maga EA (2015) Genetically engineered livestock: ethical use for food and medical models. Annu Rev Anim Biosci 3:1.1–1.17. https://doi.org/10.1146/annurev-animal-022114-110739

Garas LC, Feltrin C, Hamilton MK, Hagey JV, Murray JD, Bertolini LR, Bertolini M, Raybould HE, Maga EA (2016) Milk with and without lactoferrin can influence intestinal damage in a pig model of malnutrition. Food Funct 7:665–678

Garas LC, Cooper CA, Dawson MW, Wang J-L, Murray JD, Maga EA (2017) Young pigs consuming lysozyme transgenic goat milk are protected from clinical symptoms of enterotoxigenic E. coli infection. J. Nutrition 147:2050–2059. https://doi.org/10.3945/jn.117.251322

Golovan SP, Meidinger RG, Ajakaiye A, Cottrill M, Wiederkehr MZ, Barney DJ, Plante C, Pollard JW, Fan MZ, Anthony Hayes M, Laursen J, Peter Hjorth J, Hacker RR, Phillips JP, Forsberg CW (2001) Pigs expressing salivary phytase produce low-phosphorus manure. Nat Biotechnol 19:741–745

Gordon JW, Scangos GA, Plotkin DJ, Barbosa JA, Ruddle FH (1980) Genetic transformation of mouse embryos by microinjection of purified DNA. Proc Natl Acad Sci U S A 77:7380–7384

Guo T, Liu XF, Ding XB, Yang FF, Nie YW, An YJ, Guo H (2011) Fat-1 transgenic cattle as a model to study the function of ω-3 fatty acids. Lipids Health Dis 10:244–253. https://doi.org/10.1186/1476-511X-10-244

Hammer RE, Pursel VG, Rexroad CE Jr, Wall RJ, Bolt DJ, Ebert KM, Palmiter RD, Brinster RL (1985) Production of transgenic rabbits, sheep and pigs by microinjection. Nature 315: 680–683

Hartke JL, Monaco MH, Wheeler MB, Donovan SM (2005) Effect of a short-term fast on intestinal disaccharidase activity and villus morphology of piglets suckling insulin-like growth factor-I transgenic sows1. J An Sci 83:2404–2413. https://doi.org/10.2527/2005.83102404x

Haskell RE, Bowen RA (1995) Efficient production of transgenic cattle by retroviral infection of early embryos. Mol Reprod Dev 40:386–390

Hu S, Ni W, Sai W, Zi H, Qiao J, Wang P, Sheng J, Chen C (2013) Knockdown of myostatin expression by RNAi enhances muscle growth in transgenic sheep. PLoS One 8(3):e58521. https://doi.org/10.1371/journal.pone.0058521

Ivics Z, Hackett PB, Plasterk RH, Izsvák Z (1997) Molecular reconstruction of Sleeping Beauty, a Tc1-like transposon from fish, and its transposition in human cells. Cell 91:501–510

Jabed A, Wagner S, McCracken J, Wells DN, Laible G (2012) Targeted microRNA expression in dairy cattle directs production of β-lactoglobulin-free, high-casein milk. Proc Natl Acad Sci U S A 109:16811–16816

Jackson KA, Berg JM, Murray JD, Maga EA (2010) Evaluating the fitness of human lysozyme transgenic dairy goats: growth and reproductive traits. Transgenic Res 19:977–986

Jinek M, Chylinski K, Fonfara I, Hauer M, Doudna JA, Charpentier E (2012) A programmable dual-RNA-guided DNA endonuclease in adaptive bacterial immunity. Science 337:816–821

Krimpenfort P, Rademakers A, Eyestone W, van der Schans A, van den Broek S, Kooiman P et al (1991) Generation of transgenic dairy cattle using 'in vitro' embryo production. Bio/Technology 9:844–847

Kues WA, Niemann H (2011) Advances in farm animal transgenesis. Prev Vet Med 102:146–156

Lai L, Kang JX, Li R, Wang J, Witt WT, Yong HY, Hao Y, Wax DM, Murphy CN, Rieke A, Samuel M, Linville ML, Korte SW, Evans RW, Starzl TE, Prather RS, Dai Y (2006) Generation of cloned transgenic pigs rich in omega-3 fatty acids. Nat Biotechnol 24:435–436

Lavitrano M, Forni M, Varzi V, Pucci L, Bacci ML, Di Stefano C, Fioretti D, Zoraqi G, Moioli B, Rossi M, Lazzereschi D, Stoppacciaro A, Seren E, Alfani D, Cortesini R, Frati L (1997) Sperm-mediated gene transfer: production of pigs transgenic for a human regulator of complement activation. Transplant Proc 29:3508–3509

Lee CS, Lee DS, Fang NZ, Oh KB, Shin ST, Lee KK (2006) Integration and expression of goat b-casein/hGH hybrid gene in a transgenic goat. Reprod Dev Biol 30:293–299

Li L, Li Q, Bao Y, Li J, Chen Z, Yu X, Zhao Y, Tian Y, Li N (2014) RNAi-based inhibition of porcine reproductive and respiratory syndrome virus replication in transgenic pigs. J Biotechnol 171:17–24

Laible G (2009) Enhancing livestock through genetic engineering – recent advances and future prospects. Comp Immunol Microbiol Infect Dis 32:123–137

Liu X, Pang D, Yuan T, Li Z, Li Z, Zhang M, Ren W, Ouyang H, Tang X (2016) N-3 polyunsaturated fatty acids attenuates triglyceride and inflammatory factors level in hfat-1 transgenic pigs. Lipids Health Dis 15:89–96

Lu KH, Gordon I, Gallagher M, McGovern H (1987) Pregnancy established in cattle by transfer of embryos derived from in vitro fertilisation of oocytes matured in vitro. Vet Rec 121:259–260

Maga EA, Murray JD (1995) Mammary gland expression of transgenes and the potential for altering the properties of milk. Bio/Technology 13:1452–1457

Maga EA, Murray JD (2010) Welfare applications of genetically engineered animals for use in agriculture. J An Sci 88:1588–1591

Maga EA, Sargent RG, Zeng H, Pati S, Zarling DA, Oppenheim SM, Collette NMB, Moyer AL, Conrad-Brink JS, Rowe JD, RH BD, Anderson GB, Murray JD (2003) Increased efficiency of transgenic livestock production. Transgenic Res 12:485–496

Maga EA, Shoemaker CF, Rowe JD, BonDurant RH, Anderson GB, Murray JD (2006a) Production and processing of milk from transgenic goats expressing human lysozyme in the mammary gland. J Dairy Sci 89:518–524

Maga EA, Cullor JS, Smith W, Anderson GB, Murray JD (2006b) Human lysozyme expressed in the mammary gland of transgenic dairy goats can inhibit the growth of bacteria that cause mastitis and the cold-spoilage of milk. Foodborne Pathog Dis 3:384–392

Maga EA, Walker RL, Anderson GB, Murray JD (2006c) Consumption of milk from transgenic goats expressing human lysozyme in the mammary gland results in the modulation of intestinal microflora. Transgenic Res 15:515–519

Maga EA, Desai PT, Weimer BC, Dao N, Kültz D, Murray JD (2012) Consumption of lysozyme-rich milk can alter microbial fecal populations. Appl Environ Microbiol 78:6153–6160

Marshall KM, Hurley WL, Shanks RD, Wheeler MB (2006) Effects of suckling intensity on milk yield and piglet growth from lactation-enhanced gilts. J An Sci 84:2346–2351. https://doi.org/10.2527/jas.2005-764

McInnis EA, Kalanetra KM, Mills DA, Maga EA (2015) Analysis of raw goat milk microbiota: impact of stage of lactation and lysozyme on microbial diversity. Food Microbiol 46:121–131

McKnight RA, Shamay A, Sankaran L, Wall RJ, Hennighausen L (1992) Matrix-attachment regions can impart position-independent regulation of a tissue-specific gene in transgenic mice. Proc Natl Acad Sci U S A 89:6943–6947

Meidinger RG, Ajakaiye A, Fan MZ, Zhang J, Phillips JP, Forsberg CW (2013) Digestive utilization of phosphorus from plant-based diets in the Cassie line of transgenic Yorkshire pigs that secrete phytase in the saliva. J Anim Sci 91:1307–1320

Miller KF, Bolt DJ, Pursel VG, Hammer RE, Pinkert CA, Palmiter RD et al (1989) Expression of human or bovine growth hormone gene with a mouse metallothionein-1 promoter in transgenic swine alters the secretion of porcine growth hormone and insulin-like growth factor-I. J Endocrinol 120:481–488

Monaco MH, Gronlund DE, Bleck GT, Hurley WL, Wheeler MB, Donovan SM (2005) Mammary specific transgenic over-expression of insulin-like growth factor-I (IGF-I) increases pig milk IGF-I and IGF binding proteins, with no effect on milk composition or yield. Transgenic Res 14:761–773

Moscou MJ, Bogdanove AJ (2009) A simple cipher governs DNA recognition by TAL effectors. Science 326:1501

Murray JD, Maga EA (1999) Changing the composition and properties of milk. In: Murray JD, Anderson GB, Oberbauer AM, McGloughlin MM (eds) Transgenic animals in agriculture. CAB International, Wallingham, pp 193–208

Murray JD, Maga EA (2010) Is there a risk from not using GE animals? Transgenic Res 19:357–361

Murray JD, Maga EA (2016a) Genetically engineered livestock for agriculture: a generation after the first transgenic animal research conference. Transgenic Res 25:321–327

Murray JD, Maga EA (2016b) A new paradigm for regulating genetically engineered animals that are used as food. Proc Natl Acad Sci U S A 113:3410–3413. https://doi.org/10.1073/pnas.1602474113

Murray JD, Nancarrow CD, Marshall JT, Hazelton IG, Ward KA (1989) Production of transgenic merino sheep by microinjection of ovine metallothioneinovine growth hormone fusion genes. Reprod Fertil Dev 1:147–155

Naldini L, Bloemer U, Gallay P, Ory D, Mulligan R, Gage FH, Verma IM, Trono D (1996) In vivo gene delivery and stable transduction of nondividing cells by a lentiviral vector. Science 272:263–267

Nancarrow CD, Murray JD, Boland MP, Sutton R, Hazelton IG (1984) Effect of gonadotrophin releasing hormone in the production of single-cell embryos for pronuclear injection of foreign genes. In: Lindsay DR, Pearce DT (eds) Reproduction in sheep. Aust Acad Sci, Canberra, ACT, pp 286–288

Nancarrow CD, Marshall JTA, Clarkson JL, Murray JD, Millard RM, Shanahan CM, Wynn PC, Ward KA (1991) Expression and physiology of performance regulating genes in transgenic sheep. J Reprod Fertil Suppl 43:277–291

Noble MS, Rodriguez-Zas S, Cook JB, Bleck GT, Hurley WL, Wheeler MB (2002) Lactational performance of first-parity transgenic gilts expressing bovine alpha-lactalbumin in their milk. J Anim Sci 80:1090–1096. https://doi.org/10.2527/2002.8041090x

Nottle MB, Nagashima H, Verma PJ, Du ZT, Grupen CG et al (1999) Production and analysis of transgenic pigs containing a metallothionein porcine growth hormon gene construct. In: Murray JD, Anderson GB, Oberbauer AM, McGloughlin MM (eds) Trans-genic animals in agriculture. CABI Publishing, New York, NY, pp 145–156

Palmiter RD, Brinster RL, Hammer RE, Trumbauer ME, Rosenfeld MG, Birnberg NC, Evans RM (1982) Dramatic growth of mice that develop from eggs microinjected with metallothionein-growth hormone fusion genes. Nature 300:611–615

Parc AL, Karav S, Rouquié C, Maga EA, Bunyatratchata A, Barile D (2017) Characterization of recombinant human lactoferrin N-glycans expressed in the milk of transgenic cows. PLoS One 12(2):e0171477. https://doi.org/10.1371/journal.pone.0171477

Pinkert CA, Murray JD (1999) Transgenic farm animals. In: Murray JD, Anderson GB, Oberbauer AM, McGloughlin MM (eds) Transgenic animals in agriculture. CAB International, Wallingham, pp 1–18

Pinkert CA, Pursel VG, Miller KF, Palmiter RD, Brinster RL (1987) Production of transgenic pigs harboring growth hormone (MTbGH) or growth hormone releasing factor (MThGRF) genes. J Anim Sci 65(Suppl. 1):260 (Abstr.)

Polge EJC, Barton SC, Surani MAH, Miller JR, Wagner T, Rottman R et al (1989) Induced expression of a bovine growth hormone construct in transgenic pigs: biotechnology of growth regulation. Butterworths, London, pp 279–289

Pursel VG, Rexroad CE Jr (1993) Status of research with transgenic farm animals. J Anim Sci 71(Suppl):10–19

Pursel VG, Rexroad CE Jr, Bolt DJ, Miller KF, Wall RJ, Hammer RE et al (1987) Progress on gene transfer in farm animals. Vet Immunol Immunopathol 17:303–312

Pursel VG, Pinkert CA, Miller KF, Bolt DJ, Campbell RG, Palmiter RD, Brinster RL, Hammer RE (1989) Genetic engineering of livestock. Science 244:1281–1288

Pursel VG, Wall RJ, Solomon MB, Bolt DJ, Murray JD, Ward KA (1997) Transfer of an ovine metallothionein-ovine growth hormone fusion gene into swine. J Anim Sci 75:2208–2214

Pursel V, Wall RJ, Mitchell AD, Elsasser TH, Solomon MB, Coleman ME et al (1999) Expression of insulin-like growth factor I in skeletal muscle of transgenic swine. In: Murray JD, Anderson GB, Oberbauer AM, McGloughlin MM (eds) Transgenic animals in agriculture. CAB International, Wallingford

Pursel VG, Mitchell AD, Bee G, Elsasser TH, McMurtry JP, Wall RJ et al (2004) Growth and tissue accretion rates of swine expressing an insulin-like growth factor I transgene. Anim Biotechnol 15:33–45

Reh WA, Maga EA, Collette NMB, Moyer A, Conrad-Brink JS, Taylor SJ, DePeters EJ, Oppenheim S, Rowe JD, BonDurant RH, Anderson GB, Murray JD (2004) Hot topic: using a Stearoyl-CoA Desaturase transgene to Alter milk fatty acid composition. J Dairy Sci 87:3510–3514

Rexroad CE Jr, Hammer RE, Bolt DJ, Mayo KE, Frohman LA, Palmiter RD et al (1989) Production of transgenic sheep with growth-regulating genes. Mol Reprod Dev 1:164–169

Rexroad CE Jr, Mayo K, Bolt DJ, Elsasser TH, Miller KF, Behringer RR et al (1991) Transferrin- and albumin-directed expression of growth-related peptides in transgenic sheep. J Anim Sci 69:2995–3004

Richt JA, Kasinathan P, Hamir AN, Castilla J, Sathiyaseelan T, Vargas F, Sathiyaseelan J, Wu H, Matsushita H, Koster J, Kato S, Ishida I, Soto C, Robl JM, Kuroiwa Y (2007) Production of cattle lacking prion protein. Nat Biotechnol 25:132–138

Rocheleau CE, Downs WD, Lin R, Wittmann C, Bei Y, Cha Y-H, Ali M, Priess JR, Mello CC (1997) Wnt signaling and an APC-related gene specify endoderm in early C. elegans embryos. Cell 90:707–716

Rogers GE (1990) Improvement of wool production through genetic engineering. Trends Biotechnol 8:6

Saeki K, Matsumoto K, Kinoshita M, Suzuki I, Tasaka Y, Kano K, Taguchi Y, Mikami K, Hirabayashi M, Kashiwazaki N, Hosoi Y, Murata N, Iritani A (2004) Functional expression of a $\Delta 12$ fatty acid desaturase gene from spinach in transgenic pigs. PNAS 101:6361–6366

Scharfen EC, Mills DA, Maga EA (2007) Use of human lysozyme transgenic goat milk in cheese making: effects on lactic acid bacteria performance. J Dairy Sci 90:4084–4091

Schnieke AE, Kind AJ, Ritchie WA, Mycock K, Scott AR, Ritchie M, Wilmut I, Colman A, Campbell KHS (1997) Human factor IX transgenic sheep produced by transfer of nuclei from transfected fetal fibroblasts. Science 278:2130–2133

Shi Y, Berg JM (1995) A direct comparison of the properties of natural and designed zinc-finger proteins. Chem Biol 2:83–89

Simojoki H, Hyvönen P, Orro T, Pyörälä S (2010) High concentration of human lactoferrin in milk of rhLf-transgenic cows relieves signs of bovine experimental Staphylococcus chromogenes intramammary infection. Vet Immunol Immunopathol 136:265–271

Tan W, Carlson DF, Walton MW, Fahrenkrug SC, Hackett PB (2012) Precision editing of large animal genomes. Adv Genet 80:37–97

Tessanne K, Golding MC, Long CR, Peoples MD, Hannon G, Westhusin ME (2012) Production of transgenic calves expressing an shRNA targeting myostatin. Mol Reprod Dev 79:176–185

Thomassen EA, van Veen HA, van Berkel PH, Nuijens JH, Abrahams JP (2005) The protein structure of recombinant human lactoferrin produced in the milk of transgenic cows closely matches the structure of human milk-derived lactoferrin. Transgenic Res 14:397–405

Tong J, Wei H, Liu X, Hu W, Bi M, Wang YY, Li QY, Li N (2011) Production of recombinant human lysozyme in the milk of transgenic pigs. Transgenic Res 20:417–419

Van Eenennaam AL, Young AE (2014) Prevalence and impacts of genetically engineered feedstuffs on livestock populations1. J Anim Sci 92:4255–4278. https://doi.org/10.2527/jas.2014-8124

Vize PD, Michalska AE, Ashman R, Lloyd B, Stone BA, Quinn P et al (1988) Introduction of a porcine growth hormone fusion gene into transgenic pigs promotes growth. J Cell Sci 90:295–300

Wall RJ, Pursel VG, Hammer RE, Brinster RL (1985) Development of porcine ova that were centrifuged to permit visualization of pronuclei and nuclei. Biol Reprod 32:645–651

Wall RJ, Hawk HW, Nel N (1992) Making transgenic livestock: genetic engineering on a large scale. J Cell Biochem 49:113–120

Wang J, Yang P, Tang B, Sun X, Zhang R, Guo C, Gong G, Liu Y, Li R, Zhang L, Dai Y, Li N (2008) Expression and characterization of bioactive recombinant human alpha-lactalbumin in the milk of transgenic cloned cows. J Dairy Sci 91:4466–4476

Wang YS, He X, Du Y, Su J, Gao M, Ma Y, Hua S, Quan F, Liu J, Zhang Y (2015) Transgenic cattle produced by nuclear transfer of fetal fibroblasts carrying *Ipr1* gene at a specific locus. Theriogenology 84:608–616

Ward KA, Nancarrow CD (1991) The genetic engineering of production traits in domestic animals. Experientia 47:913

Ward KA, Franklin IR, Murray JD, Nancarrow CD, Raphael KA, Rigby NW, Byrne CR, Wilson BW, Hunt CL (1986) The direct transfer of DNA by embryo microinjection. Proc. 3rd World Congress Genetics Applied to Livestock Breeding 12:6–21. Lincoln, Nebraska.

Ward KA, Nancarrow CD, Murray JD, Shanahan CM, Byrne CR, Rigby NW, Townrow CA, Leish Z, Wilson BW, Graham NM, Wynn PC, Hunt CL, Speck PA (1990) The current status of genetic engineering in domestic animals. J Dairy Sci 73:2586–2592

Wheeler MB, Bleck GT, Donovan SM (2001) Transgenic alteration of sow milk to improve piglet growth and health. Reprod Suppl 58:313–324

Whyte JJ, Zhao J, Wells KD, Samuel MS, Whitworth KM, Walters EM, Laughlin MH, Prather RS (2011) Gene targeting with zinc finger nucleases to produce cloned eGFP knockout pigs. Mol Reprod Dev 78:2

Wiedenheft B, Sternberg SH, Doudna JA (2012) RNA-guided genetic silencing systems in bacteria and archaea. Nature 482:331–338

Wieghart M, Hoover JL, McGrane MM, Hanson RW, Rottman FM, Holtzman SH et al (1990) Production of transgenic pigs harbouring a rat phosphoenolpyruvate carboxykinase-bovine growth hormone fusion gene. J Reprod Fertil Suppl 41:89–96

Wilmut I, Schnieke AE, McWhir J, Kind AJ, Campbell KH (1997) Viable offspring derived from fetal and adult mammalian cells. Nature 385:810–813

Wu X, Ouyang H, Duan B et al (2012) Production of cloned transgenic cow expressing omega-3 fatty acids. Transgenic Res 21:537–543. https://doi.org/10.1007/s11248-011-9554-2

Yang P, Wang J, Gong G, Sun X, Zhang R, Du Z, Liu Y, Li R, Ding F, Tang B, Dai Y, Li N (2008) Cattle mammary bioreactor generated by a novel procedure of transgenic cloning for large-scale production of functional human lactoferrin. PLoS One 3:e3453

Yang B, Wang J, Tang B, Liu Y, Guo C, Yang P, Yu T, Li R, Zhao J, Zhang L, Dai Y, Li N (2011) Characterization of bioactive recombinant human lysozyme expressed in milk of cloned transgenic cattle. PLoS One 6:e17593

Zhang J, Li L, Cai Y, Xu X, Chen J, Wu Y, Yu H, Yu G, Liu S, Zhang A, Chen J, Cheng G (2008) Expression of active recombinant human lactoferrin in the milk of transgenic goats. Protein Expr Purif 57:127–135

Genetically Engineered Large Animals in Biomedicine

9

Eckhard Wolf, Alexander Kind, Bernhard Aigner, and Angelika Schnieke

Abstract

Major progress in genetic engineering and genome editing of livestock species has extended their use to biomedical applications, the most notable being tailored large animal models for translational medicine; porcine cells, tissues and organs for xenotransplantation; and production of pharmaceutical proteins in transgenic large animals. The translation of novel discoveries from basic research to clinical application is a long, often inefficient and costly process. Appropriate animal models are critical for the success of translational research. Although rodent models are widely used, they often do not accurately represent the human disease. Thus, additional animal models that more closely mimic aspects of human anatomy and physiology are required. Several genetically engineered pig models have been generated, many of which represent human disease mechanisms and phenotypes more closely than existing rodent models. In addition, genetically modified small ruminants and rabbits are interesting models for specific disease entities. Pigs are the most promising donor species for xenotransplantation. Since multiple genetic modifications are required to prevent immune

E. Wolf (✉)
Gene Centre and Department of Veterinary Sciences, Ludwig-Maximilians-Universität München, Munich, Germany

German Centre for Diabetes Research (DZD), Neuherberg, Germany
e-mail: ewolf@lmu.de

A. Kind · A. Schnieke (✉)
TUM School of Life Sciences Weihenstephan, Technische Universität München, Munich, Germany
e-mail: schnieke@wzw.tum.de; alex.kind@wzw.tum.de

B. Aigner
Gene Centre and Department of Veterinary Sciences, Ludwig-Maximilians-Universität München, Munich, Germany
e-mail: B.Aigner@gen.vetmed.uni-muenchen.de

© Springer International Publishing AG, part of Springer Nature 2018
H. Niemann, C. Wrenzycki (eds.), *Animal Biotechnology 2*,
https://doi.org/10.1007/978-3-319-92348-2_9

rejection, to overcome physiological incompatibilities of xeno-organs and to eliminate potential risk factors such as porcine endogenous retroviruses (PERV), genome editing is speeding progress in this field. Last but not least, genetic engineering of large animal species as bioreactors for the production of pharmaceutical proteins is still an interesting option, though only a few such products are on the market. In summary, genetically engineered large animals are playing an increasingly important role in biomedicine. In particular, genetically tailored large animal models may help to bridge the gap between proof-of-concept studies in rodent models and clinical trials in human patients.

9.1 Genetically Engineered Large Animal Models for Translational Medicine

Relevant animal models are critically important for discovering key disease mechanisms, identifying targets for medical intervention and developing new diagnoses and therapies. Rodents are the most popular, but in many disease areas, findings in rodents differ markedly from those in human patients, as demonstrated, for example, by Leigh syndrome (Dell'agnello et al. 2007), Huntington's disease (Ehrnhoefer et al. 2009), cystic fibrosis (Wilke et al. 2011), Duchenne muscular dystrophy (Nakamura and Takeda 2011), hereditary colon cancer (Flisikowska et al. 2013) and inflammatory processes (Seok et al. 2013). Large animals offer an alternative and can provide more relevant and predictable models of human disease. These models share close similarities with humans in size and features of anatomy, physiology and pathology (reviewed in Bähr and Wolf 2012). Importantly, tailored large animal disease models can be studied with the same clinical approach used for human patients.

The use of pigs (*Sus scrofa*) in biomedical research is steadily increasing. Pigs share more anatomical and physiological similarities with humans than do any other small or large domestic animal (reviewed in Aigner et al. 2010) and consequently often model the human situation more accurately than other species. Pigs also compare very favourably with other large domestic animal species in their reproductive performance; their relatively early sexual maturity (6 months), short generation interval (12 months), large litters (8–12 piglets) and year-round breeding all aid their practical usefulness. The availability of porcine genome data (Groenen et al. 2012) and efficient and precise methods for genetic engineering/gen(om)e editing (reviewed in Whitelaw et al. 2016) enable the replication of genetic lesions to mimic human disease mechanisms at the molecular level. Genetically engineered strains extend the range of porcine experimental models beyond spontaneous or artificially induced conditions (e.g. via surgery or chemical agents), providing far greater scope, precision, reproducibility and informative power. Genetically tailored pig models are thus ideally placed to bridge the gap between proof-of-concept models and clinical testing of new diagnostic and therapeutic procedures (Fig. 9.1). Comprehensive reviews of genetically engineered pig models have been published recently (Fan and Lai 2013; Flisikowska et al. 2014; Luo et al. 2012; Rogers 2016).

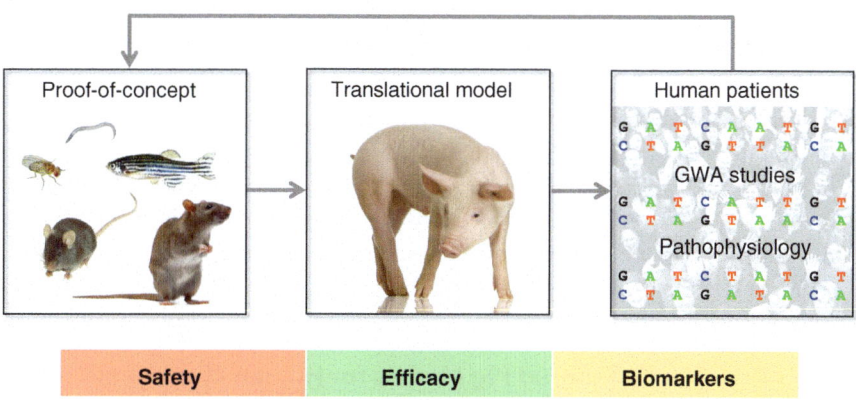

Fig. 9.1 Genetically engineered pig models as link between basic research and clinical trials in drug development

In addition to pigs, rabbits and small ruminants are increasingly used for translational research.

The rabbit (*Oryctolagus cuniculus*) is highly relevant to the study of specific diseases, such as osteoarthritis, asthma, diabetes mellitus, hypercholesterolaemia and its cardiovascular effects and reproductive biology (reviewed in Bähr and Wolf 2012). Basic reproductive techniques, including in vitro fertilisation and embryo cryopreservation, are well established in rabbits, and more advanced techniques such as transgenesis, intracytoplasmic sperm injection, nuclear transfer, gene disruption, targeted gene replacement and gene editing are now available (reviewed in Duranthon et al. 2012). Several genetically engineered rabbit models of human diseases have been generated. These include polymorphic ventricular tachycardia/ long QT syndrome type 1 (Kim et al. 2015), hyperlipidaemia (Koike et al. 2009), atherosclerosis (Niimi et al. 2016), retinitis pigmentosa (Ueno et al. 2013), congenital cataract (Yuan et al. 2016) and infection/vaccine research (Srivastava et al. 2015).

Goats (*Capra hircus*) and sheep (*Ovis aries*) are used as models for orthopaedic research and regenerative medicine. The goat is also used as a model for complete female-to-male XX sex reversal owing to a genomic mutation. Sheep are used for asthma research because key respiratory parameters, such as airflow, resistance and breathing rates, are similar to humans. Sheep are also used as models for reproductive pathology, including premature ovarian failure or polycystic ovaries (reviewed in Bähr and Wolf 2012). The generation and characterisation of a transgenic sheep model for Huntington's disease has been reported (Handley et al. 2016; Jacobsen et al. 2010; Reid et al. 2013). Transgenic goats constitutively overexpressing Toll-like receptor 2 (TLR2) have been used to study the effects of elevated TLR2 levels on inflammatory processes and clearance of bacterial infections (Deng et al. 2012). Expression of constitutively active transforming growth factor-beta1 (TGFB1) in transgenic goats results in atrial fibrosis and increased susceptibility to atrial fibrillation (Polejaeva et al. 2016).

While sheep, goats and rabbits are the species of choice for particular diseases, the main emphasis of disease modelling in larger species has been in pigs. This chapter therefore provides more detailed examples of pig models genetically engineered for specific disease areas and discusses their phenotypic characteristics and the opportunities they provide to study disease mechanisms and test therapeutic options. This is not a comprehensive list of all large animals genetically modified with the intention to produce a model. There are many more at early stage or where the disease phenotype still has to emerge. For further examples, please also see review by Flisikowska et al. (2014).

9.1.1 Genetically Tailored Pig Models for Human Monogenic Diseases

Rare monogenic diseases are an attractive market for the pharmaceutical industry, because identification of the underlying mutations provides validated targets for drug development or genetic treatment. Development of drugs for these orphan diseases frequently has higher success rates and shorter times to approval and, despite relatively small target patient populations, may generate lifetime revenues comparable to non-orphan drugs. The molecular aetiology of more than half the estimated 7,000 rare monogenic human diseases is now known, and the rate of disease gene discovery is expected to accelerate markedly with dramatic improvements in DNA sequencing technologies and associated analytical tools (reviewed in Klymiuk et al. 2016b).

Cystic fibrosis (CF) is the most frequent inherited disease in Caucasians and affects ~70,000 individuals worldwide (reviewed in Cutting 2015). This autosomal recessive disorder is caused by alterations in the gene encoding the cystic fibrosis transmembrane conductance regulator (CFTR), an epithelial anion channel. While more than 1,000 different mutations within the *CFTR* gene are associated with manifestation of CF, deletion of phenylalanine at position 508 (ΔF508) accounts for ~70% of all cases (reviewed in Fanen et al. 2014). This mutation causes aberrant folding of CFTR and subsequent degradation of most synthesised protein. If F508del-CFTR is trafficked to the cell membrane, it has reduced membrane residency and aberrant chloride channel function (reviewed in Cutting 2015; Klymiuk et al. 2016b). CF is a multi-systemic disease, affecting the airways, the gastrointestinal tract including the pancreas and hepatobiliary system and the reproductive tract. Chronic bacterial infections and persistent inflammatory processes of the lung are the main cause of morbidity and mortality (reviewed in Stoltz et al. 2015). Although numerous *Cftr* mutant mouse models have been established, they only partially reproduce the disease processes in CF patients (reviewed in Wilke et al. 2011). This is particularly true for the pathology of the respiratory tract, which is the primary cause of the patient's declining quality of life leading to death. A CFTR-deficient rat model has been reported to exhibit histological abnormalities in the ileum, increased intracellular mucus in the proximal nasal septa, reduced airway surface liquid and periciliary liquid depth and abnormal submucosal gland size

(Tuggle et al. 2014). Although the CF rat recapitulates several aspects of human CF (aberrant chloride transport, intestinal obstruction, impaired growth, malformation of the trachea, anomalous vas deferens), important hallmarks such as obstructive lung disease, dysfunction of liver and exocrine pancreas and diabetes mellitus have not been reported (reviewed in Cutting 2015; Klymiuk et al. 2016b).

Non-rodent animal models of CF have been established only recently. CFTR-deficient pigs were generated by introducing a stop codon in exon 10 (Rogers et al. 2008) or a STOP box that terminates both transcription and translation in exon 1 (Klymiuk et al. 2012a). A third CF pig model reproduced the most relevant human *CFTR* mutation, F508del in exon 10 (Ostedgaard et al. 2011). Despite the different *CFTR* mutations, each of these models had an almost identical phenotype. One major feature was almost 100% penetrance of meconium ileus, a mechanical obstruction of the gut that also occurs in human patients, but only in 10–20% of cases (reviewed in Klymiuk et al. 2016b). Neither ileostomy, i.e. surgical removal of meconium, nor intensive enema, the standard treatment for human patients, was sufficient to resolve the obstruction, and the piglets usually died at several weeks old. Stoltz et al. (2013) generated a 'gut-corrected' CF pig that expresses a *CFTR* transgene in the gut under the control of the rat fatty acid binding protein 2 gene (*Fabp2*) promoter. This transgenic rescue can extend life up to 12 months; however in-depth evaluation of CF pigs has been performed only in neonates and a limited number of ileostomised pigs. Despite these limitations, the CF pig model has already contributed tremendously to the understanding of CF pathogenesis. In particular, the availability of neonatal material, which can be seen as a 'native' tissue, has revealed new insights into early stage disease development.

Progressive obstruction of the respiratory tract is the most important cause of morbidity in CF patients. While histological examination of newborn CF pigs revealed apparently normal lung tissue (Rogers et al. 2008), the trachea had a triangular rather than circular cross section, and the cartilage appeared thicker and more discontinuous than in wild-type samples (Klymiuk et al. 2012a; Meyerholz et al. 2010). This was confirmed in human CF infants (Meyerholz et al. 2010). Similar to findings in CF patients, sinus disease developed spontaneously in older CF pigs, whereas at birth sinuses were hypoplastic, but showed no evidence of infection or inflammation. Although the lungs of newborn CF piglets showed no signs of infection, bacterial eradication was defective (Stoltz et al. 2010) and attributed to decreased pH on airway epithelia (Pezzulo et al. 2012). Furthermore, mucus detachment from submucosal glands of the airways (Hoegger et al. 2014b) and mucociliary transport of particles is impaired in CF piglets (Hoegger et al. 2014a). Analyses of epithelial tissue and cultivated cells from CF pigs revealed that the lack of CFTR caused reduced trans-cellular transport of Cl^- and HCO_3^-, but no alteration of Na^+ transport or liquid absorption (Chen et al. 2010). This was later confirmed in primary epithelial cells from human CF patients (Itani et al. 2011). This major paradigm shift was driven by interrogation of the pig model. The thickening of mucus in CF airways had previously been postulated to be a consequence of disturbed osmotic balance, but the pig model data suggested a Na^+-independent mechanism.

The established CF pig lines also provide excellent models to evaluate therapeutic strategies (Klymiuk et al. 2016b). For instance, gene therapy using viral vectors (adenovirus, adeno-associated virus, lentivirus) and non-viral vectors has not yet led to a clinically applicable therapy, but has uncovered a number of problems that limit efficacy for CF patients (reviewed in Griesenbach and Alton 2009; Prickett and Jain 2013). These include challenges with local delivery of gene therapy vectors into epithelial cells through a thickened mucus layer and immune reactions against the viral vectors. A recent clinical trial of repeated nebulisation of non-viral *CFTR* gene therapy in CF patients revealed a significant, albeit modest, treatment effect with stabilisation of lung function (Alton et al. 2015). Large CF animal models with airway and lung structures similar to human CF patients will help improve vector design and delivery strategies. For example, Cao et al. (2013) demonstrated efficient transfer of *lacZ* reporter genes and human *CFTR* expression cassettes into airway epithelia and submucosal glands of normal pigs after intra-tracheal application of aerosolised helper-dependent adenoviral vectors. In addition, intra-tracheal delivery of transfected airway epithelial cells has been suggested as a treatment for CF and proof of principle for efficient delivery shown in mice and wild-type pigs (Gui et al. 2015). It will be interesting to test these strategies in CF pigs, where gene or cell delivery may be more challenging because of pre-existing mucus and inflammation. CF pigs are also an interesting model for testing viral *CFTR* gene delivery via the coeliac artery into the pancreas, a technique recently established in wild-type piglets (Griffin et al. 2014). The F508del-CFTR pig model can be used to evaluate combinations of CFTR correctors and potentiators. CFTR correctors, such as lumacaftor, reverse the folding defect of F508del-CFTR and increase its stability. CFTR potentiators, such as ivacaftor, increase the activity of the folding-corrected F508del-CFTR (reviewed in Cutting 2015; Klymiuk et al. 2016b).

Duchenne muscular dystrophy (DMD) is a severe X-linked disease that affects 1 in 3,500 males. It is caused by loss-of-function mutations of the *DMD* gene (~2.5 Mb, 79 exons) that lead to a shift in reading frame, out-of-frame transcripts and loss of the essential muscle cytoskeletal protein dystrophin (Hoffman et al. 1987; reviewed in Klymiuk et al. 2016b). Hotspots for mutations lie in exons 3–7 and exons 45–55. DMD is characterised by progressive muscle weakness and wasting. Patients present first symptoms before 5 years old, lose ambulation around 12 years and die of respiratory or heart failure in the second to fourth decade of life (reviewed in Spurney 2011). While curative treatments are currently not available, genetic and pharmacological approaches are in different phases of clinical testing (reviewed in Fairclough et al. 2013). Animal models in different species have been instrumental in understanding the pathophysiology of DMD and developing therapeutic strategies, but have several limitations (reviewed in Nakamura and Takeda 2011; Klymiuk et al. 2016b).

The original X-linked muscular dystrophy mouse (*mdx*) occurred spontaneously in the C57BL/10 strain and has a nonsense mutation in *Dmd* exon 23. Four other strains of *mdx* mice have been identified with different mutations. A mouse lacking *Dmd* exon 52 has also been generated by gene targeting (Araki et al. 1997). Various *mdx* mouse models have been used to develop therapeutic strategies, including exon skipping, gene therapy and cell therapy. However, *mdx* mice do not develop overt muscle wasting except in the diaphragm and have a near-normal life span (Nakamura and Takeda 2011).

Mutations in the *DMD* gene have also been identified in several dog breeds, including Golden Retriever, Rottweiler, German Shorthaired Pointer and Cavalier King Charles Spaniel, with Golden Retriever muscular dystrophy (GRMD) being the most extensively examined and characterised (reviewed in McGreevy et al. 2015). GRMD is caused by a point mutation at the *DMD* intron 6 splice acceptor site, leading to skipping of exon 7 and a premature stop codon in exon 8 (Sharp et al. 1992). Dystrophin-deficient GRMD dogs have a more severe condition than *mdx* mice, but the phenotype is highly variable and animals are difficult to breed. Furthermore, the *DMD* mutation of the GRMD model does not correspond with the location of mutations occurring in most human DMD patients. Feline muscular dystrophy with dystrophin deficiency is caused by a large deletion of muscle-specific and Purkinje cell promoters of the *DMD* gene (Winand et al. 1994). The nature of the mutation and the severe and complex phenotype of this mutant have however limited its use as a model (reviewed in Nakamura and Takeda 2011). Dystrophin-deficient rat models have been generated by CRISPR/Cas9-induced deletions between *Dmd* exons 3 and 16 (Nakamura et al. 2014) and a TALEN-induced 11-bp deletion in *Dmd* exon 23 (Larcher et al. 2014). The DMD rats showed muscle weakness and histological signs of muscular dystrophy. However, no treatment studies have so far been reported, and any such findings may be difficult to extrapolate to humans due to the small size of rats. Rhesus monkeys with mutant *DMD* alleles have recently been generated by injecting CRISPR/Cas9 into fertilised oocytes (Chen et al. 2015). Although partial dystrophin depletion and hypertrophic myopathy were observed, the monkeys were mosaic, resulting in genetic and phenotypic variability, which limits their value as translational animal models.

To establish a tailored large animal model of DMD, Klymiuk et al. (2013) deleted *DMD* exon 52 in male pig cells by gene targeting using a bacterial artificial chromosome modified by recombination and generated *DMD* mutant pigs by nuclear transfer. Cloned DMD pigs lacked dystrophin in skeletal muscles (Fig. 9.2a) and showed increased serum creatine kinase levels, impaired movement and muscle weakness (Fig. 9.2b) and a maximum life expectancy of 14 weeks (reviewed in Klymiuk et al. 2016b). Pathological analysis found the skeletal muscles to be pale and moist in texture, with multifocal areas of pale discolouration. Histological examination revealed myopathy with excessive fibre size variation, numerous large rounded hypertrophic fibres, branching fibres and fibres with central nuclei, as well as scattered clusters of segmentally necrotic fibres, next to hypercontracted fibres and groups of small regenerating muscle fibres (Fig. 9.2c). These lesions were accompanied by interstitial fibrosis and mononuclear inflammatory cell infiltration, mimicking the hallmarks of the human disease. The severity and extent of these alterations progressed with age (reviewed in Klymiuk et al. 2016b). $DMD^{\Delta exon52}$ Yucatan minipigs have been developed by Exemplar Genetics, Inc., and a limited characterisation is included in their patent application WO2014117045A2.

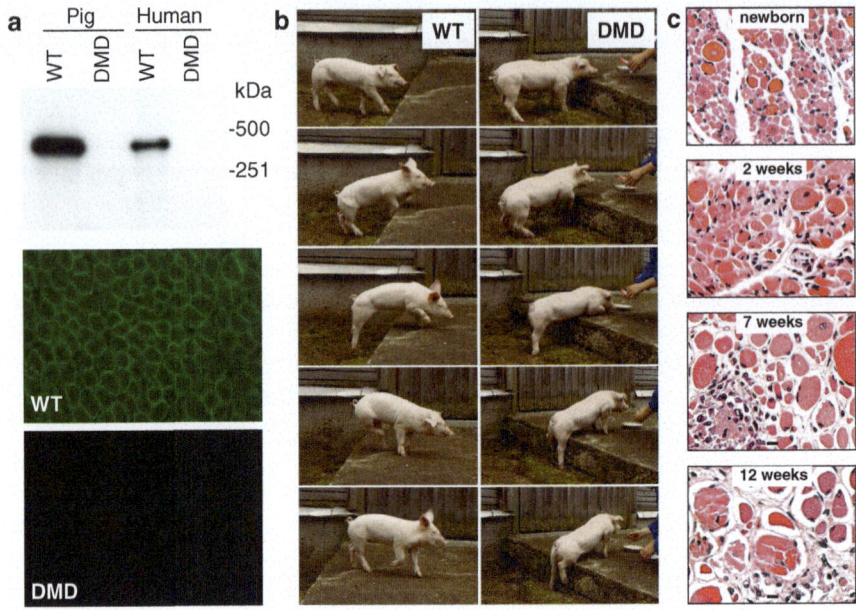

Fig. 9.2 Biochemical, clinical and pathological alterations after deletion of *DMD* exon 52. (**a**) Loss of the essential muscle protein dystrophin as demonstrated by Western blot and immunofluorescence analyses. (**b**) Characteristic clinical phenotype, including the inability of climbing a small platform. (**c**) Progressive, severe muscular dystrophy (from Klymiuk et al. 2013)

To facilitate new insights into the hierarchy of physiological derangements of dystrophic muscle, transcriptome studies were performed with skeletal muscle from young (2 days old) and older (around 3 months) DMD pigs and age-matched controls. The transcriptome changes in 3-month-old DMD pigs accorded with gene expression profiles in human DMD muscle, reflecting the processes of degeneration, regeneration, inflammation, fibrosis and impaired metabolic activity. In contrast, the transcriptome profile of 2-day-old DMD pigs showed similarities with transcriptome changes induced by acute exercise muscle injury, suggesting mechanical stress on the muscle cell membranes as an early factor in the pathogenesis of DMD (Klymiuk et al. 2013). Fröhlich and co-workers (2016) performed a label-free proteome analysis of the same set of muscle samples. The extent of proteome changes in DMD vs. wild-type muscle increased markedly with age, reflecting progression of the pathological changes. In 3-month-old DMD muscle, proteins related to muscle repair such as vimentin, nestin, desmin and tenascin C were found to be increased in abundance, whereas a large number of respiratory chain proteins were decreased, indicating serious disturbances in aerobic energy production and reduction of functional muscle tissue. The combination of proteome data for fibre-type-specific myosin heavy chain proteins and immunohistochemistry showed preferential degeneration of fast-twitch fibre types. The stage-specific proteome changes detected in the DMD pig model provide novel molecular readouts for future treatment trials.

The DMD pig appears to be a bona fide model of the human dystrophy as ascertained by the absence of the dystrophin protein, elevated serum creatine kinase, progressive muscular dystrophy and characteristic disturbances of locomotion, including the inability to climb onto a platform, which is comparable to the early difficulties of DMD patients in climbing stairs. DMD pigs exhibit the functional and pathological hallmarks of the human disease, but development is accelerated. This offers improved opportunities for early and clear-cut readouts from efficacy studies of new treatments, compared with currently available animal models. Since loss of exon 52 is a frequent mutation in human DMD, and can be treated by exon 51 skipping (reviewed in Fairclough et al. 2013), this pig model could be suitable for testing and refinement of this therapeutic strategy. To provide sufficient numbers for systematic treatment trials, female $DMD^{+/\Delta exon52}$ pigs, which have 50% male DMD offspring, have been generated (Klymiuk et al. 2016b). Alternatively, an elegant blastocyst complementation method has been used to generate chimeric boars that can transmit mutant DMD alleles and other X-linked diseases to progeny (Matsunari et al. 2018).

Genetic approaches to cure DMD include replacing the defective DMD gene, read-through of translation stop codons, exon skipping to restore the reading frame and increased expression of the utrophin ($UTRN$) gene which may compensate for loss of dystrophin (reviewed in Fairclough et al. 2013; Klymiuk et al. 2016b). Challenges for gene therapy include the large size of the DMD mRNA (14 kb) and the need to target all muscles. DMD mini- and micro-genes have been developed to overcome the size problem (reviewed in Davies 2013). The most common viral vectors used to transduce muscle cells are based on adeno-associated virus, but DMD gene delivery using this vector has resulted in immune responses against mini-dystrophin (reviewed in Davies 2013).

Exon skipping is another strategy that could work for more than 80% of DMD mutations, including most out-of-frame deletions (reviewed in Fairclough et al. 2013). This strategy aims to restore an intact reading frame of the transcript. Skipping of specific exons can be induced by intramuscular or systemic treatment with RNase H-independent antisense oligonucleotides (AONs) that hybridise to complementary sequences in, or adjacent to, the target exon. 2′-O-Methyl-phosphorothioate (2′OMe) AONs and phosphorodiamidate morpholino oligonucleotides (PMOs) have been tested in preclinical studies and clinical trials (reviewed in Fairclough et al. 2013), but failed to show clear clinical benefit. A new class of AONs made of tricyclo-DNA (tcDNA) rescued dystrophin expression in skeletal muscles and heart and to a lesser extent in the central nervous system of mdx mice (Goyenvalle et al. 2015). Improvement of several clinical parameters was reported in mdx mice treated with tcDNA AON and also in double-mutant mice that lack both dystrophin and utrophin and show a more severe phenotype than mdx mice. A disadvantage of the exon skipping approach is that only the DMD RNA is modified and the therapeutic effect is therefore transient. Recently, in vivo genome editing using the CRISPR-Cas9 system has been successfully applied in the mdx mouse model to delete the mutated exon 23 of the Dmd gene, thus restoring expression of a truncated dystrophin and achieving functional improvements (Long et al. 2016; Nelson et al. 2016; Tabebordbar et al. 2016).

Results in dystrophic mouse models are thus promising, but it would be beneficial to test the efficacy of exon skipping and gene editing strategies in a clinically severe large animal model before commencing clinical trials, because a number of key parameters cannot be easily investigated in dystrophic mouse models or DMD patients. These include (1) the best timing to initiate AON therapy or gene editing related to disease progression; (2) the amount of dystrophin required for near-normal muscle function; (3) the optimal study duration, read-outs and outcome measures; (4) the most effective systemic administration route; and (5) the optimal dosage for long-term therapy (Klymiuk et al. 2016a). The *DMD* exon 52-deficient pig model is amenable to correction by skipping of exons 51 or 53 from the *DMD* primary transcript, or by deletion of these *DMD* exons in the muscle cells using gene editing, and is thus ideally suited to test these aspects of potential treatments. The DMD pig will also be useful in studying efficacy and safety aspects of *DMD* mini-gene therapy, including potential immunological complications, and of read-through treatment strategies or cellular therapies. In comparison with the existing canine DMD models, studies in DMD pigs may also be ethically more acceptable.

9.1.2 Genetically Engineered Pig Models for Diabetes Research

Diabetes mellitus (DM) encompasses a heterogeneous group of metabolic disorders characterised by steadily increasing blood glucose levels, polydipsia, polyuria, weight loss as well as diabetic ketoacidosis and non-ketotic hyperosmolar syndrome as life-threatening consequences of metabolic decompensation (American Diabetes Association 2013). DM in humans is categorised as four aetiopathogenic classes:

1. Type 1 diabetes mellitus (T1DM; 5–10% of cases) usually leads to absolute insulin deficiency due to autoimmune-mediated or idiopathic destruction of the pancreatic beta cells.
2. Type 2 diabetes mellitus (T2DM; 90–95% of cases) is mostly caused by insulin resistance combined with inadequate compensatory insulin secretion.
3. Other types of DM, including a broad range of causes, such as genetic defects of beta-cell function or insulin action, diseases of the exocrine pancreas, endocrinopathies leading to DM, drug- or chemical-induced DM, DM caused by infections, uncommon forms of immune-mediated DM and other genetic syndromes sometimes associated with DM.
4. Gestational diabetes mellitus (GDM).

DM has emerged as a steadily increasing health problem, and the predicted future dimension of the global DM epidemic is alarming. An increase from the current 346 million people affected to over 400 million worldwide by the year 2030 has been extrapolated (reviewed in Wolf et al. 2014). Concerted research efforts are therefore required to gain insight into disease mechanisms and to expand the basis for development of preventive and therapeutic strategies. Diabetic rodent models, derived either by forward (e.g. Aigner et al. 2008) or reverse genetics (e.g. Plum et al. 2005), have traditionally been used to follow these goals, but have limitations for translational research (reviewed in Kleinert et al. 2018).

The pig is a classical animal model for diabetes research (reviewed in Renner et al. 2016b). Physiological blood glucose levels in domestic pigs vary between 70 and 115 mg/dl, the same range as humans. Pig and human pancreas and pancreatic islets are very similar, e.g. endocrine cell distribution and beta-cell content. Porcine insulin differs from human insulin in only one amino acid at position 30 of the B-chain and has been used for decades to treat diabetic patients. In both species, beta-cell mass clearly correlates with beta-cell function. While rodent beta cells have substantial proliferative capacity, beta-cell proliferation is less in humans and pigs (reviewed in Wolf et al. 2014; Renner et al. 2016b). Also, the combination of functional data from physiological tests and quantitative morphological data provides new opportunities to evaluate the relationship between beta-cell function and beta-cell mass in wild-type pigs and pig models with abnormalities in glucose homeostasis and/or beta-cell function.

The incretin hormones glucose-dependent insulinotropic polypeptide (GIP) and glucagon-like peptide-1 (GLP1) are secreted by enteroendocrine cells and are potent enhancers of insulin secretion (reviewed in Baggio and Drucker 2007; Renner et al. 2016a). Type 2 diabetic patients have diminished insulinotropic GIP activity, and it has been suggested that impaired GIP action might be involved in the early pathogenesis of type 2 diabetes mellitus (Nauck et al. 2004). To address this question, Renner et al. (2010) generated transgenic pigs expressing a dominant-negative GIP receptor (GIPR^{dn}) in the pancreatic islets. GIPR^{dn} transgenic pigs develop normally and show similar body weight gain as control littermates. However they exhibit blunted insulinotropic action of GIP; progressive deterioration of glucose control due to delayed and—at later stages—reduced insulin secretion (Fig. 9.3); and

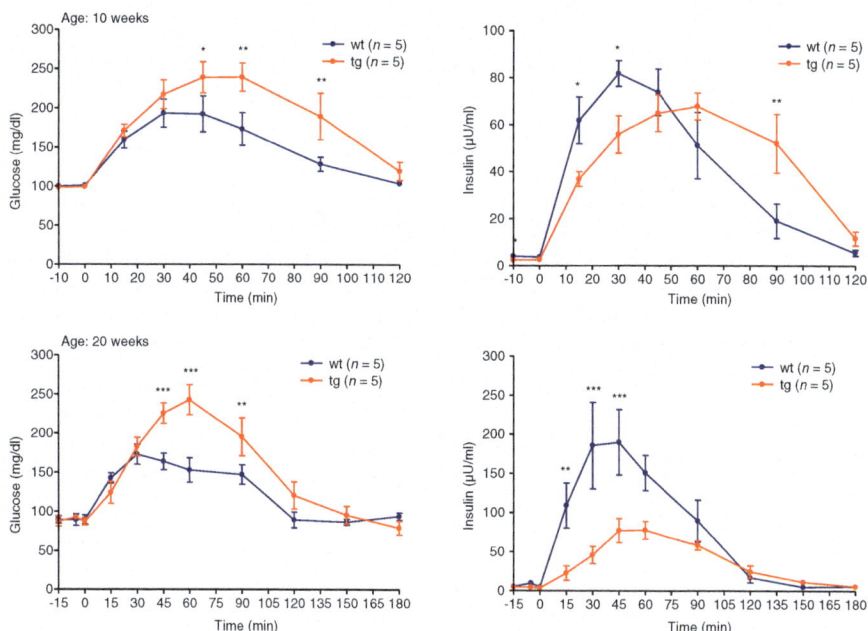

Fig. 9.3 Progressive deterioration of oral glucose tolerance in prediabetic GIPR^{dn} transgenic pigs (from Renner et al. 2010)

impairment of age-related expansion of beta-cell mass (Renner et al. 2010). GIPR^{dn} transgenic pigs thus provide a unique model to develop and test novel therapeutics and search for candidate biomarkers that predict deteriorating glucose control. In a targeted metabolomics approach, Renner et al. (2012) identified a number of plasma amino acids and lipids significantly associated with parameters of glucose homeostasis, insulin secretion and beta-cell mass. A subsequent project using the GIPR^{dn} transgenic pig model investigated whether reduced GIP function can be compensated by pharmacological stimulation of the GLP1 receptor (GLP1R) (Streckel et al. 2015). Several GLP1R agonists have been approved for the treatment of adult T2DM patients and shown to improve glycaemic control and moderately reduce body weight. GLP1R agonists also increase beta-cell mass in rodent animal models (reviewed in Renner et al. 2016a). However, since rodent pancreas has a much higher capacity for beta-cell proliferation than human pancreas (reviewed in Renner et al. 2016b), effects on beta-cell mass observed in rodent studies may not be predictive for humans. An additional question was how juvenile animals respond to treatment with a GLP1R agonist. Streckel et al. 2015) thus studied effects of the GLP1R agonist liraglutide in adolescent (2-month-old) GIPR^{dn} transgenic pigs. The animals were treated daily with liraglutide (0.6–1.2 mg per day) or placebo for 90 days. Liraglutide led to markedly reduced body weight gain (−31%) and food intake (−30%) compared to placebo, associated with reduced phosphorylation of insulin receptor beta (INSRB)/insulin-like growth factor-1 receptor beta (IGF1RB) and protein kinase B (AKT) in skeletal muscle. Absolute alpha- and beta-cell mass was reduced in liraglutide-treated animals, but alpha- and beta-cell mass-to-body weight ratios were unchanged. Liraglutide neither stimulated beta-cell proliferation in the endocrine pancreas nor acinus-cell proliferation in the exocrine pancreas, excluding both beneficial and detrimental effects on the pig pancreas. While pro-proliferative effects on endocrine or exocrine pancreas were not observed, the marked reductions in weight gain and growth observed in this study on adolescent pigs warrant special care in clinical trials with adolescent patients (Streckel et al. 2015). Another important question is whether GIPR^{dn} transgenic pigs develop clinical diabetes and/or insulin resistance after challenge with a high-carbohydrate, high-fat diet. If this was the case, the GIPR^{dn} transgenic pig model would be even more interesting for drug treatment studies and also study of the effects of bariatric surgery on glucose homeostasis. The pig is already used to develop bariatric surgical techniques, such as the laparoscopic Roux-en-Y gastric bypass (Escareno et al. 2012).

To date, 20 heterozygous missense mutations in the human insulin (*INS*) gene have been identified as common causes of insulin-deficient, permanent diabetes mellitus diagnosed predominantly in the neonatal period and referred to as mutant *INS* gene-induced diabetes of youth (MIDY) (Liu et al. 2010). One *INS* mutation, *INS*^{C96Y}, disrupts the C(B7)-C(A7) interchain disulphide bond of the insulin molecule. The widely used Akita mouse model has the corresponding mutation in the *Ins2* gene (Yoshioka et al. 1997). Mutant (pro)insulin impairs trafficking of normal proinsulin by forming high-molecular-weight complexes, and misfolded insulin accumulates in the endoplasmic reticulum (ER) causing ER stress which finally triggers beta-cell apoptosis (Liu et al. 2010). To generate a large animal model of

PNDM, Renner et al. (2013) produced transgenic pigs expressing INS^{C94Y} under the control of porcine *INS* promoter sequences. A founder with high-level transgene expression (ratio of mutant to wild-type *INS* transcripts ~0.75) developed early clinical DM. On insulin treatment the diabetic founder was fertile and produced 50% transgenic offspring after mating with wild-type females. The transgenic offspring were hyperglycaemic shortly after birth; however pancreatic beta-cell mass was unaltered, suggesting impaired insulin secretion as the major cause of diabetes at this stage. At 4.5 months old, INS^{C94Y} transgenic pigs exhibited 41% reduced body weight, 72% decreased beta-cell mass (−53% relative to body weight) and 61% lower fasting insulin levels compared with control littermates. Beta cells of 4.5-month-old INS^{C94Y} transgenic pigs showed marked reduction of insulin secretory granules and severe dilation of the ER (Fig. 9.4). Cataract development was

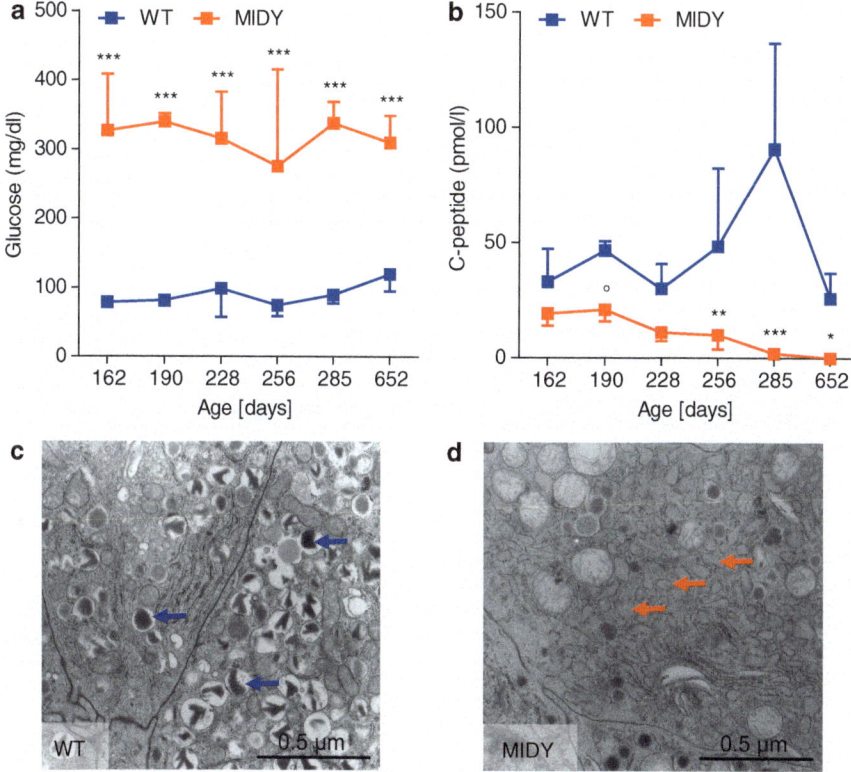

Fig. 9.4 Consequences of expression of mutant insulin C94Y in INS^{C94Y} transgenic MIDY (mutant *INS* gene-induced diabetes of youth) pigs. (**a**) Permanently elevated fasting blood glucose levels. (**b**) Decreasing plasma C-peptide concentrations indicating perturbed insulin secretion and a decrease in beta-cell mass. (**c, d**) Ultrastructural changes of beta cells from MIDY pigs (age: 4.5 months), which are indicative of ER stress. Beta cells of wild-type (WT) pigs (**c**) show multiple insulin granules (blue arrows). In beta cells from MIDY pigs, the number of insulin granules is markedly reduced, and characteristic dilations of the endoplasmic reticulum are visible (**d**; red arrows; from Renner et al. (2013)

already visible in 8-day-old INS^{C94Y} transgenic pigs and became more severe with age. Diabetes-associated pathological alterations of kidney and nervous tissue were not detected during the 1-year observation period (Renner et al. 2013). INS^{C94Y} transgenic pigs exhibit a stable diabetic phenotype without any further manipulation. In this respect this animal model is superior to chemically induced diabetes models (e.g. by streptozotocin), which may have variable outcomes (Dufrane et al. 2006), or to diabetes induced by pancreatectomy (Stump et al. 1988), which is highly invasive and requires substitution of pancreatic enzymes. Due to their exceedingly stable diabetic phenotype, INS^{C94Y} transgenic pigs are an interesting model to address a number of questions. They can be used for insulin treatment studies, to test the efficacy of gene (Callejas et al. 2013) or cell therapies (Sapir et al. 2005), and act as well-defined recipients for islet transplantation (Sakata et al. 2012). Secondary lesions of diabetes mellitus are another interesting area of research. As early as 5 months old, INS^{C94Y} transgenic pigs showed reduced capillarisation and pericyte investment in myocardium compared to age-matched controls. After experimental induction of an ischaemic lesion, the myocardium responded with increased fibrosis. Local gene therapy with thymosin B4 markedly improved capillarisation and pericyte investment in wild-type pigs, but to a lesser extent in INS^{C94Y} transgenic pigs (Hinkel et al. 2017). These findings are clinically relevant since reduced capillarisation and pericyte investment are also observed in myocardium of diabetic patients. Although diabetic neuropathy or kidney disease was not observed within the first year, long-term observation or challenge treatments, e.g. with high-protein diet, might reveal secondary lesions. As a unique resource for studying systemic consequences of chronic hyperglycaemia, the 'Munich MIDY Pig Biobank' was established from 2-year-old INS^{C94Y} transgenic pigs and littermate controls (Blutke et al. 2017; highlighted in Abbott 2015). The biobank includes samples from more than 50 different tissues and body fluids that have been recovered following the principles of systematic random sampling and preserved to enable a broad spectrum of molecular and morphological analyses (Albl et al. 2016). The retina of INS^{C94Y} transgenic pigs showed interesting diabetes-associated alterations with similarities to diabetic retinopathy in human patients (Kleinwort et al. 2017).

Effects of maternal diabetes mellitus on embryos, foetuses and offspring are another interesting research topic. Preconceptional diabetes mellitus (PCDM) in humans adversely affects pre-implantation embryo development and pregnancy outcomes (Farrell et al. 2002). The underlying mechanisms have been investigated mainly in diabetic rodent models and are only partially understood. Pig embryos are more similar to human embryos than are mouse embryos, e.g. in the timing of embryonic genome activation and in metabolic characteristics (reviewed in Wolf et al. 2014). Thus the INS^{C94Y} transgenic pig model appears to be ideally suited to study consequences of preconceptional maternal DM for oocyte, pre- and post-implantation embryonic development and foetal development. However, a major limitation of the pig model in this context is the different placental structure compared to humans (epitheliochorial vs. haemochorial).

Different autosomal dominant mutations in the hepatocyte nuclear factor-1α (*HNF1A*) gene, located on human chromosome 12q, are responsible for defects in insulin secretion and derailed glucose homeostasis and have been categorised as maturity-onset diabetes of the young 3 (MODY 3) (reviewed in Nyunt et al. 2009). To generate a large animal model for this condition, Umeyama et al. (2009) produced transgenic pigs expressing a dominant-negative human HNF1α (P291fsinsC). The expression vector included the mutant *HNF1A* cDNA under the transcriptional control of cytomegalovirus immediate early gene enhancer/porcine insulin promoter sequences and flanked by chicken β-globin insulators. Expression of transgene-derived mRNA was shown in the brain, heart, lung, liver, pancreas, spleen and kidney, and the 315-amino-acid HNF1α (P291fsinsC) protein was detected in the brain, lung, heart, pancreas, spleen and kidney. Persistent diabetes with non-fasting blood glucose levels above 200 mg/dl was observed in four longer-living transgenic pigs. Histological analysis revealed abnormal pancreatic islet morphogenesis, immature renal development and pathological alterations of the kidneys, such as glomerular hypertrophy and sclerosis (Hara et al. 2014) as well as liver alterations (Umeyama et al. 2009). Since expression of the transgene was not limited to pancreatic beta cells, it is not clear whether the kidney lesions were caused by the diabetic condition or toxic effects of the locally expressed HNF1α (P291fsinsC) protein. The latter is likely the case, since INS^{C94Y} transgenic pigs (see above) develop similar hyperglycaemia, but show no signs of diabetic nephropathy, at least before 12 months.

9.1.3 Genetically Engineered Pig Models for Dyslipidaemia Research

Monogenic disorders causing abnormal levels of plasma cholesterol and triglycerides may result in metabolic dysfunction and cardiovascular disease. Prominent examples of monogenic dyslipidaemias associated with increased LDL cholesterol in the paediatric and adolescent population include homozygous or heterozygous familial hypercholesterolaemia with mutations in the low-density lipoprotein receptor (*LDLR*) gene, familial defective apolipoprotein B resulting from *APOB* mutation and gain-of-function mutations of *PCSK9*, encoding proprotein convertase subtilisin/kexin 9 (reviewed in Rahalkar and Hegele 2008).

The mouse is currently most widely used for atherosclerosis studies. However, mice and humans differ in the quality and localisation of atherosclerotic lesions. Human lesions occur more frequently in the coronary arteries, carotids and peripheral vessels (e.g. iliac artery), while in mice the primary locations are the aortic root, aortic arch and innominate artery (reviewed in Getz and Reardon 2012). Since wild-type mice are relatively resistant to diet-induced atherosclerosis, a number of mutant strains exhibiting hypercholesterolaemia and atherosclerosis have been derived by forward and reverse genetics, with *Apoe* and *Ldlr* mutant mice being the most frequently used (reviewed in Getz and Reardon 2012). Genetically modified mouse models have provided insights into disease mechanisms and the role of signalling pathways and genetic factors during disease initiation and progression and have been extensively used to test pharmaceutical modifiers, including 3-hydroxy-3-methylglutaryl-coenzyme A reductase inhibitors (statins) and other cholesterol-lowering drugs (Zadelaar et al. 2007). However, efficacy studies of anti-atherosclerotic drugs in large animal models that mimic human pathophysiology more closely may reduce the risk of failure of anti-atherosclerotic interventions in clinical studies due to lack of efficacy.

Targeted disruption of the *LDLR* gene in Yucatan miniature pigs has provided a porcine model of hypercholesterolaemia and atherosclerosis. Davis et al. (2014) generated homozygous knockout pigs by breeding heterozygous knockouts produced by adeno-associated virus-mediated gene targeting. Phenotypic examination demonstrated significantly increased total and LDL cholesterol levels that were more pronounced in homozygous than heterozygous *LDLR* knockout pigs. These severely hypercholesterolaemic animals subsequently developed atherosclerotic lesions in the coronary arteries and abdominal aorta that resemble human atherosclerosis. High-fat, high-cholesterol diets accelerated lesion development. In humans, loss of LDLR function results in familial hypercholesterolaemia (FH) (Rader et al. 2003) consistent with the phenotype of *LDLR* knockout pigs, including the dose effect between hetero- and homozygous knockouts. This porcine model therefore offers an opportunity to evaluate new therapeutic and diagnostic tools for cardiovascular disease (Li et al. 2016).

A similar phenotype was observed in another transgenic minipig model of FH and atherosclerosis by Al-Mashhadi et al. (2013). Here, a D374Y mutant human PCSK9 was overexpressed in the liver. When fed a high-fat, high-cholesterol diet, these pigs develop severe hypercholesterolaemia and human-like atherosclerotic lesions. PCSK9 downregulates hepatic LDLR activity by increasing lysosomal degradation. With D374Y PCSK9, this inhibitory effect is enhanced, causing a severe form of autosomal dominant hypercholesterolaemia (Soutar 2011) that could be recapitulated in this transgenic pig model. Importantly, the authors demonstrated that development of atherosclerotic lesions in their minipig model could be monitored by imaging techniques, facilitating longitudinal studies of anti-atherosclerotic drug effects and testing of new atherosclerosis imaging modalities. This model was also used to study the effect of diabetes on atherosclerotic lesions (Al-Mashhadi et al. 2015).

Nevertheless, the pig has limitations as a model of human lipoprotein metabolism and atherosclerosis. These include a relatively small effect of statin treatment on plasma LDL cholesterol levels, e.g. in Yucatan minipigs, and the lack of the cholesteryl ester transport protein (CETP), hampering the testing of CETP inhibitors in pigs. This drawback may be overcome by expression of human CETP in genetically engineered pigs (Agarwala et al. 2013).

Apolipoprotein C3 (APOC3) transgenic mice have been generated to study the role of hypertriglyceridaemia as a risk factor for coronary heart disease (Ito et al. 1990). But as lipoprotein metabolism in mice is rather different to humans, the pig might be more suitable for evaluating the role of APO in cardiovascular disease. Wei et al. (2012) developed a transgenic miniature pig that expresses human APOC3. Analysis of the founder animals demonstrated expression of the transgene exclusively in the liver and intestine. Phenotypic examination showed increased overall plasma triglyceride levels and a shift towards large particle fractions in the lipoprotein profile corresponding to very low-density lipoprotein/chylomicrons. Delayed triglyceride absorbance and clearance was also detected. This represents mild to moderate hypertriglyceridaemia as often found in humans. As the lipoprotein profile of pigs and humans is very similar, this model might be useful in gaining insight into the functional role of human APOC3 in lipid metabolism and of triglycerides in atherosclerosis. Furthermore, this model may be used for evaluating drugs for hypertriglyceridaemia (Wei et al. 2012).

Transgenic pigs expressing human apolipoprotein(a) (APOA) exhibit high plasma levels of human lipoprotein(a) [LPA] and may be useful for evaluating the pharmacology and efficacy of new drugs for atherosclerosis (Shimatsu et al. 2016; Ozawa et al. 2015).

Overexpression of lipoprotein-associated phospholipase A2 (Lp-PLA2) in transgenic pigs resulted in increased postprandial plasma triglyceride levels and increased expression of pro-inflammatory genes in peripheral blood mononuclear cells. This model has been generated to study the consequences of elevated circulating Lp-PLA2 levels and to test Lp-PLA2 inhibitors (Tang et al. 2015).

9.1.4 Genetically Engineered Pig Models for Cancer Research

Animal models are crucial for the development of cancer diagnostic and therapeutic techniques. Genetically modified mice are widely used, but their small size and short life span preclude some preclinical studies. It is, for example, difficult to scale down radiological, thermal or surgical treatment of tumours or perform longitudinal studies of tumour progression and remission or longer-term response to therapy.

Mouse and human cancer biology also differ. Murine cells are more easily transformed in vitro than human cells (Rangarajan et al. 2004), and the set of genetic events required for tumorigenesis is different (Kendall et al. 2005). Mouse models may therefore not always provide the best representation of human cancers. Although pigs have not so far played a major role in this field, this is set to change. New genetically modified pig lines will provide valuable complementary resources for cancer research.

As in humans, spontaneous cancer caused by natural mutations is rare in pigs, the most common forms being lymphosarcoma in young animals (Anderson and Jarrett 1968; Stevenson and DeWitt 1973) and melanoma in adult pigs (Fisher and Olander 1978). For many years, only two spontaneous pig tumour models, based on germline mutations, were available for biomedical research. These are the Libechov and Sinclair minipigs, both predisposed to melanoma. These, however, differ from humans because the melanomas spontaneously regress at high frequency (reviewed in Flisikowski et al. 2015). The causative genetic lesions are also undefined, making it difficult to draw parallels with human melanoma.

A variety of non-transgenic strategies have been employed to experimentally generate cancers in pigs. Adam et al. (2007) reported a porcine cancer model based on autologous transplantation of primary porcine cells transduced with retroviral vectors carrying oncogenic cDNAs. These studies revealed important similarities in tumorigenesis between pig and human. However this model falls somewhat short as a representation of human cancer. The use of viral cDNA constructs does not reliably reflect the expression and regulation of endogenous genes. Tumours arising from grafted cells also differ in important respects from autochthonous tumours. Tumours arising from grafted cell lines also tend to be poor predictors of clinical efficacy, for example, anticancer drugs found to be effective on such grafts can be ineffective on real tumours (Zhou et al. 2009).

The first transgenic pigs designed to model cancer carried the v-Ha-ras oncogene directed by the mouse mammary tumour virus long terminal repeat promoter, but no phenotype was observed (Yamakawa et al. 1999). Constitutive expression of the Gli2 transcriptional activator in keratinocytes resulted in basal cell carcinoma-like lesions in young pigs, but these were euthanised due to bacterial infection before fuller investigation could be carried out (McCalla-Martin et al. 2010). More recent porcine models carrying a combination of two or three oncogenic transgenes have been generated. Schook et al. (2015) have reported pigs carrying randomly integrated Cre-inducible porcine transgenes encoding $KRAS^{G12D}$ and $TP53^{R167H}$. Although these are generally expressed at nonphysiological levels, they demonstrated transformation and tumorigenesis after Cre activation of the latent transgenes. With the aim of generating a model for colorectal cancer, Callesen et al. used transposon vectors to generate a double transgenic pig in which one vector carried a Flp-inducible oncogene cassette containing $KRAS^{G12D}$, cMYC and SV40LT and the second vector Flp recombinase controlled by the gut-specific villin promoter (Callesen et al. 2017). Unexpectedly the resulting neoplasia was a large cell neuroendocrine tumour (at the time of writing); nevertheless this is currently the only porcine cancer model showing metastatic spread to a local lymph node. A similar combination using the pancreas-specific PDX1 promoter to express FLP resulted in increased proliferation and clonal expansion of epithelial acinar cells in one of the founder animals (Berthelsen et al. 2017). Unfortunately none of the founder animals survived beyond day 45.

To more closely mimic the molecular changes in human cancer, other researchers have modified endogenous porcine genes. Mutations in the human breast cancer-associated gene 1 (BRCA1) are known to predispose to breast and/or ovarian cancer. Brca1 mutant mouse models are numerous (reviewed in Dine and Deng 2013) and have helped to elucidate the basic biological functions of this gene. However, heterozygous Brca1 mutations in mice are not sufficient to cause breast cancer even at advanced ages, while humans heterozygous for BRCA1 mutations have up to 80% increased risk of developing breast cancer (Alberg et al. 1999). The need for an alternative animal model led Luo et al. (2011) to inactivate BRCA1 in pigs. The gene was inactivated in porcine fibroblasts using adeno-associated virus, and animals were produced by nuclear transfer. These were the first gene-targeted pigs for cancer. The heterozygous founder piglets displayed no visible phenotype at birth. BRCA1 mRNA levels were found to be lower than in controls, but protein levels were unaffected. Unfortunately none of these founders survived beyond 18 days, most likely due to nuclear transfer-related problems. The same group did later report a 2-year-old sow with morphological changes in the mammary gland.

The Schnieke group is engaged in program to model human cancers in pigs. Similar genetic alterations in a relatively small set of genes are responsible for initiating many human tumour types (Futreal et al. 2004), so it should be possible to replicate a variety of cancers by combining and activating defined oncogenic mutations in chosen tissues. An initial disease focus was cancer of the colon and rectum, a serious and common condition in which the molecular basis of disease initiation is understood (Bogaert and Prenen 2014). Disruption of the Wnt pathway, often by reduced function of the tumour suppressor APC, is key to initiating the sporadic form of the disease (Powell et al. 1992) and an inherited predisposition, familial adenomatous polyposis (FAP) (Kinzler et al. 1991; Groden et al. 1991). FAP varies widely in severity, but patients typically develop hundreds of adenomatous polyps in the colon and rectum between puberty and 20 years old, almost inevitably progressing to cancer in midlife (Croner et al. 2005).

Pigs have been generated carrying missense mutations of the adenomatous polyposis coli (APC) gene at codons 1061 and 1311, orthologous to germline mutations responsible for mild and severe forms of FAP (Flisikowska et al. 2012). Examination of a 1-year-old founder and offspring animals carrying the APC^{1311} mutation (orthologous to a human severe FAP mutation, APC^{1309}) revealed polyps in the colon and rectum. This accords with the location and early onset of human FAP and contrasts with equivalent mutations in mice where polyps develop predominantly in the small intestine. Several generations of APC^{1311} pigs have been produced and animals examined up to 4 years of age. Regular colonoscopic examinations revealed that, as in humans, the severity of polyposis in individuals carrying the same APC mutation varied from mild to very severe. Histological and immunohistological analysis showed the same progression to tumours as in human FAP—aberrant crypt foci, adenomatous polyps with low- to high-grade dysplasia and carcinoma in situ up to 5 cm in diameter. Gene expression data of polyps from APC^{1311} pigs correlates with alterations in molecular pathways involved in early pathogenesis of human colorectal cancer (Flisikowska et al. 2017). Time will tell whether porcine polyps progress to invasive disease and metastasis. Examination of APC^{1061} mutant animals revealed a far milder phenotype; many were free of polyps or had one or two very small (~3 mm) polyps.

Tan et al. (2013) used TALENs to produce Landrace and Ossabaw pigs carrying knockout alleles in the APC gene. These however differ from the model described above, as they carry a truncation mutation upstream of the APC^{1061} position. No phenotype has yet been reported.

Loss of p53 function occurs in more than 50% of human cancers, so this tumour suppressor has been a natural focus for cancer modelling. Pigs have been generated with a Cre-dependent $TP53^{R167H}$ mutation, which in latent form is a knockout (Leuchs et al. 2012). Evidence so far indicates that the porcine $TP53^{R167H}$ mutation, which is orthologous to human $TP53^{R175H}$ and mouse $Trp53^{R172H}$, confers similar changes as in humans and mice. Mutant porcine p53-R167H protein accumulates in affected cells, indicating failure of normal p53 degradation (Saalfrank et al. 2016).

Analysis of older heterozygous pigs carrying the uninduced *TP53* knockout mutation revealed spontaneous osteosarcoma development, while homozygous *TP53* knockout resulted in multiple large osteosarcomas in 7–8-month-old animals, located mainly in the long bones, skull and mandible (Saalfrank et al. 2016). Osteosarcoma is a relatively rare solid tumour, but the most common primary bone cancer. It predominantly affects young people and is highly malignant, requiring aggressive surgical resection and cytotoxic chemotherapy. Five-year survival for patients with metastatic osteosarcoma is only around 30% (Mirabello et al. 2009). The initiation and progression of human osteosarcoma are not well understood. The new porcine model promises new resources to elucidate the molecular pathways and driver mutations involved and devise means of managing and treating this devastating disease.

Genetically modified mice have been used for many years to model osteosarcoma. *Trp53* inactivation in mice results in diverse cancers, with ~25% osteosarcomas in heterozygotes and ~4% osteosarcomas in homozygotes, which mainly develop lymphomas (Jacks et al. 1994). The high incidence of other tumours has motivated development of improved mouse osteosarcoma models with Cre-mediated conditional deletion of Trp53 in the osteogenic lineage, sometimes in combination with Rb1. These show highly penetrant osteosarcoma formation, but have been criticised because murine primary tumours predominantly affect the axial skeleton, while human osteosarcomas are most common in the long bones of the limbs (Guijarro et al. 2014). Rats show a similarly mixed tumour spectrum, with approximately half of heterozygous Trp53 knockout rats developing osteosarcomas, while most homozygotes develop haemangiosarcoma (van Boxtel et al. 2011).

Yucatan pigs have also been described that carry the $TP53^{R167H}$ mutation (Sieren et al. 2014). Pigs heterozygous for this mutant allele were reported to be free of tumours, while some homozygous pigs developed lymphomas and some developed osteogenic tumours.

Activating mutations in the *KRAS* proto-oncogene initiate pancreatic adenocarcinoma and non-small-cell lung cancer (reviewed in Pylayeva-Gupta et al. 2011) and occur in many other human cancers such as colorectal cancer, where *KRAS* mutations are associated with poor prognosis (Karnoub and Weinberg 2008). Efforts to devise effective therapies that target mutant *KRAS* have had little success, and there is an urgent need to develop means of early diagnosis and new targets for drug therapy. Li et al. (2015) have generated pigs with a Cre-dependent oncogenic $KRAS^{G12D}$ mutation and demonstrated the functionality of both $KRAS^{G12D}$ and $TP53^{R167H}$ in xenotransplantation experiments.

An important component of pig cancer modelling is local and tissue-specific activation of mutant oncogene(s) in a chosen tissue to mimic the spontaneous somatic events that initiate many human cancers. Such activation, e.g. by Cre-mediated recombination, can thus enable replication of diverse cancer types using the same mutant gene. In mice, Cre reporter strains provide a means of monitoring the location, pattern and extent of Cre recombination in vivo. A dual fluorescent reporter pig for Cre recombination that provides a multipurpose indicator of Cre activity has been generated (Li et al. 2014), and the Schnieke group is in the process of producing transgenic pigs with tissue-specific Cre recombinase expression and investigating local application of Cre recombinase.

The advent of efficient CRISPR/CAS9 genome editing now makes it possible to introduce oncogenic mutations into the tissue of choice in live animals, eliminating the need to generate pigs with germline modifications. Building on similar experiments in mice, Wang et al. (2017) have generated transgenic pigs carrying the CAS9 component to enable genome editing of somatic tissues in vivo. Using lentiviral vectors to introduce multiple guide RNAs, they simultaneously targeted five tumour suppressor genes (*APC, BRCA1, BRCA2, TP53, PTEN*) as well as *KRAS* in the porcine lung. Necropsy revealed multiple pulmonary adenocarcinomas, proving the efficacy of this approach. Without doubt genome editing provides an incredibly versatile tool to introduce modifications both in vivo and in vitro and can drastically reduce the time required for generating new porcine models.

The examples shown in this chapter demonstrate that genetically tailored pig models have the potential to bridge the gap between basic research in rodent models and clinical trials in human patients, providing an improved prediction of efficacy and safety of new treatments. However, compared with the large number of scientists and huge infrastructures involved in mouse genetics and phenotyping, the number of laboratories working on genetically engineered pig models is small. Thus it will be necessary to establish appropriate scientific networks and infrastructures aiming at the generation, characterisation and implementation of genetically tailored pig models in biomedical research. Such institutions would benefit multiple parties: Genetically tailored pig models offer clinicians the chance to develop and test new therapeutic treatments in human-sized animals that show the same disease mechanisms and symptoms as the affected patients. Since large animal models are expected to accelerate the development of efficacious and safe therapies, patients would also benefit. Importantly, rare diseases such as cystic fibrosis or genetic muscle diseases can be readily addressed in genetically tailored pig models, increasing the chance of therapeutic options for affected patients. Improved predictive data is likely to reduce the risk of drug failure in clinical trials and thus has enormous value for pharmaceutical and biotech companies. Healthcare systems can only cope with increased need if new disease treatments can be developed at acceptable cost and if categories of patients likely to benefit are identified. The concept of personalised or stratified medicine is thus central to all healthcare agendas. Tailored large animal models are a promising means of discovering new biomarkers and will support personalised medicine. Analogous to rodent species, attempts for increasing the genetic, environmental and experimental standardisation of the studies may further increase the success in the work with genetically modified pigs.

9.2 Genetically Modified Pigs as Donors of Cells, Tissues and Organs for Xenotransplantation

The number of donated human organs and tissues falls far short of the need (Ekser et al. 2015), a situation that threatens the life of many potential recipients. Alternative techniques such as xenotransplantation are therefore urgently needed. The pig is the preferred donor species for a number or reasons, including size, anatomical and physiological similarities with humans and efficient techniques for genetic engineering/gene editing. More than 40 different genetic modifications have been introduced into pigs to prevent immune rejection of xenografts, overcome physiological incompatibilities and reduce risk of transmitting zoonotic pathogens (reviewed in Cooper et al. 2016). The challenge of combining multiple transgenes to enable practical animal breeding, avoiding segregation of independent integration sites, can be overcome by 'combineering' and gene stacking, i.e. random or targeted placement of multiple expression cassettes at a single genomic locus (Fischer et al. 2016; Rieblinger et al. 2018).

Technical advances in the generation of genetically multi-modified pigs and new developments in the field of immunosuppression have supported significant progress in many areas of xenotransplantation, including pancreatic islets (reviewed in Klymiuk et al. 2016a; Park et al. 2015), neuronal cells (reviewed in Vadori et al. 2015) and corneas (reviewed in Kim and Hara 2015), but also vascularised organs, especially the kidney (reviewed in Iwase and Kobayashi 2015) and heart (reviewed in Mohiuddin et al. 2015). Xenotransplantation can thus be considered as a realistic future therapeutic option together with other regenerative medicine strategies, e.g. stem cells (Perkel 2016).

This chapter summarises the genetic modifications introduced into pigs to render them suitable as donors of cells, tissues and organs for xenotransplantation and also outlines recent attempts at developing human tissues within large animal hosts.

9.2.1 Genetic Modifications to Overcome Hyperacute and Acute Vascular Rejection of Pig-to-Primate Xenografts

Hyperacute rejection of pig-to primate xenografts is triggered by binding of preformed antibodies to specific xeno-antigens. Subsequent activation of the complement system cannot be controlled due to species incompatibilities between the regulators on the xenograft and the recipient's effector molecules (reviewed in Kourtzelis et al. 2015). The major xeno-antigen is galactose-α1,3-galactose (αGal) synthesised by α-1,3-galactosyltransferase (GGTA1). This enzyme is defective in humans and old-world monkeys, and αGal is consequently absent. Immunogenic contact with bacterial αGal epitopes in the intestinal tract causes humans and old-world monkeys to raise anti-αGal antibodies in early life. Other prominent xeno-antigens are N-acetylneuraminic acid (Neu5Gc, also called Hanganutziu-Deicher antigen) synthesised by cytidine monophosphate-N-acetylneuraminic acid hydroxylase (CMAH) and an Sd(a)-like glycan made by porcine β-1,4-N-acetylgalactosaminyltransferase 2 (B4GALNT2) (reviewed in Byrne et al. 2015).

A first step to overcome hyperacute rejection of pig-to-primate xenografts was the generation of transgenic donor pigs that express human complement-regulatory proteins, such as CD46 (membrane cofactor protein, MCP), CD55 (complement decay-accelerating factor, DAF) and CD59 (membrane inhibitor of reactive lysis, MIRL), singly or in combination. These attenuate complement activation and significantly prolong survival of pig-to-primate xenografts (reviewed in Cooper et al. 2016). A major step towards long-term survival of vascularised pig-to-primate xenografts was the inactivation of the porcine *GGTA1* gene. Following the first publication of GGTA1 (and consequently αGal)-deficient pigs (Phelps et al. 2003), multiple *GGTA1* knockout pig lines were generated, initially by gene targeting (reviewed in Klymiuk et al. 2010) and later by gene editing (e.g. Hauschild et al. 2011). To remove the Neu5Gc xeno-antigen, pigs lacking CMAH were generated (Kwon et al. 2013) and have been combined with GGTA1 deficiency (Lutz et al. 2013; Burlak et al. 2014; Miyagawa et al. 2015). The porcine *B4GALNT2* gene was also inactivated to eliminate an Sd(a)-like xeno-antigen (Estrada et al. 2015). The authors showed that cells from GGTA1/CMAH/B4GALNT2-deficient pigs exhibited reduced human IgM and IgG binding compared to cells lacking only GGTA1 and CMAH.

Besides preformed antibody binding to carbohydrate antigens, porcine cells elicit a humoral immune response, as after allotransplantation (reviewed in Vadori and Cozzi 2015). The risk is likely increased in pre-sensitised patients with antibodies against major histocompatibility complex (MHC) class I molecules/human leucocyte antigens (HLAs), since these may cross-react with conserved epitopes of swine MHC subclasses/swine leucocyte antigens (SLAs) (Mulder et al. 2010). MHC class I-deficient pigs have been reported; these show reduced levels of CD4$^-$ CD8$^+$ T cells in the peripheral blood, but appeared healthy and developed normally (Reyes et al. 2014). Porcine cells and tissues lacking or with reduced SLA should elicit only a weak response from the human immune system.

9.2.2 Genetic Modifications to Overcome Cellular Rejection of Pig-to-Primate Xenografts

Both innate and adaptive components of the cellular immune system contribute to xenograft rejection (reviewed in Griesemer et al. 2014). Immune cell infiltration of tissue and solid organ xenografts starts with neutrophils, followed by macrophages and T cells (reviewed in Vadori and Cozzi 2015). In addition, natural killer (NK) cells may induce endothelial cell activation in the xenograft (Dawson et al. 2000) and lyse porcine cells directly and via antibody-dependent cytotoxicity (reviewed in Weiss et al. 2009).

Cellular xenografts such as porcine islets in non-human primates are mainly rejected by CD4$^+$ T cells. Their activation can be induced by direct binding of primate T-cell receptors to SLA class 1 and class 2 molecules of porcine cells or indirectly by antigen-presenting cells (APCs) of the recipient expressing MHCs with processed xeno-antigens (reviewed in Vadori and Cozzi 2015). In addition to this T-cell receptor-mediated signal, T-cell activation is regulated by a co-stimulatory signal, which may—depending on its nature—induce and amplify an effective immune response or exhibit an inhibitory tolerogenic function. In the context of xenotransplantation, the best-studied T-cell co-stimulatory signalling complexes are CD80/CD86-CD28 and CD40-CD154, with CD28 and CD154 (=CD40L) being localised on T cells and CD80/CD86 and CD40 on APCs. The CD80/CD86-CD28 co-stimulation pathway can be blocked by systemic treatment with CTLA4-Ig (abatacept®) or LEA29Y (belatacept®), which have markedly improved the long-term outcome of allogeneic and xenogeneic tissue grafts (reviewed in Bartlett et al. 2016). These molecules can also be expressed in genetically modified donor pigs, opening the prospect of inhibiting T-cell activation locally at the graft site, thus avoiding systemic immunosuppression of the recipient and the consequent risk of infection (commented in Aikin 2012). LEA29Y expressing transgenic porcine neonatal islet cell clusters (NICCs) transplanted into immunodeficient diabetic mice normalised blood glucose levels and, in contrast to wild-type NICCs, were not rejected after the recipient mice were reconstituted with human immune cells (Fig. 9.5) (Klymiuk et al. 2012b). A subsequent study using diabetic mice with a long-term 'humanised' immune system as recipients showed that xenografted porcine islets that expressed LEA29Y survived for several months and normalised the recipients' blood glucose levels, whereas wild-type islets did not engraft in this

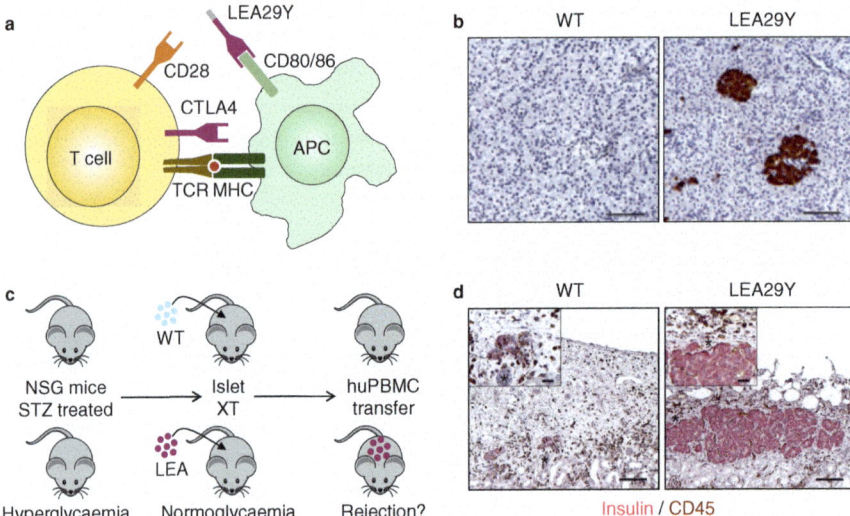

Fig. 9.5 Protection of xenografted porcine pancreatic islets against T-cell-mediated rejection by local expression of LEA29Y. (**a**) Principle of co-stimulation blockade of T cells. Activation of T cells requires interaction between the T-cell receptor and a peptide-loaded major histocompatibility complex (MHC) on an antigen-presenting cell (APC). In addition a second signal such as the interaction between CD28 und CD80/CD86 is required. The interaction of CTLA4 and CD80/CD86 blocks T-cell activation. The latter can also be achieved by the soluble molecule CTLA4-Ig or its affinity-optimised version LEA29Y. (**b**) Immunohistochemical staining of LEA29Y in pancreas sections from INS-LEA29Y transgenic pigs. (**c**) Transplantation of neonatal islet-like cell clusters (NICCs) from wild-type (WT) or INS-LEA29Y transgenic pigs (LEA29Y) into immune-deficient streptozotocin (STZ) diabetic NSG mice results in an insulin-positive cell mass that is able to normalize their blood glucose level. If the mice are subsequently reconstituted with human peripheral blood mononuclear cells (hPBMCs), the WT islets are rejected, while the LEA29Y transgenic islets are protected (from Klymiuk et al. 2012b). (**d**) Histology of the transplantation site. CD45 labels infiltrating T cells

model (Wolf-van Buerck et al. 2017). Transgenic porcine neuronal cells expressing human CTLA4-Ig have been transplanted into a non-human primate model of Parkinson's disease. Although peripheral immunosuppression was necessary for long-term survival of porcine neuronal xenografts, the transgenic cells showed improved survival and effects on locomotor functions (Aron Badin et al. 2016). Skin from transgenic pigs expressing human CTLA4-Ig under control of the human keratin 14 gene promoter exhibited remarkably prolonged survival after transplantation to rats, compared with wild-type porcine skin (Wang et al. 2015).

Transgenic pigs expressing human TNF-related apoptosis-inducing ligand (TRAIL) (Klose et al. 2005; Kemter et al. 2012), human FAS ligand (FASL) (Seol et al. 2010) and HLA-E/beta2-microglobulin have also been generated to prevent cellular rejection. Cells from the HLA-E/beta2-microglobulin pigs were effectively protected against human NK-cell-mediated cytotoxicity, depending on the level of CD94/NKG2A expression on the NK cells (Weiss et al. 2009). To control macrophage activity, human CD47 has been expressed on porcine cells to activate the 'don't eat me signal' receptor SIRPα on (human) monocytes/macrophages and to suppress phagocytic activity (reviewed in Cooper et al. 2016).

9.2.3 Genetic Modifications to Overcome Dysregulation of Coagulation and Inflammation

Dysregulation of coagulation and disordered haemostasis are frequent complications in preclinical pig-to-non-human primate xenotransplantation (reviewed in Bulato et al. 2012). Manifestations range from acute life-threatening consumptive coagulopathy with thrombocytopenia and bleeding to thrombotic microangiopathy leading to loss of the xenograft. Proposed causes include inflammation; vascular injury; innate, humoral and cellular immune responses; and in particular molecular incompatibilities between porcine and primate regulators of coagulation (Cowan and Robson 2015).

Key endothelial anticoagulant/antithrombotic proteins that have been modified/supplemented by genetic engineering of donor pigs include human thrombomodulin (THBD), endothelial protein C receptor (EPCR), tissue factor pathway inhibitor (TFPI) and ectonucleoside triphosphate diphosphohydrolase 1 (ENTPD1 alias CD39) (reviewed in Cowan and Robson 2015).

Porcine THBD binds human thrombin, but is a poor cofactor for activation of human protein C (Roussel et al. 2008). To overcome this problem, lines of transgenic pigs have been generated that express human THBD under various promoters: CMV (Petersen et al. 2009), CAGGS (Yazaki et al. 2012), human ICAM (Mohiuddin et al. 2016) and porcine THBD (Fig. 9.6) (Wuensch et al. 2014). A *GGTA1* knockout, hCD46 transgenic pig heart with the latter THBD expression vector survived for more than 900 days after heterotopic abdominal transplantation into a baboon with appropriate immunosuppression (induction with anti-thymocyte globulin and anti-CD20 antibody, maintenance with mycophenolate mofetil and intensively dosed anti-CD40 (2C10R4) antibody) (Mohiuddin et al. 2016).

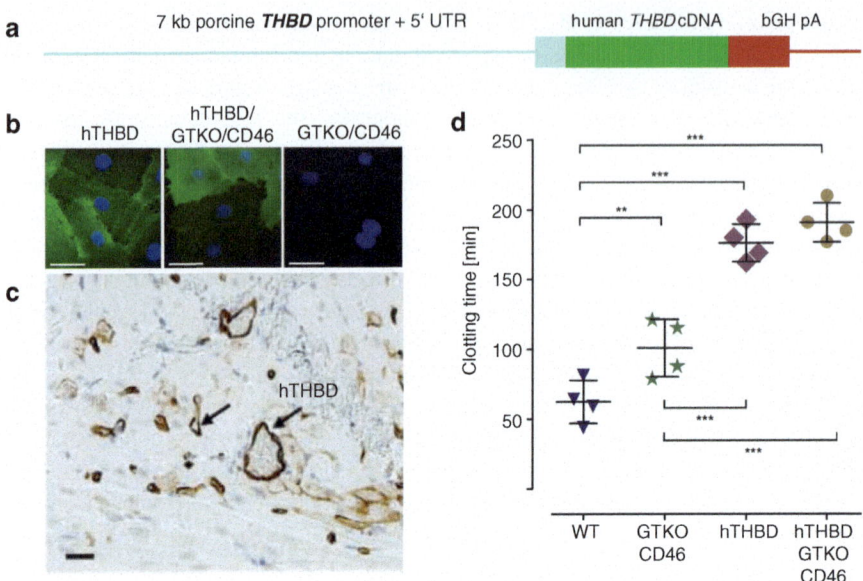

Fig. 9.6 Expression of human thrombomodulin (hTHBD) in genetically (multi-)modified pigs. (a) Expression vector with the porcine *THBD* gene promoter. (b) Immunofluorescence staining of hTHBD in transgenic porcine endothelial cells. (c) Expression of hTHBD in vascular endothelial cells of myocardium from transgenic pigs. (d) Beads covered with hTHBD-expressing endothelial cells from genetically (multi-)modified pigs prevent clotting of human blood (from Wünsch et al. 2014)

EPCR binds protein C and presents it to the THBD/thrombin complex for activation, enhancing the generation of activated protein C (Taylor et al. 2001). Expression of human EPCR in porcine aortic endothelial cells reduced human platelet aggregation almost as efficiently as human THBD (Iwase et al. 2014). The authors concluded that transgenic expression of human EPCR can enhance the effect of porcine THBD, but would have an even greater effect when co-expressed with human THBD.

Porcine TFPI is a less efficient inhibitor of human TF/factor VIIa than human TFPI (reviewed in Cowan and Robson 2015). A recent study demonstrated that Kunitz domain 1 is critical for the species incompatibility between pig TFPI and human tissue factor (TF) and that clotting can be inhibited by human TFPI-transfected porcine bone marrow mesenchymal cells (Ji et al. 2015). Transgenic pigs expressing human TFPI have been generated to overcome this incompatibility (Lee et al. 2011; Wijkstrom et al. 2015).

CD39 rapidly hydrolyses ATP and ADP to AMP. AMP is hydrolysed by ecto-5′-nucleotidase (CD73) to adenosine, an antithrombotic and cardiovascular protective mediator. Transgenic pigs expressing human CD39 under the control of the murine H-2K^b promoter showed reduced infarct size after myocardial ischaemia/reperfusion injury (Wheeler et al. 2012). Subsequently, kidneys (Le Bas-Bernardet et al. 2015;

Pintore et al. 2013) and islets (Bottino et al. 2014) of genetically multi-modified pigs (including expression of human CD39) were tested in xenotransplantation experiments in non-human primates. However specific effects of human CD39 could not be distinguished among the other genetic modifications.

Reduced expression of pro-coagulatory TF has been achieved by siRNA-mediated knockdown (Ahrens et al. 2015). Aberrant phagocytosis of human platelets during perfusion of porcine livers could be partially overcome by deleting the porcine asialoglycoprotein receptor (ASGR) (Paris et al. 2015).

In addition to modifications targeting coagulation disorders in xenotransplantation, transgenic pigs have been produced that express anti-apoptotic and anti-inflammatory proteins, such as human tumour necrosis factor-alpha-induced protein 3 (A20) (Oropeza et al. 2009) and human haem oxygenase-1 (HO-1) (Petersen et al. 2011).

9.2.4 Genetic Modifications to Decrease the Risk for Zoonoses

Xenotransplantation of porcine cells, tissues or organs carries the risk of transmitting porcine microorganisms to the human recipient and induction of zoonotic disease (Fishman et al. 2012; Mueller et al. 2011). Allotransplantation is in rare cases associated with transmission of pathogenic microorganisms such as human immunodeficiency and rabies viruses. Xenotransplants are however strictly regulated as advanced therapy medicinal products (ATMP) and subject to the highest level of safety requirements and thus may actually be safer than allografts in the future (Fishman 2014). Although a wide range of bacteria, viruses, parasites and fungi pose theoretical risks, comprehensive investigations of different pig colonies have shown that the actual number of microorganisms found in pigs is limited (e.g. Wynyard et al. 2014; Morozov et al. 2015). A complete picture of the microorganisms in individual pigs is not yet available and will strongly depend on the test methods used, but it is unlikely that many new pathogens will be identified. One reason is the thousand years of co-existence of humans and pigs, and second is that most pig donors are produced under SPF or near-SPF conditions, preventing infections from third species. Nevertheless, screening for putative zoonotic microorganisms and sterility testing is mandatory.

Porcine endogenous retroviruses (PERVs) cannot easily be eliminated, since they are integrated into the porcine genome, and thus pose the most refractory source of risk (reviewed in Denner and Tonjes 2012). PERV-A and PERV-B are able to infect some human tumour cells and to a lesser extent human primary cells. PERV-C infects only pig cells, but recombinants with PERV-A can infect human cells and are characterised by a high replication rate. Since PERV-C is not present in all pigs, it is possible to identify and select PERV-C-free animals as potential donors. Thus far, in vivo transmission of PERVs has not been documented, despite many studies, including the first clinical trials of porcine islets with more than 200 patients. Neither has PERV transmission been documented in numerous preclinical trials of pig cells and organs transplanted into non-human primates (reviewed in

Denner and Tonjes 2012). However, in almost all clinical xenotransplantation trials, no immunosuppression was applied and only a small number of cells were transplanted. Animal xenotransplantation and PERV infection experiments involve recipient species that do not have a fully functional PERV receptor, e.g. non-human primates or rats. Data on PERV infectivity in vivo are therefore not considered to be complete.

Strategies to prevent PERV transmission include (1) selection of PERV-C-free animals (Fischer et al. 2016) with low expression of PERV-A and PERV-B, (2) vaccines based on neutralising antibodies (for review, see Denner 2013) and (3) long-term reduction of PERV expression by PERV-specific siRNA (e.g. Dieckhoff et al. 2008; Ramsoondar et al. 2009). A paper reported simultaneous inactivation of up to 62 proviruses in an immortal pig kidney cell line (PK15) using the CRISPR/Cas9 system (Yang et al. 2015). This approach could be reproduced with primary pig cells, generating healthy PERV-free animals (Niu et al. 2017; commented in Denner 2017a, b). It is currently a matter of debate whether genome-wide PERV inactivation is a prerequisite for entering clinical xenotransplantation trials (Guell et al. 2017).

9.2.5 Attempts to Develop Human Organs in Animal Hosts

The group of Hiromitsu Nakauchi (Tokyo University) provided the first proof of concept for developing an allo- or xenogeneic pancreas in interspecific chimaeras (Kobayashi et al. 2010). They used mouse blastocysts with defective copies of the pancreatic and duodenal homeobox 1 (*Pdx1*) gene, which is essential for pancreas development. Embryos and foetuses developing from these blastocysts have an empty pancreas niche, and apancreatic pups die shortly after birth from severe hyperglycaemia. This lethal phenotype could be rescued by injecting embryonic stem cells (ESC) or induced pluripotent stem cells (iPSC) from *Pdx1*-intact mouse strains into the *Pdx1*-defective blastocysts. The resulting chimaeras had a pancreas entirely derived from injected pluripotent stem cells. Injection of rat iPSC into *Pdx1*-defective mouse blastocysts resulted in interspecific chimaeras with a rat-derived pancreas, demonstrating that intra- and interspecific blastocyst complementation with pluripotent stem cells can form an entire organ in a host engineered to have a free developmental niche.

As a first step towards generation of a human pancreas in an animal host, Matsunari et al. (2013) generated pancreatogenesis-disabled transgenic pigs that express *Hes1* (hairy/enhancer of split-1) under *PDX1* promoter control. Since no fully functional porcine pluripotent stem cells are available, blastocyst complementation was performed with embryonic blastomeres expressing the fluorescent marker Kusabira-Orange (K-O). The resulting chimaeric foetuses and offspring had pancreata derived entirely from the K-O-labelled cells.

The availability of specific organogenesis-impaired pig hosts opens the possibility of targeted organ generation from human pluripotent stem cells, but whether this is feasible between phylogenetically distant species remains to be seen.

Masaki et al. (2015) developed an assay to test the ability of pluripotent stem cells to form interspecies chimaeras and found that human iPSC failed to integrate into the epiblast of mouse egg-cylinder stage embryos. This they ascribed to different gastrulation mechanisms and incompatibility of ligands or adhesion molecules and proposed the use of host embryos from more closely related species. However even between rat and mouse, developmental incompatibilities may limit the success of blastocyst complementation. Usui et al. (2012) used this approach for de novo kidney formation by injecting mouse ESC or iPSC into mouse blastocysts lacking a functional *Sall1* gene. SALL1 deficiency results in impaired development of the metanephric mesenchyme (MM)-derived components of the kidney (reviewed in Kemter and Wolf 2015). Chimaeric mice generated by this approach had kidneys derived almost entirely from the injected pluripotent cells, except for the renal vascular and nervous system (Usui et al. 2012). However injection of rat iPSC into *Sall1* mutant mouse blastocysts failed to generate rat kidneys in mouse, suggesting insufficient crosstalk between ureteric bud (UB) and MM from different species (reviewed in Kemter and Wolf 2015).

A recent study evaluated the ability of different types of human PSC to contribute to chimaeras after injection into porcine and bovine blastocysts. While naïve PSC were reported to engraft in blastocysts of both host species, contribution to post-implantation pig embryos was limited. A higher (although still very low) degree of chimaerism was observed when using an intermediate PSC type (Wu et al. 2017). These observations suggest that species barriers prevent extensive post-implantation chimaerism.

In addition to biological difficulties, such use of human pluripotent stem cells raises ethical issues, for example, the possibility of unanticipated human contributions to neurons or even germ cells (Hermeren 2015). A possible solution would be stem cells with limited differentiation potential. Kobayashi et al. (2015) demonstrated that inducible expression of *Mixl1* (Mix-like protein 1) limits the differentiation potential of mouse pluripotent stem cells to derivatives of the endodermal germ layer. Such cells could still form pancreas, but not neurons or other ectoderm or mesoderm derivatives.

Another possibility to avoid ubiquitous chimaerism is 'conceptus complementation', i.e. transplantation of progenitor tissue into the appropriate foetal (or postnatal) environment (reviewed in Nagashima and Matsunari 2016). Porcine embryonic metanephroi (the progenitor structures of the definitive kidney, including the UB and MM) developed fully functional nephrons after implantation into immunosuppressed or immunodeficient mice (reviewed in Yamanaka and Yokoo 2015). The need for immunosuppression might be avoided by using genetically engineered porcine donor embryos lacking major xeno-antigens and expressing immune modulatory proteins (reviewed in Kemter and Wolf 2015). While the size of a metanephros-derived kidney tissue is primarily controlled by the metanephros donor species, growth of a porcine 'neo-kidney' in a rodent host is obviously limited. Therefore, Yokote et al. (2015) transplanted metanephroi from 30-day-old pig embryos into the omentum of syngeneic cloned recipient pigs. The grafted metanephroi showed substantial growth (5–7 mm after 3 weeks, 3 cm after 8 weeks), formation of kidney glomeruli and tubuli and production of urine. When a larger embryonic graft including the metanephros and cloaca (a common opening to the urinary and digestive tracts during development which permits formation of a urinary bladder) was used, a graft-derived neo-bladder was formed. This structure could be surgically connected to the recipient's ureter system to facilitate continuous discharge of graft-derived urine. The development of a de novo kidney with a 'stepwise peristaltic ureter (SWPU) system' in an animal with similar size as humans may pave the way to overcome the shortage of donor organs for kidney transplantation.

Metanephroi cannot only form functional kidney tissue per se but also serve as a developmental niche inducing differentiation of injected human mesenchymal stem cells (hMSCs) into all kinds of specialised cell types of the kidney (Yokoo et al. 2005). A scenario can thus be envisaged where metanephroi in pig embryos are employed as a developmental niche for hMSCs to derive human kidney progenitor tissue that can then be transplanted into patients with end-stage renal disease to form a de novo kidney (reviewed in Kemter and Wolf 2015).

9.3   Genetically Modified Large Animals for the Production of Pharmaceutical Proteins

For three decades transgenic animals have been promoted as a cost-effective method of producing biopharmaceuticals, often termed 'pharming'. The original proposal for large-scale production of pharmaceutical proteins in milk was made by Rick Lathe, John Clark and collaborators at a seminar in Edinburgh in June 1985 (Lathe et al. 1986; Clark et al. 1987). Expression in milk has developed furthest (Clark 1998), but the idea has also been extended to blood and proposed for urine and seminal fluid (Dyck et al. 2003). In 2006 the first therapeutic product produced in milk gained regulatory approval by the European Medicines Agency, recombinant human antithrombin III 'Atryn' an anticoagulant produced by rEVO biologics (formerly GTC Biotherapeutics). This was an important milestone, but the field remains very small and has now largely been overtaken by advances in rival means of production such as chemical synthesis and bulk cell culture. To date, only one other transgenic milk-derived product has gained approval, C1 esterase inhibitor 'Ruconest' produced by Pharming for the treatment of a rare disease hereditary angio-oedema, approved by the European Medicines Agency in 2010 and by the US Food and Drug Administration in 2014. Throughout its history pharming in mammals has faced considerable commercial and regulatory obstacles, and there is a long list of abandoned projects. It is however worth mentioning that animal pharming did play a central motivating role in key advances in mammalian reproductive and transgenic technology, most notably somatic cell nuclear transfer in the form of Dolly the sheep (Wilmut et al. 1997) and her transgenic and gene-targeted successors (Schnieke et al. 1997; McCreath et al. 2000), which paved the way for most of the other large animal work described in this chapter.

For completeness we should also mention that chickens are also a useful species, with transgene expression directed into egg white (Sang 2006; Lillico et al. 2007). To date one such drug has gained regulatory approval, lysosomal acid lipase 'Kanuma' produced by Alexion and approved by the US Food and Drug Administration in 2009 for the treatment of rare hereditary lysosomal acid lipase deficiency.

Acknowledgements Our studies on the development of large animal models are supported by the German Research Council, the Federal Ministry for Education and Research, the Mildred Scheel Foundation of German Cancer Aid, the Bavarian Research Council and the Mukoviszidose Institut gemeinnützige Gesellschaft für Forschung und Therapieentwicklung mbH. The authors are members of EU COST Action BM1308.

References

Abbott A (2015) Inside the first pig biobank. Nature 519(7544):397–398. https://doi. org/10.1038/519397a

Adam SJ, Rund LA, Kuzmuk KN, Zachary JF, Schook LB, Counter CM (2007) Genetic induction of tumorigenesis in swine. Oncogene 26(7):1038–1045. https://doi.org/10.1038/sj.onc.1209892

Agarwala A, Billheimer J, Rader DJ (2013) Mighty minipig in fight against cardiovascular disease. Sci Transl Med 5(166):166fs161. https://doi.org/10.1126/scitranslmed.3005369

Ahrens HE, Petersen B, Herrmann D, Lucas-Hahn A, Hassel P, Ziegler M, Kues WA, Baulain U, Baars W, Schwinzer R, Denner J, Rataj D, Werwitzke S, Tiede A, Bongoni AK, Garimella PS, Despont A, Rieben R, Niemann H (2015) siRNA mediated knockdown of tissue factor expression in pigs for xenotransplantation. Am J Transplant 15(5):1407–1414. https://doi. org/10.1111/ajt.13120

Aigner B, Rathkolb B, Herbach N, Hrabe de Angelis M, Wanke R, Wolf E (2008) Diabetes models by screen for hyperglycemia in phenotype-driven ENU mouse mutagenesis projects. Am J Physiol Endocrinol Metab 294(2):E232–E240. https://doi.org/10.1152/ajpendo.00592.2007

Aigner B, Renner S, Kessler B, Klymiuk N, Kurome M, Wunsch A, Wolf E (2010) Transgenic pigs as models for translational biomedical research. J Mol Med 88(7):653–664. https://doi. org/10.1007/s00109-010-0610-9

Aikin RA (2012) How to kill two birds with one transgenic pig. Diabetes 61(6):1348–1349. https:// doi.org/10.2337/db12-0201

Alberg AJ, Lam AP, Helzlsouer KJ (1999) Epidemiology, prevention, and early detection of breast cancer. Curr Opin Oncol 11(6):435–441

Albl B, Haesner S, Braun-Reichhart C, Streckel E, Renner S, Seeliger F, Wolf E, Wanke R, Blutke A (2016) Tissue sampling guides for porcine biomedical models. Toxicol Pathol 44(3):414–420. https://doi.org/10.1177/0192623316631023

Al-Mashhadi RH, Sorensen CB, Kragh PM, Christoffersen C, Mortensen MB, Tolbod LP, Thim T, Du Y, Li J, Liu Y, Moldt B, Schmidt M, Vajta G, Larsen T, Purup S, Bolund L, Nielsen LB, Callesen H, Falk E, Mikkelsen JG, Bentzon JF (2013) Familial hypercholesterolemia and atherosclerosis in cloned minipigs created by DNA transposition of a human PCSK9 gain-of-function mutant. Sci ransl Med 5(166):166ra161. https://doi.org/10.1126/scitranslmed.3004853

Al-Mashhadi RH, Bjorklund MM, Mortensen MB, Christoffersen C, Larsen T, Falk E, Bentzon JF (2015) Diabetes with poor glycaemic control does not promote atherosclerosis in genetically modified hypercholesterolaemic minipigs. Diabetologia 58(8):1926–1936. https://doi. org/10.1007/s00125-015-3637-1

Alton EW, Armstrong DK, Ashby D, Bayfield KJ, Bilton D, Bloomfield EV, Boyd AC, Brand J, Buchan R, Calcedo R, Carvelli P, Chan M, Cheng SH, Collie DD, Cunningham S, Davidson

HE, Davies G, Davies JC, Davies LA, Dewar MH, Doherty A, Donovan J, Dwyer NS, Elgmati HI, Featherstone RF, Gavino J, Gea-Sorli S, Geddes DM, Gibson JS, Gill DR, Greening AP, Griesenbach U, Hansell DM, Harman K, Higgins TE, Hodges SL, Hyde SC, Hyndman L, Innes JA, Jacob J, Jones N, Keogh BF, Limberis MP, Lloyd-Evans P, Maclean AW, Manvell MC, McCormick D, McGovern M, McLachlan G, Meng C, Montero MA, Milligan H, Moyce LJ, Murray GD, Nicholson AG, Osadolor T, Parra-Leiton J, Porteous DJ, Pringle IA, Punch EK, Pytel KM, Quittner AL, Rivellini G, Saunders CJ, Scheule RK, Sheard S, Simmonds NJ, Smith K, Smith SN, Soussi N, Soussi S, Spearing EJ, Stevenson BJ, Sumner-Jones SG, Turkkila M, Ureta RP, Waller MD, Wasowicz MY, Wilson JM, Wolstenholme-Hogg P, Consortium UKCFGT (2015) Repeated nebulisation of non-viral CFTR gene therapy in patients with cystic fibrosis: a randomised, double-blind, placebo-controlled, phase 2b trial. Lancet Respir Med 3:684. https://doi.org/10.1016/S2213-2600(15)00245-3

American Diabetes Association (2013) Diagnosis and classification of diabetes mellitus. Diabetes Care 36(Suppl 1):S67–S74. https://doi.org/10.2337/dc13-S067

Anderson LJ, Jarrett WF (1968) Lymphosarcoma (leukemia) in cattle, sheep and pigs in Great Britain. Cancer 22(2):398–405

Araki E, Nakamura K, Nakao K, Kameya S, Kobayashi O, Nonaka I, Kobayashi T, Katsuki M (1997) Targeted disruption of exon 52 in the mouse dystrophin gene induced muscle Degeneration similar to that observed in Duchenne muscular dystrophy. Biochem Biophys Res Commun 238(2):492–497

Aron Badin R, Vadori M, Vanhove B, Nerriere-Daguin V, Naveilhan P, Neveu I, Jan C, Leveque X, Venturi E, Mermillod P, Van Camp N, Dolle F, Guillermier M, Denaro L, Manara R, Citton V, Simioni P, Zampieri P, D'Avella D, Rubello D, Fante F, Boldrin M, De Benedictis GM, Cavicchioli L, Sgarabotto D, Plebani M, Stefani AL, Brachet P, Blancho G, Soulillou JP, Hantraye P, Cozzi E (2016) Cell therapy for Parkinson's disease: a translational approach to assess the role of local and systemic immunosuppression. Am J Transplant 16:2016. https://doi.org/10.1111/ajt.13704

Baggio LL, Drucker DJ (2007) Biology of incretins: GLP-1 and GIP. Gastroenterology 132(6):2131–2157. https://doi.org/10.1053/j.gastro.2007.03.054

Bähr A, Wolf E (2012) Domestic animal models for biomedical research. Reprod Domest Anim 47(Suppl 4):59–71. https://doi.org/10.1111/j.1439-0531.2012.02056.x

Bartlett ST, Markmann JF, Johnson P, Korsgren O, Hering BJ, Scharp D, Kay TW, Bromberg J, Odorico JS, Weir GC, Bridges N, Kandaswamy R, Stock P, Friend P, Gotoh M, Cooper DK, Park CG, O'Connell P, Stabler C, Matsumoto S, Ludwig B, Choudhary P, Kovatchev B, Rickels MR, Sykes M, Wood K, Kraemer K, Hwa A, Stanley E, Ricordi C, Zimmerman M, Greenstein J, Montanya E, Otonkoski T (2016) Report from IPITA-TTS opinion leaders meeting on the future of beta-cell replacement. Transplantation 100(Suppl 2):S1–S44. https://doi.org/10.1097/tp.0000000000001055

Berthelsen MF, Callesen MM, Ostergaard TS, Liu Y, Li R, Callesen H, Dagnaes-Hansen F, Hamilton-Dutoit S, Jakobsen JE, Thomsen MK (2017) Pancreas specific expression of oncogenes in a porcine model. Transgenic Res 26(5):603–612. https://doi.org/10.1007/s11248-017-0031-4

Blutke A, Renner S, Flenkenthaler F, Backman M, Haesner S, Kemter E, Landstrom E, Braun-Reichhart C, Albl B, Streckel E, Rathkolb B, Prehn C, Palladini A, Grzybek M, Krebs S, Bauersachs S, Bahr A, Bruhschwein A, Deeg CA, De Monte E, Dmochewitz M, Eberle C, Emrich D, Fux R, Groth F, Gumbert S, Heitmann A, Hinrichs A, Kessler B, Kurome M, Leipig-Rudolph M, Matiasek K, Ozturk H, Otzdorff C, Reichenbach M, Reichenbach HD, Rieger A, Rieseberg B, Rosati M, Saucedo MN, Schleicher A, Schneider MR, Simmet K, Steinmetz J, Ubel N, Zehetmaier P, Jung A, Adamski J, Coskun U, Hrabe de Angelis M, Simmet C, Ritzmann M, Meyer-Lindenberg A, Blum H, Arnold GJ, Frohlich T, Wanke R, Wolf E (2017) The Munich MIDY Pig Biobank - a unique resource for studying organ crosstalk in diabetes. Mol Metabol 6(8):931–940. https://doi.org/10.1016/j.molmet.2017.06.004

Bogaert J, Prenen H (2014) Molecular genetics of colorectal cancer. Ann Gastroenterol 27(1):9–14

Bottino R, Wijkstrom M, van der Windt DJ, Hara H, Ezzelarab M, Murase N, Bertera S, He J, Phelps C, Ayares D, Cooper DK, Trucco M (2014) Pig-to-monkey islet xenotransplantation using multi-transgenic pigs. Am J Transplant 14(10):2275–2287. https://doi.org/10.1111/ajt.12868

van Boxtel R, Kuiper RV, Toonen PW, van Heesch S, Hermsen R, de Bruin A, Cuppen E (2011) Homozygous and heterozygous p53 knockout rats develop metastasizing sarcomas with high frequency. Am J Pathol 179(4):1616–1622. https://doi.org/10.1016/j.ajpath.2011.06.036

Bulato C, Radu C, Simioni P (2012) Studies on coagulation incompatibilities for xenotransplantation. Methods Mol Biol 885:71–89. https://doi.org/10.1007/978-1-61779-845-0_6

Burlak C, Paris LL, Lutz AJ, Sidner RA, Estrada J, Li P, Tector M, Tector AJ (2014) Reduced binding of human antibodies to cells from GGTA1/CMAH KO pigs. Am J Transplant 14(8):1895–1900. https://doi.org/10.1111/ajt.12744

Byrne GW, McGregor CG, Breimer ME (2015) Recent investigations into pig antigen and anti-pig antibody expression. Int J Surg 23(Pt B):223–228. https://doi.org/10.1016/j.ijsu.2015.07.724

Callejas D, Mann CJ, Ayuso E, Lage R, Grifoll I, Roca C, Andaluz A, Ruiz-de Gopegui R, Montane J, Munoz S, Ferre T, Haurigot V, Zhou S, Ruberte J, Mingozzi F, High KA, Garcia F, Bosch F (2013) Treatment of diabetes and long-term survival after insulin and glucokinase gene therapy. Diabetes 62(5):1718–1729. https://doi.org/10.2337/db12-1113

Callesen MM, Arnadottir SS, Lyskjaer I, Orntoft MW, Hoyer S, Dagnaes-Hansen F, Liu Y, Li R, Callesen H, Rasmussen MH, Berthelsen MF, Thomsen MK, Schweiger PJ, Jensen KB, Laurberg S, Orntoft TF, Elverlov-Jakobsen JE, Andersen CL (2017) A genetically inducible porcine model of intestinal cancer. Mol Oncol 11(11):1616–1629. https://doi.org/10.1002/1878-0261.12136

Cao H, Machuca TN, Yeung JC, Wu J, Du K, Duan C, Hashimoto K, Linacre V, Coates AL, Leung K, Wang J, Yeger H, Cutz E, Liu M, Keshavjee S, Hu J (2013) Efficient gene delivery to pig airway epithelia and submucosal glands using helper-dependent adenoviral vectors. Mol Ther Nucleic Acids 2:e127. https://doi.org/10.1038/mtna.2013.55

Chen JH, Stoltz DA, Karp PH, Ernst SE, Pezzulo AA, Moninger TO, Rector MV, Reznikov LR, Launspach JL, Chaloner K, Zabner J, Welsh MJ (2010) Loss of anion transport without increased sodium absorption characterizes newborn porcine cystic fibrosis airway epithelia. Cell 143(6):911–923. https://doi.org/10.1016/j.cell.2010.11.029

Chen Y, Zheng Y, Kang Y, Yang W, Niu Y, Guo X, Tu Z, Si C, Wang H, Xing R, Pu X, Yang SH, Li S, Ji W, Li XJ (2015) Functional disruption of the dystrophin gene in rhesus monkey using CRISPR/Cas9. Hum Mol Genet 24(13):3764–3774. https://doi.org/10.1093/hmg/ddv120

Clark AJ (1998) The mammary gland as a bioreactor: expression, processing, and production of recombinant proteins. J Mammary Gland Biol Neoplasia 3(3):337–350

Clark AJ, Simons P, Wilmut I, Lathe R (1987) Pharmaceuticals from transgenic livestock. Trends Biotechnol 5:20–24

Cooper DK, Ekser B, Ramsoondar J, Phelps C, Ayares D (2016) The role of genetically engineered pigs in xenotransplantation research. J Pathol 238(2):288–299. https://doi.org/10.1002/path.4635

Cowan PJ, Robson SC (2015) Progress towards overcoming coagulopathy and hemostatic dysfunction associated with xenotransplantation. Int J Surg 23(Pt B):296–300. https://doi.org/10.1016/j.ijsu.2015.07.682

Croner RS, Brueckl WM, Reingruber B, Hohenberger W, Guenther K (2005) Age and manifestation related symptoms in familial adenomatous polyposis. BMC Cancer 5:24. https://doi.org/10.1186/1471-2407-5-24

Cutting GR (2015) Cystic fibrosis genetics: from molecular understanding to clinical application. Nat Rev Genet 16(1):45–56. https://doi.org/10.1038/nrg3849

Davies K (2013) The era of genomic medicine. Clin Med 13(6):594–601. https://doi.org/10.7861/clinmedicine.13-6-594

Davis BT, Wang XJ, Rohret JA, Struzynski JT, Merricks EP, Bellinger DA, Rohret FA, Nichols TC, Rogers CS (2014) Targeted disruption of LDLR causes hypercholesterolemia and atherosclerosis in Yucatan miniature pigs. PLoS One 9(4):e93457. https://doi.org/10.1371/journal.pone.0093457

Dawson JR, Vidal AC, Malyguine AM (2000) Natural killer cell-endothelial cell interactions in xenotransplantation. Immunol Res 22(2-3):165–176. https://doi.org/10.1385/ir:22:2-3:165

Dell'agnello C, Leo S, Agostino A, Szabadkai G, Tiveron C, Zulian A, Prelle A, Roubertoux P, Rizzuto R, Zeviani M (2007) Increased longevity and refractoriness to Ca(2+)-dependent neurodegeneration in Surf1 knockout mice. Hum Mol Genet 16(4):431–444. https://doi.org/10.1093/hmg/ddl477

Deng S, Yu K, Zhang B, Yao Y, Liu Y, He H, Zhang H, Cui M, Fu J, Lian Z, Li N (2012) Effects of over-expression of TLR2 in transgenic goats on pathogen clearance and role of up-regulation of lysozyme secretion and infiltration of inflammatory cells. BMC Vet Res 8:196. https://doi.org/10.1186/1746-6148-8-196

Denner J (2013) Immunising with the transmembrane envelope proteins of different retroviruses including HIV-1: a comparative study. Hum Vaccin Immunother 9(3):462–470

Denner J (2017a) Advances in organ transplant from pigs. Science 357(6357):1238–1239. https://doi.org/10.1126/science.aao6334

Denner J (2017b) Paving the path toward porcine organs for transplantation. N Engl J Med 377(19):1891–1893. https://doi.org/10.1056/NEJMcibr1710853

Denner J, Tonjes RR (2012) Infection barriers to successful xenotransplantation focusing on porcine endogenous retroviruses. Clin Microbiol Rev 25(2):318–343. https://doi.org/10.1128/CMR.05011-11

Dieckhoff B, Petersen B, Kues WA, Kurth R, Niemann H, Denner J (2008) Knockdown of porcine endogenous retrovirus (PERV) expression by PERV-specific shRNA in transgenic pigs. Xenotransplantation 15(1):36–45. https://doi.org/10.1111/j.1399-3089.2008.00442.x

Dine J, Deng CX (2013) Mouse models of BRCA1 and their application to breast cancer research. Cancer Metastasis Rev 32(1-2):25–37. https://doi.org/10.1007/s10555-012-9403-7

Dufrane D, van Steenberghe M, Guiot Y, Goebbels RM, Saliez A, Gianello P (2006) Streptozotocin-induced diabetes in large animals (pigs/primates): role of GLUT2 transporter and beta-cell plasticity. Transplantation 81(1):36–45

Duranthon V, Beaujean N, Brunner M, Odening KE, Santos AN, Kacskovics I, Hiripi L, Weinstein EJ, Bosze Z (2012) On the emerging role of rabbit as human disease model and the instrumental role of novel transgenic tools. Transgenic Res 21(4):699–713. https://doi.org/10.1007/s11248-012-9599-x

Dyck MK, Lacroix D, Pothier F, Sirard MA (2003) Making recombinant proteins in animals-different systems, different applications. Trends Biotechnol 21(9):394–399. https://doi.org/10.1016/s0167-7799(03)00190-2

Ehrnhoefer DE, Butland SL, Pouladi MA, Hayden MR (2009) Mouse models of Huntington disease: variations on a theme. Dis Model Mech 2(3-4):123–129. https://doi.org/10.1242/dmm.002451

Ekser B, Cooper DK, Tector AJ (2015) The need for xenotransplantation as a source of organs and cells for clinical transplantation. Int J Surg 23:199. https://doi.org/10.1016/j.ijsu.2015.06.066

Escareno CE, Azagury DE, Flint RS, Nedder A, Thompson CC, Lautz DB (2012) Establishing a reproducible large animal survival model of laparoscopic Roux-en-Y gastric bypass. Surg Obes Relat Dis 8(6):764–769. https://doi.org/10.1016/j.soard.2011.05.021

Estrada JL, Martens G, Li P, Adams A, Newell KA, Ford ML, Butler JR, Sidner R, Tector M, Tector J (2015) Evaluation of human and non-human primate antibody binding to pig cells lacking GGTA1/CMAH/beta4GalNT2 genes. Xenotransplantation 22(3):194–202. https://doi.org/10.1111/xen.12161

Fairclough RJ, Wood MJ, Davies KE (2013) Therapy for Duchenne muscular dystrophy: renewed optimism from genetic approaches. Nat Rev Genet 14(6):373–378. https://doi.org/10.1038/nrg3460

Fan N, Lai L (2013) Genetically modified pig models for human diseases. J Genet Genom 40(2):67–73. https://doi.org/10.1016/j.jgg.2012.07.014

Fanen P, Wohlhuter-Haddad A, Hinzpeter A (2014) Genetics of cystic fibrosis: CFTR mutation classifications toward genotype-based CF therapies. Int J Biochem Cell Biol 52:94. https://doi.org/10.1016/j.biocel.2014.02.023

Farrell T, Neale L, Cundy T (2002) Congenital anomalies in the offspring of women with type 1, type 2 and gestational diabetes. Diabet Med 19(4):322–326

Fischer K, Kraner-Scheiber S, Petersen B, Rieblinger B, Buermann A, Flisikowska T, Flisikowski K, Christan S, Edlinger M, Baars W, Kurome M, Zakhartchenko V, Kessler B, Plotzki E, Szczerbal I, Switonski M, Denner J, Wolf E, Schwinzer R, Niemann H, Kind A, Schnieke A (2016) Efficient production of multi-modified pigs for xenotransplantation by 'combineering', gene stacking and gene editing. Sci Rep 6:29081. https://doi.org/10.1038/srep29081

Fisher LF, Olander HJ (1978) Spontaneous neoplasms of pigs - a study of 31 cases. J Comp Pathol 88(4):505–517

Fishman JA (2014) Assessment of infectious risk in clinical xenotransplantation: the lessons for clinical allotransplantation. Xenotransplantation 21(4):307–308. https://doi.org/10.1111/xen.12118

Fishman JA, Scobie L, Takeuchi Y (2012) Xenotransplantation-associated infectious risk: a WHO consultation. Xenotransplantation 19(2):72–81. https://doi.org/10.1111/j.1399-3089.2012.00693.x

Flisikowska T, Merkl C, Landmann M, Eser S, Rezaei N, Cui X, Kurome M, Zakhartchenko V, Kessler B, Wieland H, Rottmann O, Schmid RM, Schneider G, Kind A, Wolf E, Saur D, Schnieke A (2012) A porcine model of familial adenomatous polyposis. Gastroenterology 143(5):1173–1175.e1171-1177. https://doi.org/10.1053/j.gastro.2012.07.110

Flisikowska T, Kind A, Schnieke A (2013) The new pig on the block: modelling cancer in pigs. Transgenic Res 22(4):673–680. https://doi.org/10.1007/s11248-013-9720-9

Flisikowska T, Kind A, Schnieke A (2014) Genetically modified pigs to model human diseases. J Appl Genet 55(1):53–64. https://doi.org/10.1007/s13353-013-0182-9

Flisikowska T, Stachowiak M, Xu H, Wagner A, Hernandez-Caceres A, Wurmser C, Perleberg C, Pausch H, Perkowska A, Fischer K, Frishman D, Fries R, Switonski M, Kind A, Saur D, Schnieke A, Flisikowski K (2017) Porcine familial adenomatous polyposis model enables systematic analysis of early events in adenoma progression. Sci Rep 7(1):6613. https://doi.org/10.1038/s41598-017-06741-8

Flisikowski K, Flisikowska T, Sikorska A, Perkowska A, Kind A, Schnieke A, Switonski M (2017) Germline gene polymorphisms predisposing domestic mammals to carcinogenesis. Vet Comp Oncol 15:289. https://doi.org/10.1111/vco.12186

Frohlich T, Kemter E, Flenkenthaler F, Klymiuk N, Otte KA, Blutke A, Krause S, Walter MC, Wanke R, Wolf E, Arnold GJ (2016) Progressive muscle proteome changes in a clinically relevant pig model of Duchenne muscular dystrophy. Sci Rep 6:33362. https://doi.org/10.1038/srep33362

Futreal PA, Coin L, Marshall M, Down T, Hubbard T, Wooster R, Rahman N, Stratton MR (2004) A census of human cancer genes. Nat Rev Cancer 4(3):177–183. https://doi.org/10.1038/nrc1299

Getz GS, Reardon CA (2012) Animal models of atherosclerosis. Arterioscler Thromb Vasc Biol 32(5):1104–1115. https://doi.org/10.1161/ATVBAHA.111.237693

Goyenvalle A, Griffith G, Babbs A, El Andaloussi S, Ezzat K, Avril A, Dugovic B, Chaussenot R, Ferry A, Voit T, Amthor H, Buhr C, Schurch S, Wood MJ, Davies KE, Vaillend C, Leumann C, Garcia L (2015) Functional correction in mouse models of muscular dystrophy using exon-skipping tricyclo-DNA oligomers. Nat Med 21(3):270–275. https://doi.org/10.1038/nm.3765

Griesemer A, Yamada K, Sykes M (2014) Xenotransplantation: immunological hurdles and progress toward tolerance. Immunol Rev 258(1):241–258. https://doi.org/10.1111/imr.12152

Griesenbach U, Alton EW (2009) Cystic fibrosis gene therapy: successes, failures and hopes for the future. Expert Rev Respir Med 3(4):363–371. https://doi.org/10.1586/ers.09.25

Griffin MA, Restrepo MS, Abu-El-Haija M, Wallen T, Buchanan E, Rokhlina T, Chen YH, McCray PB Jr, Davidson BL, Divekar A, Uc A (2014) A novel gene delivery method transduces porcine pancreatic duct epithelial cells. Gene Ther 21(2):123–130. https://doi.org/10.1038/gt.2013.62

Groden J, Thliveris A, Samowitz W, Carlson M, Gelbert L, Albertsen H, Joslyn G, Stevens J, Spirio L, Robertson M et al (1991) Identification and characterization of the familial adenomatous polyposis coli gene. Cell 66(3):589–600

Groenen MA, Archibald AL, Uenishi H, Tuggle CK, Takeuchi Y, Rothschild MF, Rogel-Gaillard C, Park C, Milan D, Megens HJ, Li S, Larkin DM, Kim H, Frantz LA, Caccamo M, Ahn H, Aken BL, Anselmo A, Anthon C, Auvil L, Badaoui B, Beattie CW, Bendixen C, Berman D, Blecha F, Blomberg J, Bolund L, Bosse M, Botti S, Bujie Z, Bystrom M, Capitanu B, Carvalho-Silva D, Chardon P, Chen C, Cheng R, Choi SH, Chow W, Clark RC, Clee C, Crooijmans RP, Dawson HD, Dehais P, De Sapio F, Dibbits B, Drou N, Du ZQ, Eversole K, Fadista J, Fairley S, Faraut T, Faulkner GJ, Fowler KE, Fredholm M, Fritz E, Gilbert JG, Giuffra E, Gorodkin J, Griffin DK, Harrow JL, Hayward A, Howe K, Hu ZL, Humphray SJ, Hunt T, Hornshoj H, Jeon JT, Jern P, Jones M, Jurka J, Kanamori H, Kapetanovic R, Kim J, Kim JH, Kim KW, Kim TH, Larson G, Lee K, Lee KT, Leggett R, Lewin HA, Li Y, Liu W, Loveland JE, Lu Y, Lunney JK, Ma J, Madsen O, Mann K, Matthews L, McLaren S, Morozumi T, Murtaugh MP, Narayan J, Nguyen DT, Ni P, Oh SJ, Onteru S, Panitz F, Park EW, Park HS, Pascal G, Paudel Y, Perez-Enciso M, Ramirez-Gonzalez R, Reecy JM, Rodriguez-Zas S, Rohrer GA, Rund L, Sang Y, Schachtschneider K, Schraiber JG, Schwartz J, Scobie L, Scott C, Searle S, Servin B, Southey BR, Sperber G, Stadler P, Sweedler JV, Tafer H, Thomsen B, Wali R, Wang J, Wang J, White S, Xu X, Yerle M, Zhang G, Zhang J, Zhang J, Zhao S, Rogers J, Churcher C, Schook LB (2012) Analyses of pig genomes provide insight into porcine demography and evolution. Nature 491(7424):393–398. https://doi.org/10.1038/nature11622

Guell M, Niu D, Kan Y, George H, Wang T, Lee IH, Wang G, Church G, Yang L (2017) PERV inactivation is necessary to guarantee absence of pig-to-patient PERVs transmission in xenotransplantation. Xenotransplantation 24(6):e12366. https://doi.org/10.1111/xen.12366

Gui L, Qian H, Rocco KA, Grecu L, Niklason LE (2015) Efficient intratracheal delivery of airway epithelial cells in mice and pigs. Am J Physiol Lung Cell Mol Physiol 308(2):L221–L228. https://doi.org/10.1152/ajplung.00147.2014

Guijarro MV, Ghivizzani SC, Gibbs CP (2014) Animal models in osteosarcoma. Front Oncol 4:189. https://doi.org/10.3389/fonc.2014.00189

Handley RR, Reid SJ, Patassini S, Rudiger SR, Obolonkin V, McLaughlan CJ, Jacobsen JC, Gusella JF, MacDonald ME, Waldvogel HJ, Bawden CS, Faull RL, Snell RG (2016) Metabolic disruption identified in the Huntington's disease transgenic sheep model. Sci Rep 6:20681. https://doi.org/10.1038/srep20681

Hara S, Umeyama K, Yokoo T, Nagashima H, Nagata M (2014) Diffuse glomerular nodular lesions in diabetic pigs carrying a dominant-negative mutant hepatocyte nuclear factor 1-alpha, an inheritant diabetic gene in humans. PLoS One 9(3):e92219. https://doi.org/10.1371/journal.pone.0092219

Hauschild J, Petersen B, Santiago Y, Queisser AL, Carnwath JW, Lucas-Hahn A, Zhang L, Meng X, Gregory PD, Schwinzer R, Cost GJ, Niemann H (2011) Efficient generation of a biallelic knockout in pigs using zinc-finger nucleases. Proc Natl Acad Sci U S A 108(29):12013–12017. https://doi.org/10.1073/pnas.1106422108

Hermeren G (2015) Ethical considerations in chimera research. Development 142(1):3–5. https://doi.org/10.1242/dev.119024

Hinkel R, Howe A, Renner S, Ng J, Lee S, Klett K, Kaczmarek V, Moretti A, Laugwitz KL, Skroblin P, Mayr M, Milting H, Dendorfer A, Reichart B, Wolf E, Kupatt C (2017) Diabetes mellitus-induced microvascular destabilization in the myocardium. J Am Coll Cardiol 69(2):131–143. https://doi.org/10.1016/j.jacc.2016.10.058

Hoegger MJ, Awadalla M, Namati E, Itani OA, Fischer AJ, Tucker AJ, Adam RJ, McLennan G, Hoffman EA, Stoltz DA, Welsh MJ (2014a) Assessing mucociliary transport of single particles in vivo shows variable speed and preference for the ventral trachea in newborn pigs. Proc Natl Acad Sci U S A 111(6):2355–2360. https://doi.org/10.1073/pnas.1323633111

Hoegger MJ, Fischer AJ, McMenimen JD, Ostedgaard LS, Tucker AJ, Awadalla MA, Moninger TO, Michalski AS, Hoffman EA, Zabner J, Stoltz DA, Welsh MJ (2014b) Impaired mucus detachment disrupts mucociliary transport in a piglet model of cystic fibrosis. Science 345(6198):818–822. https://doi.org/10.1126/science.1255825

Hoffman EP, Brown RH Jr, Kunkel LM (1987) Dystrophin: the protein product of the Duchenne muscular dystrophy locus. Cell 51(6):919–928 0092-8674(87)90579-4 [pii]

Itani OA, Chen JH, Karp PH, Ernst S, Keshavjee S, Parekh K, Klesney-Tait J, Zabner J, Welsh MJ (2011) Human cystic fibrosis airway epithelia have reduced Cl- conductance but not increased Na+ conductance. Proc Natl Acad Sci U S A 108(25):10260–10265. https://doi.org/10.1073/pnas.1106695108

Ito Y, Azrolan N, O'Connell A, Walsh A, Breslow JL (1990) Hypertriglyceridemia as a result of human apo CIII gene expression in transgenic mice. Science 249(4970):790–793

Iwase H, Kobayashi T (2015) Current status of pig kidney xenotransplantation. Int J Surg 23(Pt B):229–233. https://doi.org/10.1016/j.ijsu.2015.07.721

Iwase H, Ekser B, Hara H, Phelps C, Ayares D, Cooper DK, Ezzelarab MB (2014) Regulation of human platelet aggregation by genetically modified pig endothelial cells and thrombin inhibition. Xenotransplantation 21(1):72–83. https://doi.org/10.1111/xen.12073

Jacks T, Remington L, Williams BO, Schmitt EM, Halachmi S, Bronson RT, Weinberg RA (1994) Tumor spectrum analysis in p53-mutant mice. Curr Biol 4(1):1–7

Jacobsen JC, Bawden CS, Rudiger SR, McLaughlan CJ, Reid SJ, Waldvogel HJ, MacDonald ME, Gusella JF, Walker SK, Kelly JM, Webb GC, Faull RL, Rees MI, Snell RG (2010) An ovine transgenic Huntington's disease model. Hum Mol Genet 19(10):1873–1882. https://doi.org/10.1093/hmg/ddq063

Ji H, Li X, Yue S, Li J, Chen H, Zhang Z, Ma B, Wang J, Pu M, Zhou L, Feng C, Wang D, Duan J, Pan D, Tao K, Dou K (2015) Pig BMSCs transfected with human TFPI combat species incompatibility and regulate the human TF pathway in vitro and in a rodent model. Cell Physiol Biochem 36(1):233–249. https://doi.org/10.1159/000374067

Karnoub AE, Weinberg RA (2008) Ras oncogenes: split personalities. Nat Rev Mol Cell Biol 9(7):517–531. https://doi.org/10.1038/nrm2438

Kemter E, Wolf E (2015) Pigs pave a way to de novo formation of functional human kidneys. Proc Natl Acad Sci U S A 112(42):12905–12906. https://doi.org/10.1073/pnas.1517582112

Kemter E, Lieke T, Kessler B, Kurome M, Wuensch A, Summerfield A, Ayares D, Nagashima H, Baars W, Schwinzer R, Wolf E (2012) Human TNF-related apoptosis-inducing ligand-expressing dendritic cells from transgenic pigs attenuate human xenogeneic T cell responses. Xenotransplantation 19(1):40–51. https://doi.org/10.1111/j.1399-3089.2011.00688.x

Kendall SD, Linardic CM, Adam SJ, Counter CM (2005) A network of genetic events sufficient to convert normal human cells to a tumorigenic state. Cancer Res 65(21):9824–9828. https://doi.org/10.1158/0008-5472.can-05-1543

Kim MK, Hara H (2015) Current status of corneal xenotransplantation. Int J Surg 23(Pt B):255–260. https://doi.org/10.1016/j.ijsu.2015.07.685

Kim TY, Kunitomo Y, Pfeiffer Z, Patel D, Hwang J, Harrison K, Patel B, Jeng P, Ziv O, Lu Y, Peng X, Qu Z, Koren G, Choi BR (2015) Complex excitation dynamics underlie polymorphic ventricular tachycardia in a transgenic rabbit model of long QT syndrome type 1. Heart Rhythm 12(1):220–228. https://doi.org/10.1016/j.hrthm.2014.10.003

Kinzler KW, Nilbert MC, Su LK, Vogelstein B, Bryan TM, Levy DB, Smith KJ, Preisinger AC, Hedge P, McKechnie D et al (1991) Identification of FAP locus genes from chromosome 5q21. Science 253(5020):661–665

Kleinert M, Clemmensen C, Hofmann SM, Moore MC, Renner S, Woods SC, Huypens P, Beckers J, de Angelis MH, Schurmann A, Bakhti M, Klingenspor M, Heiman M, Cherrington AD, Ristow M, Lickert H, Wolf E, Havel PJ, Muller TD, Tschop MH (2018) Animal models of obesity and diabetes mellitus. Nat Rev Endocrinol 14:140. https://doi.org/10.1038/nrendo.2017.161

Kleinwort KJH, Amann B, Hauck SM, Hirmer S, Blutke A, Renner S, Uhl PB, Lutterberg K, Sekundo W, Wolf E, Deeg CA (2017) Retinopathy with central oedema in an INS (C94Y) transgenic pig model of long-term diabetes. Diabetologia 60(8):1541–1549. https://doi.org/10.1007/s00125-017-4290-7

Klose R, Kemter E, Bedke T, Bittmann I, Kelsser B, Endres R, Pfeffer K, Schwinzer R, Wolf E (2005) Expression of biologically active human TRAIL in transgenic pigs. Transplantation 80(2):222–230

Klymiuk N, Aigner B, Brem G, Wolf E (2010) Genetic modification of pigs as organ donors for xenotransplantation. Mol Reprod Dev 77(3):209–221. https://doi.org/10.1002/mrd.21127

Klymiuk N, Mundhenk L, Kraehe K, Wuensch A, Plog S, Emrich D, Langenmayer MC, Stehr M, Holzinger A, Kroner C, Richter A, Kessler B, Kurome M, Eddicks M, Nagashima H, Heinritzi K, Gruber AD, Wolf E (2012a) Sequential targeting of CFTR by BAC vectors generates a novel pig model of cystic fibrosis. J Mol Med 90(5):597–608. https://doi.org/10.1007/s00109-011-0839-y

Klymiuk N, van Buerck L, Bahr A, Offers M, Kessler B, Wuensch A, Kurome M, Thormann M, Lochner K, Nagashima H, Herbach N, Wanke R, Seissler J, Wolf E (2012b) Xenografted islet cell clusters from INSLEA29Y transgenic pigs rescue diabetes and prevent immune rejection in humanized mice. Diabetes 61(6):1527–1532. https://doi.org/10.2337/db11-1325

Klymiuk N, Blutke A, Graf A, Krause S, Burkhardt K, Wuensch A, Krebs S, Kessler B, Zakhartchenko V, Kurome M, Kemter E, Nagashima H, Schoser B, Herbach N, Blum H, Wanke R, Aartsma-Rus A, Thirion C, Lochmuller H, Walter MC, Wolf E (2013) Dystrophin-deficient pigs provide new insights into the hierarchy of physiological derangements of dystrophic muscle. Hum Mol Genet 22(21):4368–4382. https://doi.org/10.1093/hmg/ddt287

Klymiuk N, Ludwig B, Seissler J, Reichart B, Wolf E (2016a) Current concepts of using pigs as a source for beta-cell replacement therapy of type 1 diabetes. Curr Mol Bio Rep 2:73. https://doi.org/10.1007/s40610-016-0039-1

Klymiuk N, Seeliger F, Bohlooly YM, Blutke A, Rudmann DG, Wolf E (2016b) Tailored pig models for preclinical efficacy and safety testing of targeted therapies. Toxicol Pathol 44(3):346–357. https://doi.org/10.1177/0192623315609688

Kobayashi T, Yamaguchi T, Hamanaka S, Kato-Itoh M, Yamazaki Y, Ibata M, Sato H, Lee YS, Usui J, Knisely AS, Hirabayashi M, Nakauchi H (2010) Generation of rat pancreas in mouse by interspecific blastocyst injection of pluripotent stem cells. Cell 142(5):787–799. https://doi.org/10.1016/j.cell.2010.07.039

Kobayashi T, Kato-Itoh M, Nakauchi H (2015) Targeted organ generation using Mixl1-inducible mouse pluripotent stem cells in blastocyst complementation. Stem Cells Dev 24(2):182–189. https://doi.org/10.1089/scd.2014.0270

Koike T, Kitajima S, Yu Y, Li Y, Nishijima K, Liu E, Sun H, Waqar AB, Shibata N, Inoue T, Wang Y, Zhang B, Kobayashi J, Morimoto M, Saku K, Watanabe T, Fan J (2009) Expression of human apoAII in transgenic rabbits leads to dyslipidemia: a new model for combined hyperlipidemia. Arterioscler Thromb Vasc Biol 29(12):2047–2053. https://doi.org/10.1161/atvbaha.109.190264

Kourtzelis I, Magnusson PU, Kotlabova K, Lambris JD, Chavakis T (2015) Regulation of instant blood mediated inflammatory reaction (IBMIR) in pancreatic islet xeno-transplantation: points for therapeutic interventions. Adv Exp Med Biol 865:171–188. https://doi.org/10.1007/978-3-319-18603-0_11

Kwon DN, Lee K, Kang MJ, Choi YJ, Park C, Whyte JJ, Brown AN, Kim JH, Samuel M, Mao J, Park KW, Murphy CN, Prather RS, Kim JH (2013) Production of biallelic CMP-Neu5Ac hydroxylase knock-out pigs. Sci Rep 3:1981. https://doi.org/10.1038/srep01981

Larcher T, Lafoux A, Tesson L, Remy S, Thepenier V, Francois V, Le Guiner C, Goubin H, Dutilleul M, Guigand L, Toumaniantz G, De Cian A, Boix C, Renaud JB, Cherel Y, Giovannangeli C, Concordet JP, Anegon I, Huchet C (2014) Characterization of dystrophin deficient rats: a new model for Duchenne muscular dystrophy. PLoS One 9(10):e110371. https://doi.org/10.1371/journal.pone.0110371

Lathe R, Clark AJ, Archibald AL, Bishop JO, Simons P, Wilmut I (1986) Novel products from livestock. In: Smith C, King J, Mckay J (eds) Exploiting new technologies in animal breeding: genetic developments. Clarendon Press, Oxford, pp 91–102

Le Bas-Bernardet S, Tillou X, Branchereau J, Dilek N, Poirier N, Chatelais M, Charreau B, Minault D, Hervouet J, Renaudin K, Crossan C, Scobie L, Takeuchi Y, Diswall M, Breimer ME, Klar N, Daha MR, Simioni P, Robson SC, Nottle MB, Salvaris EJ, Cowan PJ, d'Apice AJ, Sachs DH, Yamada K, Lagutina I, Duchi R, Perota A, Lazzari G, Galli C, Cozzi E, Soulillou JP, Vanhove B, Blancho G (2015) Bortezomib, C1-inhibitor and plasma exchange do not prolong the survival of multi-transgenic GalT-KO pig kidney xenografts in baboons. Am J Transplant 15(2):358–370. https://doi.org/10.1111/ajt.12988

Lee HJ, Lee BC, Kim YH, Paik NW, Rho HM (2011) Characterization of transgenic pigs that express human decay accelerating factor and cell membrane-tethered human tissue factor pathway inhibitor. Reprod Domestic Anim 46(2):325–332. https://doi.org/10.1111/j.1439-0531.2010.01670.x

Leuchs S, Saalfrank A, Merkl C, Flisikowska T, Edlinger M, Durkovic M, Rezaei N, Kurome M, Zakhartchenko V, Kessler B, Flisikowski K, Kind A, Wolf E, Schnieke A (2012) Inactivation and inducible oncogenic mutation of p53 in gene targeted pigs. PLoS One 7(10):e43323. https://doi.org/10.1371/journal.pone.0043323

Li S, Flisikowska T, Kurome M, Zakhartchenko V, Kessler B, Saur D, Kind A, Wolf E, Flisikowski K, Schnieke A (2014) Dual fluorescent reporter pig for Cre recombination: transgene placement at the ROSA26 locus. PLoS One 9(7):e102455. https://doi.org/10.1371/journal.pone.0102455

Li S, Edlinger M, Saalfrank A, Flisikowski K, Tschukes A, Kurome M, Zakhartchenko V, Kessler B, Saur D, Kind A, Wolf E, Schnieke A, Flisikowska T (2015) Viable pigs with a conditionally-activated oncogenic KRAS mutation. Transgenic Res 24(3):509–517. https://doi.org/10.1007/s11248-015-9866-8

Li Y, Fuchimoto D, Sudo M, Haruta H, Lin QF, Takayama T, Morita S, Nochi T, Suzuki S, Sembon S, Nakai M, Kojima M, Iwamoto M, Hashimoto M, Yoda S, Kunimoto S, Hiro T, Matsumoto T, Mitsumata M, Sugitani M, Saito S, Hirayama A, Onishi A (2016) Development of human-like advanced coronary plaques in low-density lipoprotein receptor knockout pigs and justification for statin treatment before formation of atherosclerotic plaques. J Am Heart Assoc 5(4):e002779. https://doi.org/10.1161/jaha.115.002779

Lillico SG, Sherman A, McGrew MJ, Robertson CD, Smith J, Haslam C, Barnard P, Radcliffe PA, Mitrophanous KA, Elliot EA, Sang HM (2007) Oviduct-specific expression of two therapeutic proteins in transgenic hens. Proc Natl Acad Sci U S A 104(6):1771–1776. https://doi.org/10.1073/pnas.0610401104

Liu M, Hodish I, Haataja L, Lara-Lemus R, Rajpal G, Wright J, Arvan P (2010) Proinsulin misfolding and diabetes: mutant INS gene-induced diabetes of youth. Trends Endocrinol Metab 21(11):652–659. https://doi.org/10.1016/j.tem.2010.07.001

Long C, Amoasii L, Mireault AA, McAnally JR, Li H, Sanchez-Ortiz E, Bhattacharyya S, Shelton JM, Bassel-Duby R, Olson EN (2016) Postnatal genome editing partially restores dystrophin expression in a mouse model of muscular dystrophy. Science 351(6271):400–403. https://doi.org/10.1126/science.aad5725

Luo Y, Li J, Liu Y, Lin L, Du Y, Li S, Yang H, Vajta G, Callesen H, Bolund L, Sorensen CB (2011) High efficiency of BRCA1 knockout using rAAV-mediated gene targeting: developing a pig model for breast cancer. Transgenic Res 20(5):975–988. https://doi.org/10.1007/s11248-010-9472-8

Luo Y, Lin L, Bolund L, Jensen TG, Sorensen CB (2012) Genetically modified pigs for biomedical research. J Inherit Metab Dis 35(4):695–713. https://doi.org/10.1007/s10545-012-9475-0

Lutz AJ, Li P, Estrada JL, Sidner RA, Chihara RK, Downey SM, Burlak C, Wang ZY, Reyes LM, Ivary B, Yin F, Blankenship RL, Paris LL, Tector AJ (2013) Double knockout pigs deficient in N-glycolylneuraminic acid and galactose alpha-1,3-galactose reduce the humoral barrier to xenotransplantation. Xenotransplantation 20(1):27–35. https://doi.org/10.1111/xen.12019

Masaki H, Kato-Itoh M, Umino A, Sato H, Hamanaka S, Kobayashi T, Yamaguchi T, Nishimura K, Ohtaka M, Nakanishi M, Nakauchi H (2015) Interspecific in vitro assay for the chimera-forming ability of human pluripotent stem cells. Development 142(18):3222–3230. https://doi.org/10.1242/dev.124016

Matsunari H, Nagashima H, Watanabe M, Umeyama K, Nakano K, Nagaya M, Kobayashi T, Yamaguchi T, Sumazaki R, Herzenberg LA, Nakauchi H (2013) Blastocyst complementation generates exogenic pancreas in vivo in apancreatic cloned pigs. Proc Natl Acad Sci U S A 110(12):4557–4562. https://doi.org/10.1073/pnas.1222902110

Matsunari H, Watanabe M, Nakano K, Enosawa S, Umeyama K, Uchikura A, Yashima S, Fukuda T, Klymiuk N, Kurome M, Kessler B, Wuensch A, Zakhartchenko V, Wolf E, Hanazono Y, Nagaya M, Umezawa A, Nakauchi H, Nagashima H (2018) Modeling lethal X-linked genetic disorders in pigs with ensured fertility. Proc Natl Acad Sci U S A 115(4):708–713. https://doi.org/10.1073/pnas.1715940115

McCalla-Martin AC, Chen X, Linder KE, Estrada JL, Piedrahita JA (2010) Varying phenotypes in swine versus murine transgenic models constitutively expressing the same human Sonic hedgehog transcriptional activator, K5-HGLl2 Delta N. Transgenic Res 19(5):869–887. https://doi.org/10.1007/s11248-010-9362-0

McCreath KJ, Howcroft J, Campbell KH, Colman A, Schnieke AE, Kind AJ (2000) Production of gene-targeted sheep by nuclear transfer from cultured somatic cells. Nature 405(6790):1066–1069. https://doi.org/10.1038/35016604

McGreevy JW, Hakim CH, McIntosh MA, Duan D (2015) Animal models of Duchenne muscular dystrophy: from basic mechanisms to gene therapy. Dis Model Mech 8(3):195–213. https://doi.org/10.1242/dmm.018424

Meyerholz DK, Stoltz DA, Namati E, Ramachandran S, Pezzulo AA, Smith AR, Rector MV, Suter MJ, Kao S, McLennan G, Tearney GJ, Zabner J, McCray PB Jr, Welsh MJ (2010) Loss of cystic fibrosis transmembrane conductance regulator function produces abnormalities in tracheal development in neonatal pigs and young children. Am J Respir Crit Care Med 182(10):1251–1261. https://doi.org/10.1164/rccm.201004-0643OC

Mirabello L, Troisi RJ, Savage SA (2009) Osteosarcoma incidence and survival rates from 1973 to 2004: data from the Surveillance, Epidemiology, and End Results Program. Cancer 115(7):1531–1543. https://doi.org/10.1002/cncr.24121

Miyagawa S, Matsunari H, Watanabe M, Nakano K, Umeyama K, Sakai R, Takayanagi S, Takeishi T, Fukuda T, Yashima S, Maeda A, Eguchi H, Okuyama H, Nagaya M, Nagashima H (2015) Generation of alpha1,3-galactosyltransferase and cytidine monophospho-N-acetylneuraminic acid hydroxylase gene double-knockout pigs. J Reprod Dev 61(5):449–457. https://doi.org/10.1262/jrd.2015-058

Mohiuddin MM, Reichart B, Byrne GW, McGregor CG (2015) Current status of pig heart xenotransplantation. Int J Surg 23(Pt B):234–239. https://doi.org/10.1016/j.ijsu.2015.08.038

Mohiuddin MM, Singh AK, Corcoran PC, Thomas ML 3rd, Clark T, Lewis BG, Hoyt RF, Eckhaus M, Pierson RN 3rd, Belli AJ, Wolf E, Klymiuk N, Phelps C, Reimann KA, Ayares D, Horvath KA (2016) Chimeric 2C10R4 anti-CD40 antibody therapy is critical for long-term survival of GTKO.hCD46.hTBM pig-to-primate cardiac xenograft. Nature Commun 7:11138. https://doi.org/10.1038/ncomms11138

Morozov VA, Morozov AV, Rotem A, Barkai U, Bornstein S, Denner J (2015) Extended microbiological characterization of Göttingen minipigs in the context of xenotransplantation: detection and vertical transmission of hepatitis E virus. PLoS One 10(10):e0139893. https://doi.org/10.1371/journal.pone.0139893

Mueller NJ, Takeuchi Y, Mattiuzzo G, Scobie L (2011) Microbial safety in xenotransplantation. Curr Opin Organ Transplant 16(2):201–206. https://doi.org/10.1097/MOT.0b013e32834486f6

Mulder A, Kardol MJ, Arn JS, Eijsink C, Franke ME, Schreuder GM, Haasnoot GW, Doxiadis II, Sachs DH, Smith DM, Claas FH (2010) Human monoclonal HLA antibodies reveal interspecies crossreactive swine MHC class I epitopes relevant for xenotransplantation. Mol Immunol 47(4):809–815. https://doi.org/10.1016/j.molimm.2009.10.004

Nagashima H, Matsunari H (2016) Growing human organs in pigs-a dream or reality? Theriogenology 86(1):422–426. https://doi.org/10.1016/j.theriogenology.2016.04.056

Nakamura A, Takeda S (2011) Mammalian models of Duchenne Muscular Dystrophy: pathological characteristics and therapeutic applications. J Biomed Biotechnol 2011:184393. https://doi.org/10.1155/2011/184393

Nakamura K, Fujii W, Tsuboi M, Tanihata J, Teramoto N, Takeuchi S, Naito K, Yamanouchi K, Nishihara M (2014) Generation of muscular dystrophy model rats with a CRISPR/Cas system. Sci Rep 4:5635. https://doi.org/10.1038/srep05635

Nauck MA, Baller B, Meier JJ (2004) Gastric inhibitory polypeptide and glucagon-like peptide-1 in the pathogenesis of type 2 diabetes. Diabetes 53(Suppl 3):S190–S196

Nelson CE, Hakim CH, Ousterout DG, Thakore PI, Moreb EA, Castellanos Rivera RM, Madhavan S, Pan X, Ran FA, Yan WX, Asokan A, Zhang F, Duan D, Gersbach CA (2016) In vivo genome editing improves muscle function in a mouse model of Duchenne muscular dystrophy. Science 351(6271):403–407. https://doi.org/10.1126/science.aad5143

Niimi M, Yang D, Kitajima S, Ning B, Wang C, Li S, Liu E, Zhang J, Eugene Chen Y, Fan J (2016) ApoE knockout rabbits: a novel model for the study of human hyperlipidemia. Atherosclerosis 245:187–193. https://doi.org/10.1016/j.atherosclerosis.2015.12.002

Niu D, Wei HJ, Lin L, George H, Wang T, Lee IH, Zhao HY, Wang Y, Kan Y, Shrock E, Lesha E, Wang G, Luo Y, Qing Y, Jiao D, Zhao H, Zhou X, Wang S, Wei H, Guell M, Church GM, Yang L (2017) Inactivation of porcine endogenous retrovirus in pigs using CRISPR-Cas9. Science 357(6357):1303–1307. https://doi.org/10.1126/science.aan4187

Nyunt O, Wu JY, McGown IN, Harris M, Huynh T, Leong GM, Cowley DM, Cotterill AM (2009) Investigating maturity onset diabetes of the young. Clin Biochem Rev 30(2):67–74

Oropeza M, Petersen B, Carnwath JW, Lucas-Hahn A, Lemme E, Hassel P, Herrmann D, Barg-Kues B, Holler S, Queisser AL, Schwinzer R, Hinkel R, Kupatt C, Niemann H (2009) Transgenic expression of the human A20 gene in cloned pigs provides protection against apoptotic and inflammatory stimuli. Xenotransplantation 16(6):522–534. https://doi.org/10.1111/j.1399-3089.2009.00556.x

Ostedgaard LS, Meyerholz DK, Chen JH, Pezzulo AA, Karp PH, Rokhlina T, Ernst SE, Hanfland RA, Reznikov LR, Ludwig PS, Rogan MP, Davis GJ, Dohrn CL, Wohlford-Lenane C, Taft PJ, Rector MV, Hornick E, Nassar BS, Samuel M, Zhang Y, Richter SS, Uc A, Shilyansky J, Prather RS, McCray PB Jr, Zabner J, Welsh MJ, Stoltz DA (2011) The DeltaF508 mutation causes CFTR misprocessing and cystic fibrosis-like disease in pigs. Sci Transl Med 3(74):74ra24. https://doi.org/10.1126/scitranslmed.3001868

Ozawa M, Himaki T, Ookutsu S, Mizobe Y, Ogawa J, Miyoshi K, Yabuki A, Fan J, Yoshida M (2015) Production of cloned miniature pigs expressing high levels of human apolipoprotein(a) in plasma. PLoS One 10(7):e0132155. https://doi.org/10.1371/journal.pone.0132155

Paris LL, Estrada JL, Li P, Blankenship RL, Sidner RA, Reyes LM, Montgomery JB, Burlak C, Butler JR, Downey SM, Wang ZY, Tector M, Tector AJ (2015) Reduced human platelet uptake by pig livers deficient in the asialoglycoprotein receptor 1 protein. Xenotransplantation 22(3):203–210. https://doi.org/10.1111/xen.12164

Park CG, Bottino R, Hawthorne WJ (2015) Current status of islet xenotransplantation. Int J Surg 23(Pt B):261–266. https://doi.org/10.1016/j.ijsu.2015.07.703

Perkel JM (2016) Xenotransplantation makes a comeback. Nat Biotechnol 34(1):3–4. https://doi.org/10.1038/nbt0116-3

Petersen B, Ramackers W, Tiede A, Lucas-Hahn A, Herrmann D, Barg-Kues B, Schuettler W, Friedrich L, Schwinzer R, Winkler M, Niemann H (2009) Pigs transgenic for human thrombomodulin have elevated production of activated protein C. Xenotransplantation 16(6):486–495. https://doi.org/10.1111/j.1399-3089.2009.00537.x

Petersen B, Ramackers W, Lucas-Hahn A, Lemme E, Hassel P, Queisser AL, Herrmann D, Barg-Kues B, Carnwath JW, Klose J, Tiede A, Friedrich L, Baars W, Schwinzer R, Winkler M, Niemann H (2011) Transgenic expression of human heme oxygenase-1 in pigs confers resistance against xenograft rejection during ex vivo perfusion of porcine kidneys. Xenotransplantation 18(6):355–368. https://doi.org/10.1111/j.1399-3089.2011.00674.x

Pezzulo AA, Tang XX, Hoegger MJ, Ramachandran S, Moninger TO, Karp PH, Wohlford-Lenane CL, Haagsman HP, van Eijk M, Banfi B, Horswill AR, Stoltz DA, McCray PB Jr, Welsh MJ, Zabner J (2012) Reduced airway surface pH impairs bacterial killing in the porcine cystic fibrosis lung. Nature 487(7405):109–113. https://doi.org/10.1038/nature11130

Phelps CJ, Koike C, Vaught TD, Boone J, Wells KD, Chen SH, Ball S, Specht SM, Polejaeva IA, Monahan JA, Jobst PM, Sharma SB, Lamborn AE, Garst AS, Moore M, Demetris AJ, Rudert WA, Bottino R, Bertera S, Trucco M, Starzl TE, Dai Y, Ayares DL (2003) Production of alpha 1,3-galactosyltransferase-deficient pigs. Science 299(5605):411–414. https://doi.org/10.1126/science.1078942

Pintore L, Paltrinieri S, Vadori M, Besenzon F, Cavicchioli L, De Benedictis GM, Calabrese F, Cozzi E, Nottle MB, Robson SC, Cowan PJ, Castagnaro M (2013) Clinicopathological findings in non-human primate recipients of porcine renal xenografts: quantitative and qualitative evaluation of proteinuria. Xenotransplantation 20(6):449–457. https://doi.org/10.1111/xen.12063

Plum L, Wunderlich FT, Baudler S, Krone W, Bruning JC (2005) Transgenic and knockout mice in diabetes research: novel insights into pathophysiology, limitations, and perspectives. Physiology 20:152–161. https://doi.org/10.1152/physiol.00049.2004

Polejaeva IA, Ranjan R, Davies CJ, Regouski M, Hall J, Olsen AL, Meng Q, Rutigliano HM, Dosdall DJ, Angel NA, Sachse FB, Seidel T, Thomas AJ, Stott R, Panter KE, Lee PM, Van Wettere AJ, Stevens JR, Wang Z, Macleod RS, Marrouche NF, White KL (2016) Increased susceptibility to atrial fibrillation secondary to atrial fibrosis in transgenic goats expressing transforming growth factor-beta1. J Cardiovasc Electrophysiol 27:1220. https://doi.org/10.1111/jce.13049

Powell SM, Zilz N, Beazer-Barclay Y, Bryan TM, Hamilton SR, Thibodeau SN, Vogelstein B, Kinzler KW (1992) APC mutations occur early during colorectal tumorigenesis. Nature 359(6392):235–237. https://doi.org/10.1038/359235a0

Prickett M, Jain M (2013) Gene therapy in cystic fibrosis. Transl Res 161(4):255–264. https://doi.org/10.1016/j.trsl.2012.12.001

Pylayeva-Gupta Y, Grabocka E, Bar-Sagi D (2011) RAS oncogenes: weaving a tumorigenic web. Nat Rev Cancer 11(11):761–774. https://doi.org/10.1038/nrc3106

Rader DJ, Cohen J, Hobbs HH (2003) Monogenic hypercholesterolemia: new insights in pathogenesis and treatment. J Clin Invest 111(12):1795–1803. https://doi.org/10.1172/JCI18925

Rahalkar AR, Hegele RA (2008) Monogenic pediatric dyslipidemias: classification, genetics and clinical spectrum. Mol Genet Metab 93(3):282–294. https://doi.org/10.1016/j.ymgme.2007.10.007

Ramsoondar J, Vaught T, Ball S, Mendicino M, Monahan J, Jobst P, Vance A, Duncan J, Wells K, Ayares D (2009) Production of transgenic pigs that express porcine endogenous retrovirus small interfering RNAs. Xenotransplantation 16(3):164–180. https://doi.org/10.1111/j.1399-3089.2009.00525.x

Rangarajan A, Hong SJ, Gifford A, Weinberg RA (2004) Species- and cell type-specific requirements for cellular transformation. Cancer Cell 6(2):171–183. https://doi.org/10.1016/j.ccr.2004.07.009

Reid SJ, Patassini S, Handley RR, Rudiger SR, McLaughlan CJ, Osmand A, Jacobsen JC, Morton AJ, Weiss A, Waldvogel HJ, MacDonald ME, Gusella JF, Bawden CS, Faull RL, Snell RG (2013) Further molecular characterisation of the OVT73 transgenic sheep model of Huntington's disease identifies cortical aggregates. J Huntington's Dis 2(3):279–295. https://doi.org/10.3233/jhd-130067

Renner S, Fehlings C, Herbach N, Hofmann A, von Waldthausen DC, Kessler B, Ulrichs K, Chodnevskaja I, Moskalenko V, Amselgruber W, Goke B, Pfeifer A, Wanke R, Wolf E (2010) Glucose intolerance and reduced proliferation of pancreatic beta-cells in transgenic pigs with impaired glucose-dependent insulinotropic polypeptide function. Diabetes 59(5):1228–1238. https://doi.org/10.2337/db09-0519

Renner S, Romisch-Margl W, Prehn C, Krebs S, Adamski J, Goke B, Blum H, Suhre K, Roscher AA, Wolf E (2012) Changing metabolic signatures of amino acids and lipids during the prediabetic period in a pig model with impaired incretin function and reduced beta-cell mass. Diabetes 61(8):2166–2175. https://doi.org/10.2337/db11-1133

Renner S, Braun-Reichhart C, Blutke A, Herbach N, Emrich D, Streckel E, Wunsch A, Kessler B, Kurome M, Bahr A, Klymiuk N, Krebs S, Puk O, Nagashima H, Graw J, Blum H, Wanke R, Wolf E (2013) Permanent neonatal diabetes in INS^{C94Y} transgenic pigs. Diabetes 62(5):1505–1511. https://doi.org/10.2337/db12-1065

Renner S, Blutke A, Streckel E, Wanke R, Wolf E (2016a) Incretin actions and consequences of incretin-based therapies: lessons from complementary animal models. J Pathol 238(2):345–358. https://doi.org/10.1002/path.4655

Renner S, Dobenecker B, Blutke A, Zols S, Wanke R, Ritzmann M, Wolf E (2016b) Comparative aspects of rodent and nonrodent animal models for mechanistic and translational diabetes research. Theriogenology 86(1):406–421. https://doi.org/10.1016/j.theriogenology.2016.04.055

Reyes LM, Estrada JL, Wang ZY, Blosser RJ, Smith RF, Sidner RA, Paris LL, Blankenship RL, Ray CN, Miner AC, Tector M, Tector AJ (2014) Creating class I MHC-null pigs using guide RNA and the Cas9 endonuclease. J Immunol 193(11):5751–5757. https://doi.org/10.4049/jimmunol.1402059

Rieblinger B, Fischer K, Kind A, Saller BS, Baars W, Schuster M, Wolf-van Buerck L, Schaffler A, Flisikowska T, Kurome M, Zakhartchenko V, Kessler B, Flisikowski K, Wolf E, Seissler J, Schwinzer R, Schnieke A (2018) Strong xenoprotective function by single-copy transgenes placed sequentially at a permissive locus. Xenotransplantation 25:e12382. https://doi.org/10.1111/xen.12382

Rogers CS (2016) Genetically engineered livestock for biomedical models. Transgenic Res 25(3):345–359. https://doi.org/10.1007/s11248-016-9928-6

Rogers CS, Stoltz DA, Meyerholz DK, Ostedgaard LS, Rokhlina T, Taft PJ, Rogan MP, Pezzulo AA, Karp PH, Itani OA, Kabel AC, Wohlford-Lenane CL, Davis GJ, Hanfland RA, Smith TL, Samuel M, Wax D, Murphy CN, Rieke A, Whitworth K, Uc A, Starner TD, Brogden KA, Shilyansky J, McCray PB Jr, Zabner J, Prather RS, Welsh MJ (2008) Disruption of the CFTR gene produces a model of cystic fibrosis in newborn pigs. Science 321(5897):1837–1841. https://doi.org/10.1126/science.1163600

Roussel JC, Moran CJ, Salvaris EJ, Nandurkar HH, d'Apice AJ, Cowan PJ (2008) Pig thrombomodulin binds human thrombin but is a poor cofactor for activation of human protein C and TAFI. Am J Transplant 8(6):1101–1112. https://doi.org/10.1111/j.1600-6143.2008.02210.x

Saalfrank A, Janssen KP, Ravon M, Flisikowski K, Eser S, Steiger K, Flisikowska T, Muller-Fliedner P, Schulze E, Bronner C, Gnann A, Kappe E, Bohm B, Schade B, Certa U, Saur D, Esposito I, Kind A, Schnieke A (2016) A porcine model of osteosarcoma. Oncogenesis 5:e210. https://doi.org/10.1038/oncsis.2016.21

Sakata N, Yoshimatsu G, Tsuchiya H, Egawa S, Unno M (2012) Animal models of diabetes mellitus for islet transplantation. Exp Diabetes Res 2012:256707. https://doi.org/10.1155/2012/256707

Sang H (2006) Transgenesis sunny-side up. Nat Biotechnol 24(8):955–956. https://doi.org/10.1038/nbt0806-955

Sapir T, Shternhall K, Meivar-Levy I, Blumenfeld T, Cohen H, Skutelsky E, Eventov-Friedman S, Barshack I, Goldberg I, Pri-Chen S, Ben-Dor L, Polak-Charcon S, Karasik A, Shimon I, Mor E, Ferber S (2005) Cell-replacement therapy for diabetes: generating functional insulin-producing tissue from adult human liver cells. Proc Natl Acad Sci U S A 102(22):7964–7969. https://doi.org/10.1073/pnas.0405277102

Schnieke AE, Kind AJ, Ritchie WA, Mycock K, Scott AR, Ritchie M, Wilmut I, Colman A, Campbell KH (1997) Human factor IX transgenic sheep produced by transfer of nuclei from transfected fetal fibroblasts. Science 278(5346):2130–2133

Schook LB, Collares TV, Hu W, Liang Y, Rodrigues FM, Rund LA, Schachtschneider KM, Seixas FK, Singh K, Wells KD, Walters EM, Prather RS, Counter CM (2015) A genetic porcine model of cancer. PLoS One 10(7):e0128864. https://doi.org/10.1371/journal.pone.0128864

Seok J, Warren HS, Cuenca AG, Mindrinos MN, Baker HV, Xu W, Richards DR, McDonald-Smith GP, Gao H, Hennessy L, Finnerty CC, Lopez CM, Honari S, Moore EE, Minei JP, Cuschieri J, Bankey PE, Johnson JL, Sperry J, Nathens AB, Billiar TR, West MA, Jeschke MG, Klein MB, Gamelli RL, Gibran NS, Brownstein BH, Miller-Graziano C, Calvano SE, Mason PH, Cobb JP, Rahme LG, Lowry SF, Maier RV, Moldawer LL, Herndon DN, Davis RW, Xiao W, Tompkins RG, Inflammation, Host Response to Injury LSCRP (2013) Genomic responses in mouse models poorly mimic human inflammatory diseases. Proc Natl Acad Sci U S A 110(9):3507–3512. https://doi.org/10.1073/pnas.1222878110

Seol JG, Kim SH, Jin D, Hong SP, Yoo JY, Choi KM, Park YC, Yun YJ, Park KW, Heo JY (2010) Production of transgenic cloned miniature pigs with membrane-bound human Fas ligand (FasL) by somatic cell nuclear transfer. Nat Preced http://hdlhandlenet/10101/npre201045391

Sharp NJ, Kornegay JN, Van Camp SD, Herbstreith MH, Secore SL, Kettle S, Hung WY, Constantinou CD, Dykstra MJ, Roses AD et al (1992) An error in dystrophin mRNA processing in golden retriever muscular dystrophy, an animal homologue of Duchenne muscular dystrophy. Genomics 13(1):115–121

Shimatsu Y, Horii W, Nunoya T, Iwata A, Fan J, Ozawa M (2016) Production of human apolipoprotein(a) transgenic NIBS miniature pigs by somatic cell nuclear transfer. Exp Anim 65(1):37–43. https://doi.org/10.1538/expanim.15-0057

Sieren JC, Meyerholz DK, Wang XJ, Davis BT, Newell JD Jr, Hammond E, Rohret JA, Rohret FA, Struzynski JT, Goeken JA, Naumann PW, Leidinger MR, Taghiyev A, Van Rheeden R, Hagen

J, Darbro BW, Quelle DE, Rogers CS (2014) Development and translational imaging of a TP53 porcine tumorigenesis model. J Clin Invest 124(9):4052–4066. https://doi.org/10.1172/jci75447

Soutar AK (2011) Unexpected roles for PCSK9 in lipid metabolism. Curr Opin Lipidol 22(3):192–196. https://doi.org/10.1097/MOL.0b013e32834622b5

Spurney CF (2011) Cardiomyopathy of Duchenne muscular dystrophy: current understanding and future directions. Muscle Nerve 44(1):8–19. https://doi.org/10.1002/mus.22097

Srivastava R, Khan AA, Huang J, Nesburn AB, Wechsler SL, BenMohamed L (2015) A herpes simplex virus type 1 human asymptomatic CD8+ T-cell epitopes-based vaccine protects against ocular herpes in a "humanized" HLA transgenic rabbit model. Invest Ophthalmol Vis Sci 56(6):4013–4028. https://doi.org/10.1167/iovs.15-17074

Stevenson RG, DeWitt WF (1973) An unusual case of lymphosarcoma in a pig. Canadian Vet J 14(6):139–141

Stoltz DA, Meyerholz DK, Pezzulo AA, Ramachandran S, Rogan MP, Davis GJ, Hanfland RA, Wohlford-Lenane C, Dohrn CL, Bartlett JA, Nelson GA, Chang EH, Taft PJ, Ludwig PS, Estin M, Hornick EE, Launspach JL, Samuel M, Rokhlina T, Karp PH, Ostedgaard LS, Uc A, Starner TD, Horswill AR, Brogden KA, Prather RS, Richter SS, Shilyansky J, McCray PB Jr, Zabner J, Welsh MJ (2010) Cystic fibrosis pigs develop lung disease and exhibit defective bacterial eradication at birth. Sci Transl Med 2(29):29ra31. https://doi.org/10.1126/scitranslmed.3000928

Stoltz DA, Rokhlina T, Ernst SE, Pezzulo AA, Ostedgaard LS, Karp PH, Samuel MS, Reznikov LR, Rector MV, Gansemer ND, Bouzek DC, Alaiwa MH, Hoegger MJ, Ludwig PS, Taft PJ, Wallen TJ, Wohlford-Lenane C, McMenimen JD, Chen JH, Bogan KL, Adam RJ, Hornick EE, Nelson GA, Hoffman EA, Chang EH, Zabner J, McCray PB Jr, Prather RS, Meyerholz DK, Welsh MJ (2013) Intestinal CFTR expression alleviates meconium ileus in cystic fibrosis pigs. J Clin Invest 123(6):2685–2693. https://doi.org/10.1172/JCI68867

Stoltz DA, Meyerholz DK, Welsh MJ (2015) Origins of cystic fibrosis lung disease. N Engl J Med 372(4):351–362. https://doi.org/10.1056/NEJMra1300109

Streckel E, Braun-Reichhart C, Herbach N, Dahlhoff M, Kessler B, Blutke A, Bahr A, Ubel N, Eddicks M, Ritzmann M, Krebs S, Goke B, Blum H, Wanke R, Wolf E, Renner S (2015) Effects of the glucagon-like peptide-1 receptor agonist liraglutide in juvenile transgenic pigs modeling a pre-diabetic condition. J Transl Med 13:73. https://doi.org/10.1186/s12967-015-0431-2

Stump KC, Swindle MM, Saudek CD, Strandberg JD (1988) Pancreatectomized swine as a model of diabetes mellitus. Lab Anim Sci 38(4):439–443

Tabebordbar M, Zhu K, Cheng JK, Chew WL, Widrick JJ, Yan WX, Maesner C, Wu EY, Xiao R, Ran FA, Cong L, Zhang F, Vandenberghe LH, Church GM, Wagers AJ (2016) In vivo gene editing in dystrophic mouse muscle and muscle stem cells. Science 351(6271):407–411. https://doi.org/10.1126/science.aad5177

Tan W, Carlson DF, Lancto CA, Garbe JR, Webster DA, Hackett PB, Fahrenkrug SC (2013) Efficient nonmeiotic allele introgression in livestock using custom endonucleases. Proc Natl Acad Sci U S A 110(41):16526–16531. https://doi.org/10.1073/pnas.1310478110

Tang X, Wang G, Liu X, Han X, Li Z, Ran G, Li Z, Song Q, Ji Y, Wang H, Wang Y, Ouyang H, Pang D (2015) Overexpression of porcine lipoprotein-associated phospholipase A2 in swine. Biochem Biophys Res Commun 465(3):507–511. https://doi.org/10.1016/j.bbrc.2015.08.048

Taylor FB Jr, Peer GT, Lockhart MS, Ferrell G, Esmon CT (2001) Endothelial cell protein C receptor plays an important role in protein C activation in vivo. Blood 97(6):1685–1688

Tuggle KL, Birket SE, Cui X, Hong J, Warren J, Reid L, Chambers A, Ji D, Gamber K, Chu KK, Tearney G, Tang LP, Fortenberry JA, Du M, Cadillac JM, Bedwell DM, Rowe SM, Sorscher EJ, Fanucchi MV (2014) Characterization of defects in ion transport and tissue development in cystic fibrosis transmembrane conductance regulator (CFTR)-knockout rats. PLoS One 9(3):e91253. https://doi.org/10.1371/journal.pone.0091253

Ueno S, Koyasu T, Kominami T, Sakai T, Kondo M, Yasuda S, Terasaki H (2013) Focal cone ERGs of rhodopsin Pro347Leu transgenic rabbits. Vision Res 91:118–123. https://doi.org/10.1016/j.visres.2013.08.006

Umeyama K, Watanabe M, Saito H, Kurome M, Tohi S, Matsunari H, Miki K, Nagashima H (2009) Dominant-negative mutant hepatocyte nuclear factor 1alpha induces diabetes in transgenic-cloned pigs. Transgenic Res 18(5):697–706. https://doi.org/10.1007/s11248-009-9262-3

Usui J, Kobayashi T, Yamaguchi T, Knisely AS, Nishinakamura R, Nakauchi H (2012) Generation of kidney from pluripotent stem cells via blastocyst complementation. Am J Pathol 180(6):2417–2426. https://doi.org/10.1016/j.ajpath.2012.03.007

Vadori M, Cozzi E (2015) The immunological barriers to xenotransplantation. Tissue Antigens 86(4):239–253. https://doi.org/10.1111/tan.12669

Vadori M, Aron Badin R, Hantraye P, Cozzi E (2015) Current status of neuronal cell xenotransplantation. Int J Surg 23(Pt B):267–272. https://doi.org/10.1016/j.ijsu.2015.09.052

Wang Y, Yang HQ, Jiang W, Fan NN, Zhao BT, Ou-Yang Z, Liu ZM, Zhao Y, Yang DS, Zhou XY, Shang HT, Wang LL, Xiang PY, Ge LP, Wei H, Lai LX (2015) Transgenic expression of human cytoxic T-lymphocyte associated antigen4-immunoglobulin (hCTLA4Ig) by porcine skin for xenogeneic skin grafting. Transgenic Res 24(2):199–211. https://doi.org/10.1007/s11248-014-9833-9

Wang K, Jin Q, Ruan D, Yang Y, Liu Q, Wu H, Zhou Z, Ouyang Z, Liu Z, Zhao Y, Zhao B, Zhang Q, Peng J, Lai C, Fan N, Liang Y, Lan T, Li N, Wang X, Wang X, Fan Y, Doevendans PA, Sluijter JPG, Liu P, Li X, Lai L (2017) Cre-dependent Cas9-expressing pigs enable efficient in vivo genome editing. Genome Res 27(12):2061–2071. https://doi.org/10.1101/gr.222521.117

Wei J, Ouyang H, Wang Y, Pang D, Cong NX, Wang T, Leng B, Li D, Li X, Wu R, Ding Y, Gao F, Deng Y, Liu B, Li Z, Lai L, Feng H, Liu G, Deng X (2012) Characterization of a hypertriglyceridemic transgenic miniature pig model expressing human apolipoprotein CIII. FEBS J 279(1):91–99. https://doi.org/10.1111/j.1742-4658.2011.08401.x

Weiss EH, Lilienfeld BG, Muller S, Muller E, Herbach N, Kessler B, Wanke R, Schwinzer R, Seebach JD, Wolf E, Brem G (2009) HLA-E/human beta2-microglobulin transgenic pigs: protection against xenogeneic human anti-pig natural killer cell cytotoxicity. Transplantation 87(1):35–43. https://doi.org/10.1097/TP.0b013e318191c784

Wheeler DG, Joseph ME, Mahamud SD, Aurand WL, Mohler PJ, Pompili VJ, Dwyer KM, Nottle MB, Harrison SJ, d'Apice AJ, Robson SC, Cowan PJ, Gumina RJ (2012) Transgenic swine: expression of human CD39 protects against myocardial injury. J Mol Cell Cardiol 52(5):958–961. https://doi.org/10.1016/j.yjmcc.2012.01.002

Whitelaw CB, Sheets TP, Lillico SG, Telugu BP (2016) Engineering large animal models of human disease. J Pathol 238(2):247–256. https://doi.org/10.1002/path.4648

Wijkstrom M, Bottino R, Iwase H, Hara H, Ekser B, van der Windt D, Long C, Toledo FG, Phelps CJ, Trucco M, Cooper DK, Ayares D (2015) Glucose metabolism in pigs expressing human genes under an insulin promoter. Xenotransplantation 22(1):70–79. https://doi.org/10.1111/xen.12145

Wilke M, Buijs-Offerman RM, Aarbiou J, Colledge WH, Sheppard DN, Touqui L, Bot A, Jorna H, de Jonge HR, Scholte BJ (2011) Mouse models of cystic fibrosis: phenotypic analysis and research applications. J Cyst Fibrosis 10(Suppl 2):S152–S171. https://doi.org/10.1016/s1569-1993(11)60020-9

Wilmut I, Schnieke AE, McWhir J, Kind AJ, Campbell KH (1997) Viable offspring derived from fetal and adult mammalian cells. Nature 385(6619):810–813. https://doi.org/10.1038/385810a0

Winand NJ, Edwards M, Pradhan D, Berian CA, Cooper BJ (1994) Deletion of the dystrophin muscle promoter in feline muscular dystrophy. Neuromuscul Disord 4(5-6):433–445

Wolf E, Braun-Reichhart C, Streckel E, Renner S (2014) Genetically engineered pig models for diabetes research. Transgenic Res 23(1):27–38. https://doi.org/10.1007/s11248-013-9755-y

Wolf-van Buerck L, Schuster M, Oduncu FS, Baehr A, Mayr T, Guethoff S, Abicht J, Reichart B, Klymiuk N, Wolf E, Seissler J (2017) LEA29Y expression in transgenic neonatal porcine islet-like cluster promotes long-lasting xenograft survival in humanized mice without immunosuppressive therapy. Sci Rep 7(1):3572. https://doi.org/10.1038/s41598-017-03913-4

Wu J, Platero-Luengo A, Sakurai M, Sugawara A, Gil MA, Yamauchi T, Suzuki K, Bogliotti YS, Cuello C, Morales Valencia M, Okumura D, Luo J, Vilarino M, Parrilla I, Soto DA, Martinez CA, Hishida T, Sanchez-Bautista S, Martinez-Martinez ML, Wang H, Nohalez A, Aizawa E, Martinez-Redondo P, Ocampo A, Reddy P, Roca J, Maga EA, Esteban CR, Berggren WT, Nunez Delicado E, Lajara J, Guillen I, Guillen P, Campistol JM, Martinez EA, Ross PJ, Izpisua Belmonte JC (2017) Interspecies chimerism with mammalian pluripotent stem cells. Cell 168(3):473–486.e415. https://doi.org/10.1016/j.cell.2016.12.036

Wuensch A, Baehr A, Bongoni AK, Kemter E, Blutke A, Baars W, Haertle S, Zakhartchenko V, Kurome M, Kessler B, Faber C, Abicht JM, Reichart B, Wanke R, Schwinzer R, Nagashima H, Rieben R, Ayares D, Wolf E, Klymiuk N (2014) Regulatory sequences of the porcine THBD gene facilitate endothelial-specific expression of bioactive human thrombomodulin in single- and multitransgenic pigs. Transplantation 97(2):138–147. https://doi.org/10.1097/TP.0b013e3182a95cbc

Wynyard S, Nathu D, Garkavenko O, Denner J, Elliott R (2014) Microbiological safety of the first clinical pig islet xenotransplantation trial in New Zealand. Xenotransplantation 21(4):309–323. https://doi.org/10.1111/xen.12102

Yamakawa H, Nagai T, Harasawa R, Yamagami T, Takahashi J, Ishikawa K, Nomura N, Nagashima H (1999) Production of transgenic pig carrying MMTV/v-Ha-ras. J Reprod Dev 45(2):111–118

Yamanaka S, Yokoo T (2015) Current bioengineering methods for whole kidney regeneration. Stem Cells Int 2015:724047. https://doi.org/10.1155/2015/724047

Yang L, Guell M, Niu D, George H, Lesha E, Grishin D, Aach J, Shrock E, Xu W, Poci J, Cortazio R, Wilkinson RA, Fishman JA, Church G (2015) Genome-wide inactivation of porcine endogenous retroviruses (PERVs). Science 350(6264):1101–1104. https://doi.org/10.1126/science.aad1191

Yazaki S, Iwamoto M, Onishi A, Miwa Y, Hashimoto M, Oishi T, Suzuki S, Fuchimoto D, Sembon S, Furusawa T, Liu D, Nagasaka T, Kuzuya T, Ogawa H, Yamamoto K, Iwasaki K, Haneda M, Maruyama S, Kobayashi T (2012) Production of cloned pigs expressing human thrombomodulin in endothelial cells. Xenotransplantation 19(2):82–91. https://doi.org/10.1111/j.1399-3089.2012.00696.x

Yokoo T, Ohashi T, Shen JS, Sakurai K, Miyazaki Y, Utsunomiya Y, Takahashi M, Terada Y, Eto Y, Kawamura T, Osumi N, Hosoya H (2005) Human mesenchymal stem cells in rodent whole-embryo culture are reprogrammed to contribute to kidney tissues. Proc Natl Acad Sci U S A 102(9):3296–3300. https://doi.org/10.1073/pnas.0406878102

Yokote S, Matsunari H, Iwai S, Yamanaka S, Uchikura A, Fujimoto E, Matsumoto K, Nagashima H, Kobayashi E, Yokoo T (2015) Urine excretion strategy for stem cell-generated embryonic kidneys. Proc Natl Acad Sci U S A 112(42):12980–12985. https://doi.org/10.1073/pnas.1507803112

Yoshioka M, Kayo T, Ikeda T, Koizumi A (1997) A novel locus, Mody4, distal to D7Mit189 on chromosome 7 determines early-onset NIDDM in nonobese C57BL/6 (Akita) mutant mice. Diabetes 46(5):887–894

Yuan L, Sui T, Chen M, Deng J, Huang Y, Zeng J, Lv Q, Song Y, Li Z, Lai L (2016) CRISPR/Cas9-mediated GJA8 knockout in rabbits recapitulates human congenital cataracts. Sci Rep 6:22024. https://doi.org/10.1038/srep22024

Zadelaar S, Kleemann R, Verschuren L, de Vries-Van der Weij J, van der Hoorn J, Princen HM, Kooistra T (2007) Mouse models for atherosclerosis and pharmaceutical modifiers. Arterioscler Thromb Vasc Biol 27(8):1706–1721. https://doi.org/10.1161/ATVBAHA.107.142570

Zhou BB, Zhang H, Damelin M, Geles KG, Grindley JC, Dirks PB (2009) Tumour-initiating cells: challenges and opportunities for anticancer drug discovery. Nat Rev Drug Discov 8(10):806–823. https://doi.org/10.1038/nrd2137

Stem Cells and Cell Conversion in Livestock

10

Fulvio Gandolfi and Tiziana A. L. Brevini

Abstract

The main drive to study stem cells is their possible use as therapeutic agents. Within veterinary medicine, a direct medicinal use of stem cells is reserved to companion species. Domestic ungulates like ruminants and pig are often used for preclinical research.

A stem cell is an unspecialized cell type able to undergo asymmetrical divisions: one cell is identical to its mother; the other begins its transformation toward one or more cell types capable of specific functions.

Physiologically, small populations of stem cells are present in each organ, and their function is to counteract the physiological wear and tear. These are named organ-specific stem cells and can be isolated from any animal species as well as in humans.

Embryonic stem cells are not a physiological cell type and are derived from early embryos or can be generated artificially (induced pluripotent cells) by inducing a somatic cell to overexpress four specific pluripotency-related genes. They can proliferate indefinitely if kept undifferentiated or can give rise to any other cell type when cultured in the appropriate conditions or transplanted back into an embryo. However, as opposed to organ-specific stem cells, pluripotent stem cells have so far been difficult to obtain in any species other than humans and laboratory rodents.

In order to circumvent the lack of pluripotent cells in livestock species as well as their inherent susceptibility to culture-induced alterations and tumorigenic

F. Gandolfi
Department of Anatomy of Domestic Animals, University of Milan, Milan, Italy
e-mail: fulvio.gandolfi@unimi.it

T. A. L. Brevini (✉)
Laboratory of Biomedical Embryology, UniStem, Centre for Stem Cell Research,
Università degli Studi di Milano, Milan, Italy
e-mail: tiziana.brevini@unimi.it

© Springer International Publishing AG, part of Springer Nature 2018
H. Niemann, C. Wrenzycki (eds.), *Animal Biotechnology 2*,
https://doi.org/10.1007/978-3-319-92348-2_10

transformation, novel techniques of cell conversions have been developed that work effectively with no species-specific limitations. Epigenetic mechanisms are used to enhance cell plasticity so that the exposure to adequate culture conditions can transform easily accessible dermal fibroblasts into a wide range of different cell types. Their lack of permanent pluripotency makes them promising candidates for safe therapeutic applications in all species including livestock.

10.1   What Is a Stem Cell?

A stem cell is an unspecialized cell type defined by its capacity to undergo asymmetrical division when required by the physiological or experimental circumstances. The products of an asymmetrical division are two cells: one is identical to its mother cell; the other is different because it has begun its transformation toward one or more cell types capable of specific functions. This ensures the conservation of a stem cell population and, at the same time, enables the generation of new specialized cells.

In the early phases of mammalian embryonic development, we identify three germ layers, the endoderm, mesoderm, and ectoderm; each one forms a different set of specific organs and tissue types. Stem cells are classified according to their potency that can span from unipotency, when only a single-cell type can be generated, to multipotency, when a stem cell can originate to all or many cells of a single germ layer.

When a stem cell can differentiate into cells that arise from all three germ layers, it is defined as pluripotent.

In nature, pluripotency is limited to the epiblast, a transient tissue that exists only for a brief period of embryonic development, before giving origin to the three germ layers. Therefore, the epiblast is not a kind of stem cell because it lacks the property of asymmetric division and stable pluripotent cells are not a physiological component of the body but are created only in vitro (Smith 2001).

The main drive to study stem cells is their possible use as therapeutic agents. Within veterinary medicine, a direct medicinal use of stem cells is reserved to companion species, like dogs, cats, and horses. Domestic ungulates like ruminants and pigs are often used for preclinical research. Their use in regenerative medicine is crucial as intermediate models between laboratory rodents and humans, providing an important step for the translation of basic research into clinical applications. However, whereas the derivation of organ-specific stem cells has been successful in livestock species (Spencer et al. 2011), pluripotent stem cells have so far been difficult to obtain.

10.2   Different Properties of Adult and Embryonic Stem Cells

Every organ has its specific stem cell population, neatly located in specialized areas, called niches. Niches ensure that stem cells receive the necessary signals for regulating proliferation proportionally to cell loss, in order to maintain a functional equilibrium, defined as tissue homeostasis.

The skin, intestine, and bone marrow undergo a high rate of wearing out and, consequently, of renewal, whereas the skeletal muscle, heart, and brain typically are more stable, and their cells have a longer life span. It is not always easy to identify the stem cells of every adult organ, and their number decreases with age; however, adult stem cells have been isolated and studied in several livestock species.

The domestic pig is one of the best models for the study of human diseases, because of its well-known similarities in terms of anatomy, physiology, metabolism, and organ development with humans. As described in Chap. 9, the creation of humanized pigs and the improvement of preclinical disease models by targeted genetic engineering have further expanded the role of this species. In this context the derivation of adult stem cells further strengthens the pig as a relevant and powerful biomedical model.

Research on cardiac regenerative medicine is a typical example. The evident differences in coronary architecture and in the extent of vessel variations between mouse and human hearts severely limit the clinical relevance of the experiments performed in rodents. Differences can also be appreciated at the cellular level, as indicated by the higher capillary density and the larger cross-sectional area of the myocytes in human, in comparison to the mouse. On the contrary, the coronary anatomy and the subendocardial to epicardial collateral network of the swine heart are very similar to those of the human. Therefore, the pig has emerged as useful large animal model in cardiovascular research, bridging the gap between classical rodent models and humans.

Recent studies demonstrated the presence of resident cardiac progenitors that, although more abundant in early postnatal life, persist and assure local remodeling in the adult organism as well. These cells were isolated from three different heart regions, including the aorta, ventricle, and atrium, respectively, indicating that this subpopulation of committed, but still proliferating, cardiac progenitor cells is not confined to a specific region within the organ, but rather evenly distributed to several areas (Fig. 10.1). Interestingly enough, these cells seem to be more abundant in the outer layers of the organ, indicating the epicardium as a possible site of origin for resident stem or progenitor cell populations in the pig.

Mesenchymal stem cells (MSCs) are another example of somatic stem cells extensively studied and used in regenerative medicine. Previously known as "stromal stem cells," the term "mesenchymal stem cells" was coined in 1991 by Arnold Caplan and became widely adopted. However, not long ago Caplan himself

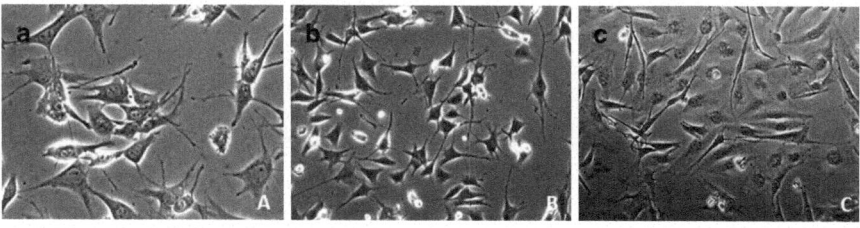

Fig. 10.1 Cardiac progenitor cells isolated from explants of the aorta (**a**), ventricle (**b**), and atrium (**c**) of a healthy adult pig heart

proposed the alternative term "medicinal signaling cells" based on the observation that MSCs get activated in an inflammatory environment and express their "medicinal" functions—which are primarily immunomodulatory and trophic. Therefore, in a way, MSC's biological function has nothing to do with "stemness." The concept behind the renaming is that we should focus on what cells can do therapeutically rather than on what they can differentiate into. Furthermore, since the experimental proof that MSCs, or even their subsets, fulfill the stem cell definition is still lacking at the single-cell level, the term "stem cell" seems inappropriate. However, despite the question whether or not MSCs qualify as "stem cells" remains legitimate, the term "mesenchymal stem cells" has gained such global usage that professionals have not adopted yet the new "medicinal signaling cells" name.

The first source reported to contain MSCs was the stromal compartment of the bone marrow. For this reason, the bone marrow is currently the best investigated origin of MSCs in domestic animals. MSCs derived from the bone marrow (BM-MSCs) are typically multipotent since they can differentiate into the bone, cartilage, and adipose tissue (Fig. 10.2).

Alternatively and, in some cases, less invasive sources of MSCs or MSC-like cells are the adult adipose tissue and amniotic fluid as well as the fetal blood, liver, bone marrow, and lung.

MSC's potency of differentiation varies depending on their origin. For example, the capability of bone marrow MSCs (BM-MSCs) to differentiate into the cartilage is higher than that of MSCs derived from the adipose tissue, but inferior to that isolated from the umbilical cord blood. Despite BM-MSCs represent the most commonly investigated cell type for application in human and veterinary regenerative medicine, it is important to remember that BM-MSCs have a relatively limited potential in vitro proliferation ability. Their plasticity and growth decline with increasing donor age and in vitro passage number.

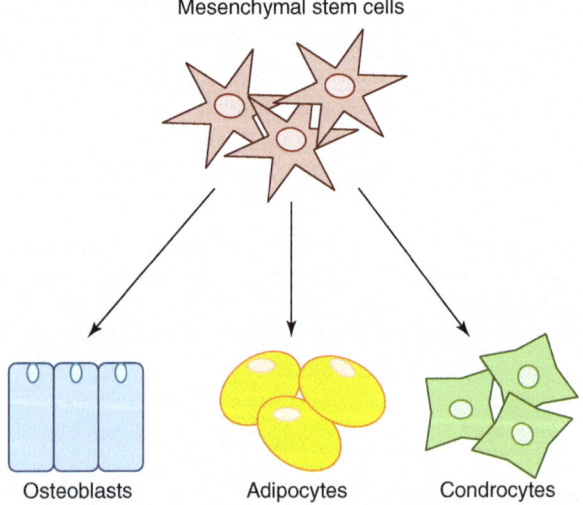

Fig. 10.2 MSCs derived from the bone marrow and multipotent differentiation capacity. BM-MSCs can generate multiple mesoderm-type cell lineages, such as osteoblasts, adipocytes, and chondrocytes

The adipose tissue is another source of MSC frequently used in humans and domestic animals. The successful and efficient recovery of adipose-derived MSC in domestic animals indicates that fat is an effective MSC source for clinical application. Furthermore, these cells display higher proliferation rate and lower senescence compared to MSCs from other sources.

All MSCs secrete soluble factors that have beneficial effects on the regeneration of injured tissues. They also inhibit apoptosis, limit pathologic fibrotic remodeling, stimulate proliferation and differentiation of endogenous progenitor cells, decrease inflammatory oxidative stress, and modulate immune reactions (Fig. 10.3).

MSCs of domestic animals are not only used as models for human therapies but are being increasingly used for the treatment of a number of diseases, including arthritis, atopic dermatitis, and tendon injury.

One major application of MSCs in livestock is repairing damaged tendons and ligaments, because MSCs are physiologically present in these structures. The local injection of MSCs in far greater numbers than normally present within tendon tissue would have the potential for regenerating or repairing the tendon. Several studies on horse tendinopathies revealed that cells remain close to the injection site and that both autologous and allogeneic MSCs do not stimulate an undesirable immune response from the host. However, it is still unclear whether the major contribution of MSCs to the healing process is their differentiation into tenocytes or the supply of growth factors, which stimulate residing cells within the tendon. A combination of the two mechanisms may also occur.

However, the collection of the bone marrow and adipose tissue in several animal species, like horse, requires an invasive procedure. To overcome the invasive

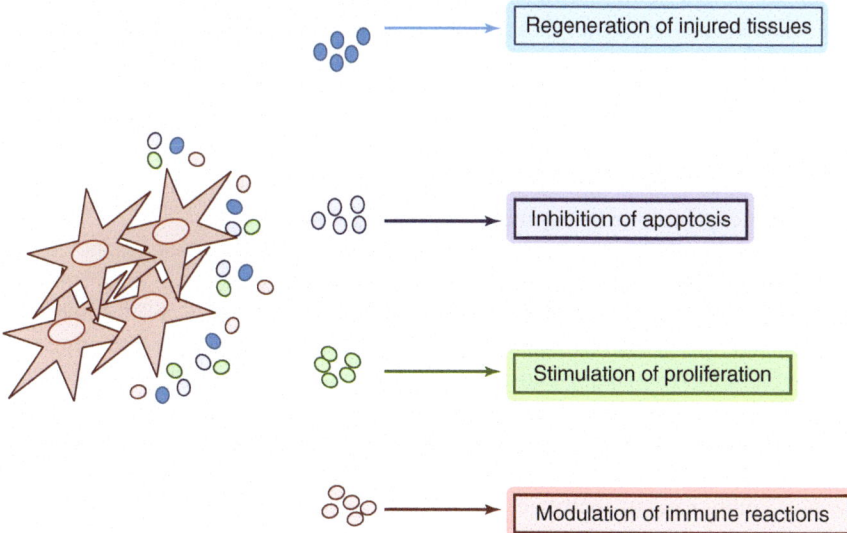

Fig. 10.3 MSCs secrete soluble factors that have beneficial effects on regeneration of injured tissues, inhibit apoptosis, stimulate proliferation, and modulate immune reactions

collection of the bone marrow and adipose tissue, progenitor cells derived from extra-fetal sources, such as the umbilical cord and amnion, could represent alternative candidates.

An attractive alternative source of MSCs is the amnion, the membrane that limits the fluid-filled cavity where the embryo develops during pregnancy, which is part of the placenta. The first amnion-mesenchymal cells (AMCs) were derived and characterized in the horse (Lange-Consiglio et al. 2012). AMCs and BM-MSCs both exhibit adult stromal cell-specific gene and protein expression, but AMCs have higher and quicker differentiation ability than MSCs.

Since amnion is discarded at birth, collecting amniotic cells is noninvasive and low cost. Adding this to their rapid proliferation and greater differentiation potential makes AMCs a potentially useful cell type for therapy.

Indeed, comparative studies that employed cryopreserved heterologous AMCs and fresh autologous BM-MSCs in spontaneous equine tendon lesions in vivo showed that AMCs were well tolerated by patients and provided beneficial effects (Lange-Consiglio et al. 2013). A distinct advantage of stored AMCs is that they can be administered at a much shorter interval from the injury than BM-MSC because fresh autologous cells require prolonged in vitro culture to reach the required number. On the contrary, the short interval between AMC thawing and their injection at the selected site enables them to act before the physiological repair mechanisms leave any permanent structural change within the injured tendon. The regenerated tissue is more elastic and therefore functionally closer or even identical to a normal tendon.

In summary, tissue-specific stem cells are a physiological component of the human and animal body that replace worn-out cells with new ones, maintaining the functional equilibrium, known as tissue homeostasis, that characterizes a healthy status. These cells are identical in humans and animals, and a large body of evidence indicates that the cells can be effectively used for therapeutic purposes in several circumstances.

However, tissue-specific stem cells have some notable limitations: their number decreases with age; they are difficult to identify or reach in certain organs, like the central nervous system; and they have a relatively limited life span in vitro which limits their capability to be expanded in very large numbers.

These limitations can be overcome by pluripotent stem cells that, as opposed to tissue-specific stem cells, are not a physiological component of the organisms but were originally derived from the mouse blastocyst, an early stage of embryonic development that is reached within 4 days after fertilization prior to implanting into the uterine wall. A small population of cells, the inner cell mass, is isolated from the rest of the blastocyst and adapted to grow in vitro under specific culture conditions. These cells become able to proliferate indefinitely as an unspecialized cell population. However, if transplanted back into a blastocyst, they are able to integrate into the host embryo and to contribute to all tissues including the gametes. In vitro, differentiation is achieved by changing culture conditions, in order to obtain the desired cell type. In this way, it is possible to obtain cell populations whose physiological stem cells have not been properly identified, e.g., endocrine pancreatic cells or liver or heart cells, or are difficult to reach in living individuals, e.g., neural stem cells. Furthermore, their unlimited capacity to proliferate in vitro enables the possibility of obtaining an equally unlimited number of differentiated cells.

10.3 Pluripotent Stem Cells Are Not Equal in Different Species

Whereas tissue-specific stem cells are physiological cell types normally residing in the organism, embryonic stem cells (ESCs) are the equivalent of the epiblast, a transient component of early embryos that normally evolves into other more differentiated tissues. Therefore, ESCs are not a permanent part of the embryo, and their establishment as a cell line requires the artificial adaptation of the epiblast to specialized culture conditions. Consequently, ESCs substantially differ among species possibly reflecting differences in embryonic development and the requirement of specific culture conditions.

Truly pluripotent, so-called bona fide, embryonic stem cells (ESCs) have been derived from the laboratory mouse (Evans and Kaufman 1981) and rats (Buehr et al. 2008; Li et al. 2008). Mouse and primate ESCs share some major properties such as unlimited replication in vitro (self-renewal); expression of core pluripotency genes such as OCT4, SOX2, and NANOG; and capacity to differentiate in vitro into any of the different tissues that make the body. If ESCs are injected into mice deprived of their immune system, they form a teratoma, a benign tumor that comprises several differentiated tissues like the cartilage, muscle, epidermis, teeth, etc. and that is considered a proof of pluripotency.

Whereas ESCs have been originally derived from embryos, it is now possible to obtain cell lines with the same properties simply by overexpression of four key genes, Oct4, Sox2, Kfl4, and c-Myc, the so-called Yamanaka factors, named after the Nobel Laureate Shinya Yamanaka who first established this technology (Takahashi and Yamanaka 2006). This "artificial" form of ESC is called induced pluripotent stem cells (iPSCs), and as for the original ESCs, they have been obtained and are best characterized in mouse, primates, and humans. Since ESC and iPSC for all practical purpose are indistinguishable from each other, they are now collectively referred to as pluripotent stem cells (PSCs).

Interestingly, however, substantial differences have been found between primate and mouse PSC. These begin in the culture medium which requires the presence of a set of special additives (LIF and BMP4) to maintain the undifferentiated state of mouse ESC (mESC). They are different from those required for primate ESC (activin A and FGF2) (Ying et al. 2003; Xu et al. 2005, 2008). Furthermore, mouse ESC can be derived only from a small number of "permissive" strains, whereas primate ESCs show no limitations related to genetic background. Mouse and primate ESCs differ also in their morphology. Small, compact, and domed colonies are typically formed by mESC as opposed to primate ESCs that grow in larger, flat colonies. Mouse ESC colonies are propagated after dissociation to single cells, but the same treatment would rapidly kill primate ESC, whose colonies need to be detached from the feeder layer and fragmented mechanically. Typically, mESC grow vigorously and can easily adapt to culture conditions that enable the derivation of lines from single cells. This is much more difficult with primate cells, which grow at a slower pace and respond poorly to single-cell culture conditions (Nichols and Smith 2009, 2011). The derivation of cell lines from a single cell is important because it is the only way to prove that each cell of a colony is truly pluripotent. Furthermore, the ability to survive and proliferate starting from single cells is crucial for mouse ESC's ability to form chimeras. If we inject mESC into an early embryo at the blastocyst stage, they mix up with its cells. Upon transfer of this embryo into a surrogate mother, a chimera is born which is a mouse whose tissues are made partly by the recipient blastocyst and partly by the injected ESC. When ESCs are of good quality, chimerism involves all tissues and organs including germ cells and gametes. This is the only experimental proof that a mESC line is truly pluripotent.

The generation of chimeras has never been achieved with nonhuman primate PSC. Recent results confirmed that rhesus monkey ESCs are unable to be integrated into host blastocysts, but chimera formation was achieved for the first time from the aggregation of several four-cell embryos (Tachibana et al. 2012). This suggests that the inability of primate ESCs to form chimeras may not be linked to a lack of pluripotency but to their inability to be dissociated to a single cell, thus explaining their inability to mix with other cells and form chimeras.

These differences have gone largely unexplained until pluripotent cell lines were derived from the epiblast of postimplantation mouse embryos (E5.5–E7.5) as opposed to standard preimplantation embryos (E3.5 or earlier) (Brons et al. 2007; Tesar et al. 2007). These cell lines were named epiblast stem cells (EpiSC). The major property of EpiSC is to share the main characteristics that differentiate human and nonhuman primate ESC from mESC. All this has provided a biological explanation for the startling differences between rodent and primate ESC. It is now clear that they derive from two different stages of embryonic development: one has been defined as naïve epiblast and can be found in the mouse preimplantation blastocyst; the other is defined as primed epiblast and is found in primate preimplantation blastocysts and mouse postimplantation embryos (Nichols and Smith 2009, 2011). The notion that primate ESCs are the equivalent of mouse EpiSC rather than of mouse ESC is now largely accepted.

It has also been observed that it is possible to derive mouse EpiSC from mESC simply by exposing them to the appropriate growth factor combination (activin A/ FGF2), therefore indicating that EpiSC are the "physiological" evolution of ESC (Guo et al. 2009). Therefore, it can be hypothesized that cells isolated from primate preimplantation blastocysts, presumably originating from naïve epiblast, as it occurs in the mouse, spontaneously progress to the primed epiblast stage in vitro, before giving rise to stable cell lines that, despite the fact that have been named ESC, are actually EpiSC. The opposite is also possible so that when primed human stem cells are cultured in a specifically formulated medium called naïve human stem cell medium (NHSM), they revert to their naïve form and acquire all the properties previously exclusive of mouse ESC (Gafni et al. 2013). These include the ability to form cross-species chimeric mouse embryos.

With a different modification of the culture medium, a third and novel type of PSC has recently been isolated: the region-selective pluripotent stem cells (rsPSCs) (Wu et al. 2015). These cells maintain the more developmentally advanced state of the primed ESC together with their positive property of being poised for rapid and efficient differentiation. In addition, they possess a high cloning efficiency and a more robust growth rate that lead to their ability to integrate in all three germ layers in chimeric embryos.

10.4 Livestock Pluripotent Stem Cells Are Difficult to Obtain

As opposed to mouse and primates, it is very difficult, if not outright impossible, to derive embryonic stem cells in livestock species. For a detailed summary of the research results obtained, you are referred to some recent reviews (Brevini et al. 2010; Koh and Piedrahita 2014; Kumar et al. 2015; Soto and Ross 2016).

In most cases, cell lines in these species are defined as ES-like and show several major deficiencies, ranging from short life in culture to lack of controlled pluripotency or of the ability to form chimeras (Talbot and Blomberg le 2008). Despite the extensive research activity, it is still unclear why it is not possible to derive truly pluripotent ESC from embryos of these species.

As we illustrated before, ESCs originate from the epiblast either naïve or primed. This leads to the question: Is the lack of domestic animals' ESC due to the lack of appropriate culture conditions or the epiblast from these species inherently different that "suspending" it in vitro may not be possible?

The process of epiblast formation is known in detail especially in mouse. During the first embryonic divisions, all blastomeres are developmentally equivalent and totipotent and all express the transcription factor OCT4. The first differentiation process consists in the generation of trophectoderm (TE) and inner cell mass (ICM) cells from their unique totipotent blastomere precursors. This is marked by the restriction of OCT4 expression to ICM cells, which is caused by its repression by CDX2. The result is that TE cells express CDX2 and ICM cells express OCT4. ICM cells will then undergo a further differentiation leading to the formation of the hypoblast, which will lose OCT4 expression, and of the epiblast that will retain it. The latter is the pluripotent tissue that will originate all three germ layers in vivo or will originate both primate and rodent ESC when cultured in vitro.

Mouse epiblast differentiation and Oct4 restriction to this tissue are completed by E3.5. By E5.5 mouse embryos are embedded into the uterine wall. Human embryos go through the same changes but at a slower pace (Rossant 2011) with OCT4 restriction to the epiblast completed by E6 and implantation taking place at E7–9.

When we examined the distribution of OCT4 in bovine embryos, we soon realized that it is not as tightly restricted to ICM as described in mouse and human embryos but it was ubiquitously expressed also in expanded blastocysts (van van Eijk et al. 1999). When observations were extended to later-stage embryos, it was determined that OCT4 restriction to the epiblast is completed only by E11 in bovine (Berg et al. 2011) and E8–9 in pig (Hall et al. 2009) embryos.

Based on this different timing, attempts have been performed using day 10–12.5 elongated pig blastocysts, this time using the knowledge that late, or primed, epiblast responds better to FGF2 than to LIF (Alberio et al. 2010). Indeed results were encouraging with cell lines showing a robust self-renewal and the ability to differentiate into precursor cells derived from all three germ layers as well as into trophectoderm and germ cell precursors. This indicates that pig cells, and possibly other ungulates, respond to culture conditions similar to primate embryos.

Since it is possible to convert mouse and human primed ESC into the naïve state simply exposing them to the suitable culture medium EpiSC to ESC culture medium (Bao et al. 2009; Gafni et al. 2013), it will be interesting to see if pig EpiSC will show the same plasticity and will provide a reliable source of pig naïve PSC.

However, performing this experiment may be difficult, since it is unclear which are the culture conditions required for pig naïve ESC. The most recent papers describing putative pig ESC either used both LIF and FGF2 (Brevini et al. 2010) or a very rich mixture of LIF, FGF2, activin, and EGF (Vassiliev et al. 2010). Therefore, the respective role of each of these molecules is unclear.

10.5 What Happens When We Do Not Use an Embryo?

Given the possibility that the specific morphological and functional characteristic of domestic ungulate preimplantation embryos may have a profound influence on the possibility to derive ungulate ESC lines, it was interesting to see whether the forced induction of pluripotency thanks to the iPS technology has made it possible to test if bypassing the embryo as a starting material could allow to obtain bona fide pluripotent stem cells in ungulates.

Indeed iPS have been obtained in a wide range of domestic ungulates which include pig (Esteban et al. 2009; Ezashi et al. 2009; Wu et al. 2009; West et al. 2010; Montserrat et al. 2011), sheep (Bao et al. 2011; Li et al. 2011a, b; Liu et al. 2012), cow (Sumer et al. 2011), and horse (Nagy et al. 2011) with some variations in the results. In some instances, however, expression of the exogenous pluripotency genes was not downregulated or was artificially maintained. In the first case, this made it difficult to induce teratoma formation. In the latter, the absence of expression induced a rapid differentiation in pig (Esteban et al. 2009; Wu et al. 2009), sheep (Li et al. 2011a, b), and cow (Sumer et al. 2011) cell lines. More importantly, the ability of livestock iPSCs to generate chimeras was very low, and even lower was their ability to contribute to the germ line (West et al. 2011). The results are consistent with the fact that most of these cell lines show the characteristics of the primed type.

The recent developments of new media able to convert primed cell lines into the naïve type or to confer higher clonal and chimeric properties to primed lines give us hope that further developments may be achieved in livestock species as well.

However, recent results suggest that it is possible that these species harbor a yet undefined "third" stage that differs from both the naïve and primed epiblast whose nature could prevent its stable conversion into a pluripotent cell line.

Somatic cells transformed into iPSC maintain the culture requirements typical of the species of origin with mouse cells giving rise to the naïve type of stem cells and human cells originating cell lines with the characteristics of the primed state (Telugu et al. 2010). This indicates that, once a somatic cell is reprogrammed to pluripotency, it follows the default behavior that is typical of the epiblast of its species.

It was therefore interesting to see which are the properties of ungulate iPSC and, as a consequence, which is the typical default behavior. But once again the picture is unclear. If we consider pig iPSC, we see that, in one case, cell lines show a slow proliferation rate and inability to form chimeras, reminiscent of an EpiSC model (Ezashi et al. 2009), whereas in another, a rapid proliferation rate was accompanied by the chimera generation, typical of mouse naïve ESC (West et al. 2010). In this case chimeric cells were detected also in the germ line of 4.7% of the offspring. However, the results were not robust since only two second-generation piglets inherited the marker gene but both died at birth or shortly thereafter (West et al. 2011).

In many cases proliferation speed of pig (Montserrat et al. 2011), sheep (Liu et al. 2012), and cow (Sumer et al. 2011) iPS cells was not reported, and cells were propagated in media supplemented with both LIF and FGF2, thereby making it impossible to classify them either as naïve or primed. However, morphology was described as similar to that of human ESC, but chimera formation, at least at the blastocyst stage, was reported for sheep iPS (Liu et al. 2012). Therefore, the picture is unclear but suggests that in most case ungulate iPS, so far, belong mainly to the EpiSC/primed category.

Altogether, these results suggest that true LIF-dependent naïve/ESC equivalent to those of mouse cannot be obtained in ungulates, possibly due to some inherent characteristic of their epiblast. It will be interesting, in the future, to see if other methods can be developed to reprogram ungulate EpiSC into naïve ESC working on the epigenome.

10.6   Epigenetics and Stem Cell Research

The current lack of true pluripotent cell lines in livestock and domestic species leaves unsolved the problem of finding a source of cells for regenerative medicine of those organs whose specific stem cells are not available or are difficult to access.

In addition, irrespective of the species, true pluripotency is unphysiological and inherently labile and makes cells prone to culture-induced alterations and tumorigenic transformations. Furthermore, although various methodologies have been established, the efficiency of iPSC induction remains low. A serious concern is also the integration of transgenes that severely limits their use in clinical studies (Okita et al. 2007) due to the problems related to residual DNA and chromosomal disruptions that may result in harmful genetic alterations (Kim et al. 2009).

In order to circumvent these limitations, several approaches were suggested, in an attempt to promote cell reprogramming without the requirement for exogenous transcription factors. In this line, it has been shown that small-molecule compounds can be used instead of some of the reprogramming genes and are able to modulate the epigenetic state of the target cells through the activation and/or inhibition of specific differentiation signaling pathways (Huangfu et al. 2008; Ichida et al. 2009; Li et al. 2011a, b; Hou et al. 2013). Since epigenetic mechanisms exert a key role in somatic cell reprogramming, small-molecule inhibitors of epigenetic-modifying enzymes were selected.

Huangfu et al. reported that valproic acid (VPA), which is an inhibitor of histone deacetylating enzymes, allows efficient induction of human and murine iPSCs and greatly improves reprogramming (Huangfu et al. 2008). Mouse adult fibroblasts could be reprogrammed using a chemical combination of VPA, CHIR99021, 616452, and tranylcypromine (TCP), in the presence only of Oct-4, and without the use of any other transcription factors (Li et al. 2011a, b).

A recent study also reports that seven small-molecule compounds, namely, VPA, CHIR99021, 616452, TCP, Forskolin (FSK), 2-methyl-5-hydroxytryptamine (2-Me-5HT), and D4476, can reactivate endogenous pluripotency programs without introduction of exogenous genes and generate iPSCs from murine somatic cells at a frequency up to 0.2% (Hou et al. 2013).

Consistent with these findings, Moschidou et al. demonstrated that the use of a low growth factor medium in combination with VPA reverts 82% of amniotic fluid cells to a pluripotent state. These cells displayed high transcriptional identity with ESC, formed embryoid bodies (EB), differentiated into the three germ layers, and generated teratomas (Moschidou et al. 2012). Similarly, endogenous high plasticity transcription factor genes were reactivated in adult human dermal fibroblasts exposed to VPA and in the absence of any transgenes (Rim et al. 2012).

Altogether, these findings represent a remarkable progress, since with these novel approaches the use of retroviruses and/or lentiviruses vectors as well as the insertion of transgenes is avoided. However, it is important to highlight that all the cells described above are characterized by a stable pluripotent state that is nonphysiological and makes them prone to error, leading to an elevated risk of malignant transformation. This suggests great caution for their use in regenerative medicine (Kim et al. 2009).

10.7 Epigenetic Erasing and Writing: A Novel Way to Direct Cell Differentiation

The first paper reporting the ability of a small molecule to induce dedifferentiation of mouse C2C12 myoblasts was published in 2004 (Chen et al. 2004). The results obtained demonstrated that reversine, a 2,6-disubstituted purine, could increase cell plasticity, inducing lineage-committed myoblasts to become multipotent mesenchymal progenitor cells. The extremely powerful effect of this compound was subsequently demonstrated in 3T3E1 osteoblasts (Chen et al. 2007), human primary skeletal myoblasts (Chen et al. 2007), and murine and human dermal fibroblasts (Anastasia et al. 2006). In all these cell types, reversine was able to increase plasticity and address toward a progenitor-like state.

The growing understanding of the mechanisms driving epigenetic controls of cell differentiation and phenotype definition, together with the huge development of epigenetic chemistry, has increased our knowledge about the use of epigenetic modifiers, such as writers and erasers (Fig. 10.4). The first catalyze modifications either on DNA, RNA, or histone proteins by the addition of chemical groups. This group includes histone methyltransferases (HMTs), histone acetyltransferases (HATs), and DNA methyltransferases (DNMTs). In contrast to the previous enzymes, erasers remove the structural modifications introduced by the writers. They comprise histone deacetylases (HDACs) and demethylases (for details see the Chap. 3 by N. Beaujean).

There is growing evidence that cell plasticity may be increased through the use of erasers. For instance, recent experiments demonstrated that brief exposure to a demethylating agent can push cells to a less committed state, increasing their permissivity for a short window of time, sufficient to readdress cells toward a different cell type (Harris et al. 2011; Pennarossa et al. 2013, 2014; Brevini et al. 2014; Mirakhori et al. 2015; Chandrakanthan et al. 2016). The starting hypothesis is based on the observation that the processes associated with differentiation are driven by several mechanisms. Among these, DNA methylation plays a fundamental role during both early embryonic development and cell lineage specification, causing silencing of a large fraction of the genome and subsequent expression of genes essential for the maintenance of the differentiated and tissue-specific phenotype.

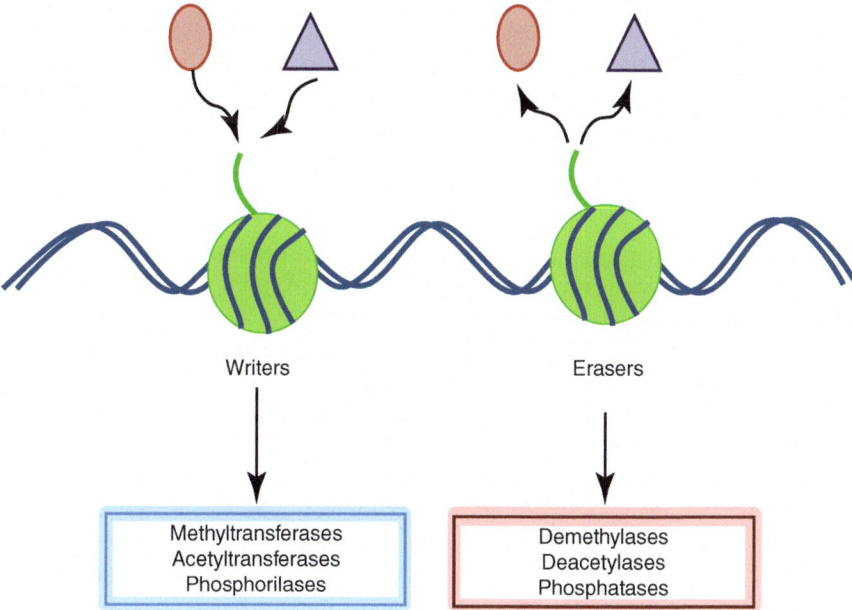

Fig. 10.4 Epigenetic modifiers: writers and erasers. The writers are enzymes able to add molecules on DNA, RNA, or histone tails. The erasers remove the structural modifications introduced by the writers

In particular, 5-azacytidine (5-aza-CR), a well-characterized DNMT inhibitor, has been used in order to remove the epigenetic "blocks" that are responsible for tissue specification. This drug is known to directly inhibit methylation in newly synthesized DNA, exerting a block on DNMT functions (Stresemann and Lyko 2008). These features give 5-aza-CR a very powerful erasing ability, resulting in DNA hypomethylation, gene expression modification, and reactivation of silent genes in eukaryotic cells (Jones and Taylor 1981; Taylor and Jones 1982; Jones et al. 1983; Jones 1985a, b; Glover et al. 1986).

In agreement with these findings, human mesenchymal stem cells (MSCs) and skin fibroblasts were transformed into hematopoietic cells after an incubation with 5-aza-CR, granulocyte-macrophage colony-stimulating factor (GM-CSF), and stem cell factor (SCF) (Harris et al. 2011). Moreover, adult skin fibroblasts and granulosa cells, derived from different species, namely, human (Pennarossa et al. 2013; Brevini et al. 2014), porcine (Pennarossa et al. 2014), and dog (Brevini et al. 2016), were converted into a different cell type belonging to the same embryonic lineage or to a different one.

The "highly permissive state" obtained by cells, after 5-aza-CR erasing, was demonstrated by a decrease in global DNA methylation and was accompanied by significant phenotype changes with increased nuclear volume and highly decondensed chromatin (Pennarossa et al. 2013, 2014; Brevini et al. 2014; Manzoni et al. 2016). These morphological features are distinctive of highly plastic cells that display loosely packed chromatin, in order to maintain genes in a potentially open state and prepare them for future expression (Tamada et al. 2006).

Differently from iPSC, the erasing process is transient and does not drive converted cells in a stable irreversible pluripotent state, greatly reducing the risk of error and malignant transformation (Pennarossa et al. 2013, 2014; Brevini et al. 2014).

Once cells enter into the "higher plasticity window," they can be easily directed toward a different phenotype if they are exposed to specific differentiation stimuli (Fig. 10.5).

In particular, human, porcine, canine, and murine skin fibroblasts were converted toward pancreatic lineage (Pennarossa et al. 2013, 2014, 2018; Brevini et al. 2014). At the end of the epigenetic conversion, they expressed the main hormones and glucose sensor genes specific of the pancreatic tissue. Furthermore, cell functionality was also demonstrated using severe combined immunodeficient (SCID) mice whose β-cells had been selectively destroyed with streptozotocin (Pennarossa et al. 2013, 2014).

The possibility to apply epigenetic conversion to different cell types was further proved using granulosa cells as starting cell population and converting them into muscle cells (Brevini et al. 2014) and human foreskin fibroblasts that were readdressed to neural progenitor-like cells (Mirakhori et al. 2015). Another study demonstrated the conversion of human and murine fibroblasts into proliferating chemical-induced neural progenitor cells (ciNPC), using a cocktail containing inhibitors of histone deacetylation, glycogen synthase kinase, and TGF-β pathway under physiological hypoxic conditions (5% O_2) (Cheng et al. 2014).

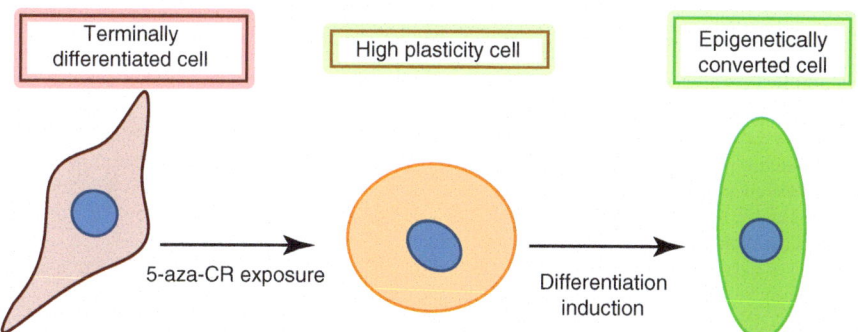

Fig. 10.5 Epigenetic-converted cells are generated through exposure to 5-aza-CR. This agent induces a transient high plasticity state that allows cells to differentiate toward a new cell lineage

Further experiments described the epigenetic conversion of human skin fibroblasts into mature Schwann cells through the use of the HDAC inhibitor VPA (Thoma et al. 2014) with neuro-supportive and myelination capacity and with the expression of proteins specific of the peripheral nervous system. Recently, a combination of 5-aza-CR and PDGF has also been shown to successfully convert somatic cells into regenerative multipotent stem cells (Chandrakanthan et al. 2016).

Conclusions

Bona fide pluripotent stem cells from ungulates are not here yet but are much closer than ever before. Above all a much better-defined conceptual framework has emerged, thanks to the recognition of the difference between naïve and primed epiblast stage and between ESC and EpiSC. Within this framework, it looks as if ungulate cell lines belong to EpiSC type, and as such they can sustain a robust self-renewal and form teratomas, and their derivation is not restricted by the genotype but is not prone to chimera formation. However, this does not solve all the problems, and ungulate cell lines do not behave exactly as primate lines, essentially because not only their pre- and peri-implantation development is substantially different but also because differences in the molecular mechanism of cell fate specification are emerging.

Building on this rapidly expanding knowledge is leading the scientific community very close to the so far elusive goal to widen the range of species where pluripotency can be captured in a stable cell line. In parallel unexpected progress have been obtained using the growing knowledge of the epigenetic mechanisms controlling cell differentiation and commitments. Novel approaches have proved successfully in controlling and erasing differentiation, paving the way to alternative protocols for the generation of stem cells in ungulates.

Acknowledgments Carraresi Foundation and European Foundation for the Study of Diabetes (EFSD). The authors are members of the COST Actions FA1201 Epiconcept: Epigenetics and Periconception Environment, BM1308 Sharing Advances on Large Animal Models (SALAAM), CM1406 Epigenetic Chemical Biology (EPICHEM), and CA16119 In vitro 3-D total cell guidance and fitness (CellFit). A special thanks to Dr. G. Pennarossa, University of Milan, for the help in the preparation of the text and images.

References

Alberio R, Croxall N, Allegrucci C (2010) Pig epiblast stem cells depend on activin/nodal signaling for pluripotency and self-renewal. Stem Cells Dev 19:1627–1636

Anastasia L, Sampaolesi M, Papini N, Oleari D, Lamorte G, Tringali C, Monti E, Galli D, Tettamanti G, Cossu G, Venerando B (2006) Reversine-treated fibroblasts acquire myogenic competence in vitro and in regenerating skeletal muscle. Cell Death Differ 13:2042–2051

Bao S, Tang F, Li X, Hayashi K, Gillich A, Lao K, Surani MA (2009) Epigenetic reversion of post-implantation epiblast to pluripotent embryonic stem cells. Nature 461:1292–1295

Bao L, He L, Chen J, Wu Z, Liao J, Rao L, Ren J, Li H, Zhu H, Qian L, Gu Y, Dai H, Xu X, Zhou J, Wang W, Cui C, Xiao L (2011) Reprogramming of ovine fibroblasts to pluripotency via drug-inducible expression of defined factors. Cell Res 21:600–608

Berg DK, Smith CS, Pearton DJ, Wells DN, Broadhurst R, Donnison M, Pfeffer PL (2011) Trophectoderm lineage determination in cattle. Dev Cell 20:244–255

Brevini TA, Pennarossa G, Attanasio L, Vanelli A, Gasparrini B, Gandolfi F (2010) Culture conditions and signalling networks promoting the establishment of cell lines from parthenogenetic and biparental pig embryos. Stem Cell Rev 6:484–495

Brevini TA, Pennarossa G, Rahman MM, Paffoni A, Antonini S, Ragni G, deEguileor M, Tettamanti G, Gandolfi F (2014) Morphological and molecular changes of human granulosa cells exposed to 5-azacytidine and addressed toward muscular differentiation. Stem Cell Rev 10:633

Brevini TA, Pennarossa G, Acocella F, Brizzola S, Zenobi A, Gandolfi F (2016) Epigenetic conversion of adult dog skin fibroblasts into insulin-secreting cells. Vet J 211:52

Brons IG, Smithers LE, Trotter MW, Rugg-Gunn P, Sun B, Chuva de Sousa Lopes SM, Howlett SK, Clarkson A, Ahrlund-Richter L, Pedersen RA, Vallier L (2007) Derivation of pluripotent epiblast stem cells from mammalian embryos. Nature 448:191–195

Buehr M, Meek S, Blair K, Yang J, Ure J, Silva J (2008) Capture of authentic embryonic stem cells from rat blastocysts. Cell 135:1287–1298

Chandrakanthan V, Yeola A, Kwan JC, Oliver RA, Qiao Q, Kang YC, Zarzour P, Beck D, Boelen L, Unnikrishnan A, Villanueva JE, Nunez AC, Knezevic K, Palu C, Nasrallah R, Carnell M, Macmillan A, Whan R, Yu Y, Hardy P, Grey ST, Gladbach A, Delerue F, Ittner L, Mobbs R, Walkley CR, Purton LE, Ward RL, Wong JW, Hesson LB, Walsh W, Pimanda JE (2016) PDGF-AB and 5-Azacytidine induce conversion of somatic cells into tissue-regenerative multipotent stem cells. Proc Natl Acad Sci U S A 113:E2306

Chen S, Zhang Q, Wu X, Schultz PG, Ding S (2004) Dedifferentiation of lineage-committed cells by a small molecule. J Am Chem Soc 126:410–411

Chen S, Takanashi S, Zhang Q, Xiong W, Zhu S, Peters EC, Ding S, Schultz PG (2007) Reversine increases the plasticity of lineage-committed mammalian cells. Proc Natl Acad Sci 104:10482–10487

Cheng L, Hu W, Qiu B, Zhao J, Yu Y, Guan W, Wang M, Yang W, Pei G (2014) Generation of neural progenitor cells by chemical cocktails and hypoxia. Cell Res 24:665–679

van Eijk MJ, van Rooijen MA, Modina S, Scesi L, Folkers G, van Tol HT, Bevers MM, Fisher SR, Lewin HA, Rakacolli D, Galli C, de Vaureix C, Trounson AO, Mummery CL, Gandolfi F (1999) Molecular cloning, genetic mapping, and developmental expression of bovine POU5F1. Biol Reprod 60:1093–1103

Esteban MA, Xu J, Yang J, Peng M, Qin D, Li W, Jiang Z, Chen J, Deng K, Zhong M, Cai J, Lai L, Pei D (2009) Generation of induced pluripotent stem cell lines from Tibetan miniature pig. J Biol Chem 284:17634–17640

Evans MJ, Kaufman MH (1981) Establishment in culture of pluripotential cells from mouse embryos. Nature 292:154–156

Ezashi T, Telugu BP, Alexenko AP, Sachdev S, Sinha S, Roberts RM (2009) Derivation of induced pluripotent stem cells from pig somatic cells. Proc Natl Acad Sci U S A 106:10993–10998

Gafni O, Weinberger L, Mansour AA, Manor YS, Chomsky E, Ben-Yosef D, Kalma Y, Viukov S, Maza I, Zviran A, Rais Y, Shipony Z, Mukamel Z, Krupalnik V, Zerbib M, Geula S, Caspi I,

Schneir D, Shwartz T, Gilad S, Amann-Zalcenstein D, Benjamin S, Amit I, Tanay A, Massarwa R, Novershtern N, Hanna JH (2013) Derivation of novel human ground state naive pluripotent stem cells. Nature 504:282–286

Glover TW, Coyle-Morris J, Pearce-Birge L, Berger C, Gemmill RM (1986) DNA demethylation induced by 5-azacytidine does not affect fragile X expression. Am J Hum Genet 38:309–318

Guo G, Yang J, Nichols J, Hall JS, Eyres I, Mansfield W, Smith A (2009) Klf4 reverts developmentally programmed restriction of ground state pluripotency. Development 136:1063–1069

Hall VJ, Christensen J, Gao Y, Schmidt MH, Hyttel P (2009) Porcine pluripotency cell signaling develops from the inner cell mass to the epiblast during early development. Dev Dyn 238:2014–2024

Harris DM, Hazan-Haley I, Coombes K, Bueso-Ramos C, Liu J, Liu Z, Li P, Ravoori M, Abruzzo L, Han L, Singh S, Sun M, Kundra V, Kurzrock R, Estrov Z (2011) Transformation of human mesenchymal cells and skin fibroblasts into hematopoietic cells. PLoS One 6:e21250

Hou P, Li Y, Zhang X, Liu C, Guan J, Li H, Zhao T, Ye J, Yang W, Liu K, Ge J, Xu J, Zhang Q, Zhao Y, Deng H (2013) Pluripotent stem cells induced from mouse somatic cells by small-molecule compounds. Science 341:651–654

Huangfu D, Maehr R, Guo W, Eijkelenboom A, Snitow M, Chen AE (2008) Induction of pluripotent stem cells by defined factors is greatly improved by small-molecule compounds. Nat Biotechnol 26:795–797

Ichida JK, Blanchard J, Lam K, Son EY, Chung JE, Egli D, Loh KM, Carter AC, Di Giorgio FP, Koszka K, Huangfu D, Akutsu H, Liu DR, Rubin LL, Eggan K (2009) A small-molecule inhibitor of tgf-Beta signaling replaces sox2 in reprogramming by inducing nanog. Cell Stem Cell 5:491–503

Jones PA (1985a) Altering gene expression with 5-azacytidine. Cell 40:485–486

Jones PA (1985b) Effects of 5-azacytidine and its 2′-deoxyderivative on cell differentiation and DNA methylation. Pharmacol Ther 28:17–27

Jones PA, Taylor SM (1981) Hemimethylated duplex DNAs prepared from 5-azacytidine-treated cells. Nucleic Acids Res 9:2933–2947

Jones PA, Taylor SM, Wilson VL (1983) Inhibition of DNA methylation by 5-azacytidine. Recent Results Cancer Res 84:202–211

Kim D, Kim CH, Moon JI, Chung YG, Chang MY, Han BS, Ko S, Yang E, Cha KY, Lanza R, Kim KS (2009) Generation of human induced pluripotent stem cells by direct delivery of reprogramming proteins. Cell Stem Cell 4:472–476

Koh S, Piedrahita JA (2014) From "ES-like" cells to induced pluripotent stem cells: a historical perspective in domestic animals. Theriogenology 81:103–111

Kumar D, Talluri TR, Anand T, Kues WA (2015) Induced pluripotent stem cells: mechanisms, achievements and perspectives in farm animals. World J Stem Cells 7:315–328

Lange-Consiglio A, Corradetti B, Bizzaro D, Magatti M, Ressel L, Tassan S, Parolini O, Cremonesi F (2012) Characterization and potential applications of progenitor-like cells isolated from horse amniotic membrane. J Tissue Eng Regen Med 6:622–635

Lange-Consiglio A, Tassan S, Corradetti B, Meucci A, Perego R, Bizzaro D, Cremonesi F (2013) Investigating the efficacy of amnion-derived compared with bone marrow-derived mesenchymal stromal cells in equine tendon and ligament injuries. Cytotherapy 15:1011–1020

Li P, Tong C, Mehrian-Shai R, Jia L, Wu N, Yan Y, Maxson RE, Schulze EN, Song H, Hsieh CL, Pera MF, Ying QL (2008) Germline competent embryonic stem cells derived from rat blastocysts. Cell 135:1299–1310

Li Y, Cang M, Lee AS, Zhang K, Liu D (2011a) Reprogramming of sheep fibroblasts into pluripotency under a drug-inducible expression of mouse-derived defined factors. PLoS One 6:e15947

Li Y, Zhang Q, Yin X, Yang W, Du Y, Hou P, Ge J, Liu C, Zhang W, Zhang X, Wu Y, Li H, Liu K, Wu C, Song Z, Zhao Y, Shi Y, Deng H (2011b) Generation of iPSCs from mouse fibroblasts with a single gene, Oct4, and small molecules. Cell Res 21:196–204

Liu J, Balehosur D, Murray B, Kelly JM, Sumer H, Verma PJ (2012) Generation and characterization of reprogrammed sheep induced pluripotent stem cells. Theriogenology 77(338-346):e331

Manzoni EF, Pennarossa G, deEguileor M, Tettamanti G, Gandolfi F, Brevini TA (2016) 5-azacytidine affects TET2 and histone transcription and reshapes morphology of human skin fibroblasts. Sci Rep 6:37017

Mirakhori F, Zeynali B, Kiani S, Baharvand H (2015) Brief azacytidine step allows the conversion of suspension human fibroblasts into neural progenitor-like cells. Cell J 17:153–158

Montserrat N, Bahima EG, Batlle L, Hafner S, Rodrigues AM, Gonzalez F, Izpisua Belmonte JC (2011) Generation of pig iPS cells: a model for cell therapy. J Cardiovasc Transl Res 4:121–130

Moschidou D, Mukherjee S, Blundell MP, Drews K, Jones GN, Abdulrazzak H, Nowakowska B, Phoolchund A, Lay K, Ramasamy TS, Cananzi M, Nettersheim D, Sullivan M, Frost J, Moore G, Vermeesch JR, Fisk NM, Thrasher AJ, Atala A, Adjaye J, Schorle H, De Coppi P, Guillot PV (2012) Valproic acid confers functional pluripotency to human amniotic fluid stem cells in a transgene-free approach. Mol Ther 20:1953–1967

Nagy K, Sung HK, Zhang P, Laflamme S, Vincent P, Agha-Mohammadi S, Woltjen K, Monetti C, Michael IP, Smith LC, Nagy A (2011) Induced pluripotent stem cell lines derived from equine fibroblasts. Stem Cell Rev 7:693

Nichols J, Smith A (2009) Naive and primed pluripotent states. Cell Stem Cell 4:487–492

Nichols J, Smith A (2011) The origin and identity of embryonic stem cells. Development 138:3–8

Okita K, Ichisaka T, Yamanaka S (2007) Generation of germline-competent induced pluripotent stem cells. Nature 448:313

Pennarossa G, Maffei S, Campagnol M, Tarantini L, Gandolfi F, Brevini TA (2013) Brief demethylation step allows the conversion of adult human skin fibroblasts into insulin-secreting cells. Proc Natl Acad Sci U S A 110:8948–8953

Pennarossa G, Maffei S, Campagnol M, Rahman MM, Brevini TA, Gandolfi F (2014) Reprogramming of pig dermal fibroblast into insulin secreting cells by a brief exposure to 5-aza-cytidine. Stem Cell Rev 10:31–43

Pennarossa G, Santoro R, Manzoni E, Pesce M, Gandolfi F, Brevini T (2018) Epigenetic erasing and pancreatic differentiation of dermal fibroblasts into insulin-producing cells are boosted by the use of low-stiffness substrate. Stem Cell Rev Rep In press

Rim JS, Strickler KL, Barnes CW, Harkins LL, Staszkiewicz J, Gimble JM, Leno GH, Eilertsen KJ (2012) Temporal epigenetic modifications differentially regulate ES cell-like colony formation and maturation. Stem Cell Discov 2:45–57

Rossant J (2011) Developmental biology: a mouse is not a cow. Nature 471:457–458

Smith AG (2001) Embryo-derived stem cells: of mice and men. Annu Rev Cell Dev Biol 17:435–462

Soto DA, Ross PJ (2016) Pluripotent stem cells and livestock genetic engineering. Transgenic Res 25:289–306

Spencer ND, Gimble JM, Lopez MJ (2011) Mesenchymal stromal cells: past, present, and future. Vet Surg 40:129–139

Stresemann C, Lyko F (2008) Modes of action of the DNA methyltransferase inhibitors azacytidine and decitabine. Int J Cancer 123:8–13

Sumer H, Liu J, Malaver-Ortega LF, Lim ML, Khodadadi K, Verma PJ (2011) NANOG is a key factor for induction of pluripotency in bovine adult fibroblasts. J Anim Sci 89:2708–2716

Tachibana M, Ma H, Sparman ML, Lee HS, Ramsey CM, Woodward JS, Sritanaudomchai H, Masterson KR, Wolff EE, Jia Y, Mitalipov SM (2012) X-chromosome inactivation in monkey embryos and pluripotent stem cells. Dev Biol 371:146–155

Takahashi K, Yamanaka S (2006) Induction of pluripotent stem cells from mouse embryonic and adult fibroblast cultures by defined factors. Cell 126:663–676

Talbot NC, Blomberg le A (2008) The pursuit of ES cell lines of domesticated ungulates. Stem Cell Rev 4:235–254

Tamada H, Van Thuan N, Reed P, Nelson D, Katoku-Kikyo N, Wudel J, Wakayama T, Kikyo N (2006) Chromatin decondensation and nuclear reprogramming by nucleoplasmin. Mol Cell Biol 26:1259–1271

Taylor SM, Jones PA (1982) Changes in phenotypic expression in embryonic and adult cells treated with 5-azacytidine. J Cell Physiol 111:187–194

Telugu BP, Ezashi T, Roberts RM (2010) The promise of stem cell research in pigs and other ungulate species. Stem Cell Rev 6:31–41

Tesar PJ, Chenoweth JG, Brook FA, Davies TJ, Evans EP, Mack DL (2007) New cell lines from mouse epiblast share defining features with human embryonic stem cells. Nature 448:196–199

Thoma EC, Merkl C, Heckel T, Haab R, Knoflach F, Nowaczyk C, Flint N, Jagasia R, Jensen Zoffmann S, Truong HH, Petitjean P, Jessberger S, Graf M, Iacone R (2014) Chemical conversion of human fibroblasts into functional Schwann cells. Stem Cell Rep 3:539–547

Vassiliev I, Vassilieva S, Beebe LF, McIlfatrick SM, Harrison SJ, Nottle MB (2010) Development of culture conditions for the isolation of pluripotent porcine embryonal outgrowths from in vitro produced and in vivo derived embryos. J Reprod Dev 56:546–551

West FD, Terlouw SL, Kwon DJ, Mumaw JL, Dhara SK, Hasneen K, Dobrinsky JR, Stice SL (2010) Porcine induced pluripotent stem cells produce chimeric offspring. Stem Cells Dev 19:1211

West FD, Uhl EW, Liu Y, Stowe H, Lu Y, Yu P, Gallegos-Cardenas A, Pratt SL, Stice SL (2011) Brief report: chimeric pigs produced from induced pluripotent stem cells demonstrate germline transmission and no evidence of tumor formation in young pigs. Stem Cells 29:1640–1643

Wu Z, Chen J, Ren J, Bao L, Liao J, Cui C, Rao L, Li H, Gu Y, Dai H, Zhu H, Teng X, Cheng L, Xiao L (2009) Generation of pig induced pluripotent stem cells with a drug-inducible system. J Mol Cell Biol 1:46–54

Wu J, Okamura D, Li M, Suzuki K, Luo C, Ma L, He Y, Li Z, Benner C, Tamura I, Krause MN, Nery JR, Du T, Zhang Z, Hishida T, Takahashi Y, Aizawa E, Kim NY, Lajara J, Guillen P, Campistol JM, Esteban CR, Ross PJ, Saghatelian A, Ren B, Ecker JR, Izpisua Belmonte JC (2015) An alternative pluripotent state confers interspecies chimaeric competency. Nature 521:316–321

Xu RH, Peck RM, Li DS, Feng X, Ludwig T, Thomson JA (2005) Basic FGF and suppression of BMP signaling sustain undifferentiated proliferation of human ES cells. Nat Methods 2:185–190

Xu XQ, Graichen R, Soo SY, Balakrishnan T, Rahmat SN, Sieh S, Tham SC, Freund C, Moore J, Mummery C, Colman A, Zweigerdt R, Davidson BP (2008) Chemically defined medium supporting cardiomyocyte differentiation of human embryonic stem cells. Differentiation 76:958–970

Ying QL, Nichols J, Chambers I, Smith A (2003) BMP induction of Id proteins suppresses differentiation and sustains embryonic stem cell self-renewal in collaboration with STAT3. Cell 115:281–292

The Legal Governance of Animal Biotechnologies

11

Nils Hoppe

Abstract

This chapter seeks to give an overview of governance approaches to animal bio-technologies. It will identify common themes and trends. The law is in most cases a domestic affair and it would be outside the remit of this chapter if it sought to analyse the legal situation in any number of states. The German and the European regulatory frameworks will be used by way of an illustration how institutions and normative concepts are connected and interact, but this will be put in the greater context of understanding regulation. The aim of this chapter is, therefore, to give the reader an overview over the structure of the governance landscape, the issues and challenges and the approaches to deal with these. Nevertheless, the difficulties in encompassing the reality of how we use animals, and need to use animals, and our moral perception of animals are a common theme which we need to be aware of to understand how the regulation described in this chapter functions and how norms and institutions interact. The chapter will initially outline the different frameworks and institutions which play a role in the legal governance of animal biotechnologies. Afterwards, it will look at how governance works in this field, before discussing how the law approaches animal biotechnologies. Finally, the chapter will take a look at issue after the market introduction of an animal biotechnology before drawing some conclusions.

N. Hoppe
CELLS—Centre for Ethics and Law in the Life Sciences, Leibniz Universität Hannover, Hannover, Germany
e-mail: nils.hoppe@cells.uni-hannover.de

© Springer International Publishing AG, part of Springer Nature 2018
H. Niemann, C. Wrenzycki (eds.), *Animal Biotechnology 2*,
https://doi.org/10.1007/978-3-319-92348-2_11

11.1   Introduction

Law is a reflection of societal attitudes to relationships. Animal biotechnology, understood as the 'application of scientific and engineering principles to the processing or production of materials by animals to provide goods or services' (USDA 2017), gives rise to a number of relationships between institutions, individuals and the subject matter of the technology: animals. As animals are not simply inert objects but living things, it is this relationship which causes a great deal of legal brow furrowing, especially where society's perception of this relationship changes dramatically.

Where our attitude towards a certain activity, relationship or tangible (commoditised) object changes over time, the law changes along with it (though, admittedly, sometimes rather slowly). This is generally true for all manner of drastic—and overdue—societal changes, and it is clear that the societal perception of animals in society has fluctuated dramatically in this way since the first half of the twentieth century. Kellert and Westervelt showed in 1983 that public debate surrounding animals was strongly utilitarian during the war years, especially during World War I (Kellert and Westervelt 1983). The contemporaneous context of society dictated that under harsh wartime conditions we could not afford to view animals as anything other than means to an end. Such utilitarian attitudes to animals then declined again between the end of the war and 1976 (which was the end of Kellert and Westervelt's analysis period). They conclude that:

> …[m]ajor increases in the aesthetic and humanistic attitudes in the urban newspapers intimates the growth of a more appreciative and emotional perspective toward animals among city residents. Additionally, a pronounced increase in the ecologistic attitude in the urban areas suggests a more protection-attitude toward wildlife and natural habitats among urban residents. Significant contrasts with the rural areas on these attitude dimensions intimates major urban/rural differences in the decades ahead. (Kellert and Westervelt 1983)

The empirical gist of their analysis can be summarised as this: the attention we pay to animals' welfare is secondary to that which we pay to our own welfare. We use animals to further our own welfare to a point and make the animals' welfare contingent on this. Whether, and if so, to what extent, this is morally defensible is a subject matter of contentious debate, and many positions in this debate are plausible and justifiable, even if some of these positions are often articulated poorly or aggressively.

When looking at legal governance aspects of animal biotechnologies, we are therefore—to a certain extent—concerned with the moral justifiability of activities involving animals. By and large, however, the law deals with phenomena in the real world, and the evidence on how *we actually treat animals* naturally interests us more than the discussion of how *we should treat animals*.

It is trite to write that the domestication of animals is one of the most important cultural achievements of the last 13,000 years (Diamond 2002). It is not so trite to posit that the social effects of domestication are manifold (Twine 2010), and a legally significant aspect of this domestication is the commodification of the animal and with it the legislative categorisation of animals as commodities that can be owned, sold, bought, used, destroyed, etc. Indeed, a first look at German civil law (*Bürgerliches Gesetzbuch, BGB*) reveals that there is an assumption that animals are constructive commodities of some sort, Section 90a BGB:

> Animals are not commodities. They are protected by specific legislation. The law in relation to commodities can be applied to them unless the law provides otherwise. (My translation.)

This is the crux of the major regulatory challenge many legislators have struggled with in relation to animal biotechnologies: whilst history bequeaths a strong, utilitarian property approach towards animals, developing societal attitudes demand that we increasingly see them as imbued with a moral status which forbids some of the activities we would ordinarily associate with regular commodities. This balancing act is a common theme that we can find in numerous instruments and norms dealing with the relationship of people and institutions in the context of the use of animals (and in other jurisdictions, too. See, e.g. the Netherlands (Brom and Schroten 1993)).

11.1.1 Strength of Norms

A particularly interesting example is, since 2002, the provision of fundamental protection of animals in the German constitution (*Grundgesetz, GG*), Art. 20a GG:

> Mindful of its responsibility for future generations, the state protects [...] animals within the framework of constitutional order by way of legislation and on the basis of the law through the executive and judiciary. (My translation.)

Whilst this does not, prima facie, sound particularly contentious, it is its positioning in the normative hierarchy that carries the hidden message here: to provide for the protection of animals in a constitutional norm means that the welfare of animals can be balanced against other constitutional norms (e.g. that of scientific freedom, Art. 5(3) GG). In other words, other constitutional rights and freedoms do not automatically trump the welfare of animals any more. This was a dramatic change in the way the law treated the use of animals within the German legal landscape (and has had significant knock-on effects, as we will see later on). I will address hierarchies of norms below, but it is worth noting the importance of understanding these hierarchies in the context of animal biotechnologies at this point. Where a rule in relation to how we treat animals is read, the reader needs to be able to contextualise the value of that rule: is it a general item of guidance, or a constitutional norm of exceptional weight?

Away from such (unusual) constitutional norms, the issue in Germany is somewhat compounded by the use of quasi-religious language in section 1 of the German Animal Protection Act (*Tierschutzgesetz, TierSchG*). Rather unusually, the provision has this wording, s. 1 TierSchG:

The purpose of this Act is to protect the life and welfare of animals, **based on our human responsibility for animals as fellow creatures.** […] (My translation, my emphasis; the German term 'Geschöpf' smacks significantly more of 'creation' than the English term 'creature'..)

Both of these norms show the difficulty legislators have in encoding our extensive history of using animals as means to an end (and its continued necessity in modern society) in a way which can be reconciled with our appreciation of the special moral status of animals as, if I may paraphrase, fellow living things (rather than 'creatures'). The German legislation is a prominent example but by no means the only context in which we encounter this issue. The last years have seen a number of attempts to free non-human primates (NHP) from captivity on the basis of seventeenth-century legal doctrine (*habeas corpus*). The doctrine of habeas corpus provides that a prisoner is entitled to be produced at court to have the legitimacy of their detention tested. These cases involving primates do not generally survive an initial test, and certainly not an appeal, but there are first instance decisions which have been significantly less than critical of the assertion that human rights and freedoms should equally apply to NHP. Instead of an outright refusal to hear such cases, a court recently issued an order requiring a New York university to explain why the chimpanzees Hercules and Leo should not be freed immediately (*Nonhuman Rights Project Inc., on behalf of Hercules and Leo v. Stony Brook University 152736/2015 (Manhattan Supreme Court), 20 April 2015*). Even more recently, the animal rights organisation PETA (People for the Ethical Treatment of Animals) attempted to assert copyright on behalf of a crested macaque who had famously taken a selfie using a camera left by a wildlife photographer (reportedly bankrupting the reporter in the process of litigation (Wong 2017)). It is difficult to see how such legal action is anything other than a disruptive abuse of process; from the perspective of how the law works, it is futile and a waste of resources to attempt to legally equate humans with animals (rather than work towards a better normative framework for the protection of animals as sui generis categories of protectable entities). What it may be an indication of is this: there is sufficient uncertainty in the law governing our use of animals to enable some actors to try to blur the boundaries between humans, animals and commodities. Nonetheless, it is clear that some form of firm regulation in this field is desirable (as is commonly the case in the science context). Where there is strong, detailed and harmonised regulation, standards are better and the scientific output improves (or so the theory goes). Where this is not the case, scientific quality may suffer as a result (Mohr et al. 2016).

11.1.2 Aims and Objectives

Two things are important to note. Firstly, this chapter seeks to give an overview of governance approaches to animal biotechnologies. It will identify common themes and trends. Law is, in the end, always a domestic affair and it would be outside the remit of this chapter (and well outside the remit of my competence) if I sought to analyse the legal situation in any number of states. I will use the German and the European regulatory frameworks by way of an illustration how institutions and normative concepts are connected and interact but will put this in the greater context of understanding regulation. The aim of this chapter is, therefore, to give the reader an overview over the structure of the governance landscape, the issues and challenges and the approaches to deal with these. Secondly, I have already alluded to the fact that much of the regulatory difficulty in this context stems from some sort of moral status we recognise in animals which makes them different to other commodities. This means that much of our analysis of this part of governance is merely a faint echo of the much more substantive and challenging discussion of the ethics of animal biotechnologies, which is in the chapter from Lanzerath in this book.

Nevertheless, the difficulties in encompassing the reality of how we use animals, and need to use animals, and our moral perception of animals are a common theme which we need to be aware of to understand how the regulation described in this chapter functions and how norms and institutions interact. I will initially outline the different frameworks and institutions which play a role in the legal governance of animal biotechnologies. Afterwards, we will look at how governance works in this field, before discussing how the law treats the market introduction of the relevant products. Finally, we will take a look at post-marketing issue in relation to these biotechnologies before drawing some conclusions.

11.2 Normative Frameworks

It is worth spending some moments considering how legal frameworks in this context work. It is clear that legal norms are delicately placed cogs in intricate hierarchies of norms: otherwise we would not be able to set them off against each other where they collide. We have already seen that Germany has enshrined animal welfare in a constitutional norm, which is at the top of the normative hierarchy domestically. Where an ordinary item of primary legislation collides with this constitutional norm, it is the constitutional norm which survives the collision. Within a domestic legal system, there are constitutional norms (right at the top) and secondary legislation (such as orders outlining the way executive agencies expect the law to be interpreted), as well as local and regional guidelines and semi-private codes of conduct at the bottom of the stack. These hierarchies become broader at the bottom: whilst there is generally only one constitution in a state, there may be several executive agencies, for example at county/regional level, or for different purposes, or as emanations of different ministries or departments. The same is true for the courts charged with interpreting the law where there is disagreement. There are many first instance courts in different locations and with different specialisations. The further a case travels to the top of a legal system, the fewer the courts with jurisdiction are and the weightier their judgement. Where we seek legal certainty in relation to animal biotechnologies, we need to look to the top of these hierarchies, norms and courts, to find it.

11.2.1 Supranational Sources of Law

Outside the domestic hierarchy, there are supranational and international instruments, and it is important to understand their place and to realise that these frameworks do not automatically supersede the domestic law (but often inform its interpretation in court). The European Union, as one of the most important creators of supranational law in this area, provides for a number of different instruments that mean different things in terms of their application and power. Starting from the bottom of the EU hierarchy of norms, there are:

- *Recommendations and opinions* (which are non-binding)
- *Decisions* (which are binding only for those to whom they are specifically addressed)
- *Directives* (which are binding on member states as to their intended result, but have to be implemented in domestic law)
- *Regulations* (which are binding in all member states and directly applicable)

The EU's hierarchy of norms is finely balanced: it contains soft mechanisms which can be used to carefully nudge protagonists in the right direction, as well as hard mechanisms (such as regulations) which create a legal reality regardless of the current statutory situation in each member state. The choice of mechanism also depends on how much competence the EU has in a particular regulatory area. In essence, the EU must not provide legislation in areas where there is no *treaty competence* or where the member state is in a significantly better place to regulate locally. Treaty competence means that the international treaties which establish the European Union contain explicit areas in which the EU can regulate for all member states. For example, article 13 of Title II of the Treaty of Lisbon (TFEU) states:

> In formulating and implementing the Union's agriculture, fisheries, transport, internal market, research and technological development and space policies, the Union and the Member States shall, since animals are sentient beings, pay full regard to the welfare requirements of animals, while respecting the legislative or administrative provisions and customs of the Member States relating in particular to religious rites, cultural traditions and regional heritage.

This, together with Article 114 TFEU (which contains the EU's treaty competence for environmental protection), is the basis for passing the EU's main instrument for regulating the use of animals for scientific purposes (Directive 2010/63/EU). The directive outlines the provisions member states had to implement in order to protect animals which are used in scientific contexts. It is worth remembering that a directive by itself is relatively unhelpful (with very few exceptions) when it comes to understanding the law in a certain context. The directive can only provide a vague idea of what the European legislator wanted to achieve. The implementation is to be found in domestic law, and this is where many of the issues we can identify as problems arise—through unclear or half-hearted implementation of EU directives in domestic law.

Outside the European Union's legislative framework, there are international treaties which are either bilateral (i.e. between two sovereign states) or multilateral (i.e. between many states), agreeing on certain common norms. The *European Convention for the Protection of Vertebrate Animals Used for Experimental and Other Scientific Purposes* (ETS No. 123) is such an international treaty (established by the Council of Europe), and it is very important to remember that the Council of Europe and the European Union are entirely separate entities. The Council of Europe has a small number of treaties (or conventions) in relation to the welfare of animals: farm animals (ETS No. 87), pets (ETS No. 125), animals destined for slaughter (ETS No. 102) and the international transport of animals (ETS No. 065; ETS No. 103). Each of these conventions has its own list of signatories and ratifications and it is by no means the case that all treaties are accepted by all states. It is also rather problematic to enforce non-adherence to a convention in relation to animals: the Council of Europe only provides one forum for enforcing convention rights—the European Court of Human Rights in Strasbourg. The court's jurisprudence regularly deals with animal rights organisations' petitions in relation to the organisations' convention rights, but quite appropriately, not with petitions on behalf of animals.

11.2.2 Non-governmental Organisations

Remaining at the level of international organisations (such as the Council of Europe), the World Trade Organisation provides a number of agreements which impact on animal biotechnologies. TRIPS, the agreement on Trade-Related Aspects of Intellectual Property Rights, regulates the intellectual property exploitation of biotechnologies in an international context. Additionally, there are agreements which seek to eliminate unnecessary product requirements where these can be a barrier to trade (e.g. the Agreement on Technical Barriers on Trade, TBT, and the Agreement on Sanitary and Phytosanitary Measures, SPS). SPS prescribes that regulatory frameworks in relation to animal health must be necessary to achieve the required level of protection only, non-arbitrary or non-discriminatory, science-based and must not lead to undue delay. The World Organisation for Animal Health (OIE), based in Paris, is concerned (amongst other things) with the sanitary safety of animals and animal products which are traded internationally. Whilst these agreements often contain relevant rules and norms, it is rare that we can refer to them directly when making a case for, or against, a certain kind of activity in the area of animal biotechnologies: we always have to go back to domestic law to find the relevant national norm. Where it is clear that the national norm is in conflict with the state's treaty or agreement obligations, it is not possible to disregard the domestic norm and choose the treaty norm. Instead, the state in question will have to be reminded of its international obligations and the long and arduous process of changing domestic law has to be started.

There is, to a certain extent, a palpable trend towards normative convergence across different jurisdictions (not just in the setting of the EU's desirable convergence and harmonisation efforts). This is by and large a phenomenon which can be attributed to the power of the markets (i.e. the ability to globally market a product, which means that some sort of common standard is helpful), as well as the international interconnectedness of research. If we take a look across the Atlantic at the legislative instruments in the United States, we can identify similar but not quite identical norms as those in EU member states. The general direction of governance is similar, but the conditions for working in the context of animal biotechnologies differ, making it an attractive proposition to 'forum shop' when it comes to the question where certain work is carried out (i.e. to seek out the jurisdiction with the least impediment to the animal biotechnology in question). It is additionally worth noting that we do not have to look all the way to the United States to identify a good amount of fragmentation. We have already seen that, whilst the international norms and norm-giving bodies we have identified are informative, it is in most cases (with the notable exception of EU Regulations) usually domestic law which we have to turn to in order to understand the legal regulation of animal biotechnologies. Looking again at our example of German law, the state's international obligations have percolated into significantly fragmented and unsatisfactory collection of norms at federal as well as regional level. This has diminished legal certainty in relation to the ability to use animals for biotechnological purposes and has made the process of research in this area enormously difficult.

11.2.3 Legal Approaches

There are a number of different legal approaches to governing animal biotechnologies. This is a result of a number of competing sets of norms in this rather diverse context. The methods deployed in this field can be used to modify and breed livestock for the provision of food or for experimental purposes, including the creation of specifically modified organisms for xenotransplantation research. These diverse uses touch on normative areas such as agriculture, consumer protection and food safety, genetically modified organisms (GMO), animal welfare and—potentially—in vitro fertilisation (IVF) and embryo protection legislation. When assessing the normative landscape in relation to a certain use, it is important to remember the discussion in relation to the hierarchies of norms. It may very well be that there is a convention in relation to a certain use case, but the member state in which the technology is used may simply not have signed up to the convention. It is worth reiterating what was discussed at the outset, namely, that most legal frameworks provide for a fundamental understanding of how the legislator envisages how animals are used (i.e. in some cases not at all) and then opens individual doors with very specific legislation. We will look at these areas of regulation in the next section.

11.3 International and Supranational Governance

The majority of European instruments for the governance of animal use in biotechnology, in particular in the context of scientific research, have an interestingly linear provenance. The germ cell of this lineage can be said to be ETS No. 123 of 1986, the *European Convention for the Protection of Vertebrate Animals Used for Experimental and Other Scientific Purposes*. Subsequent legislators, domestic and supranational, used this Convention as a mould for casting own norms. The Convention entered into force in 1991 with four ratifications. By 2016, it was signed and ratified by 21 states and the European Union. The preamble of the Convention sets the tone for this and subsequent legislation:

> [...] Recognising that man has a moral obligation to respect all animals and to have due consideration for their capacity for suffering and memory;
>
> Accepting nevertheless that man in his quest for knowledge, health and safety has a need to use animals where there is a reasonable expectation that the result will be to extend knowledge or be to the overall benefit of man or animal, just as he uses them for food, clothing and as beasts of burden;
>
> Resolved to limit the use of animals for experimental and other scientific purposes, with the aim of replacing such use wherever practical, in particular by seeking alternative measures and encouraging the use of these alternative measures;
>
> Desirous to adopt common provisions in order to protect animals used in those procedures which may possibly cause pain, suffering, distress or lasting harm and to ensure that where unavoidable they shall be kept to a minimum [...]

The tenor of this preamble can be found, in an abbreviated form, in the preamble of the main EU directive, which we will return to again below. It is worth briefly looking at the different areas which the Convention regulates in order to understand the interplay between international and supranational law here.

The Convention applies to the experimental use of animals only, which is defined as any activity which is aimed at avoiding and preventing illness, such as testing drugs and products (Article 2 a i), to diagnose and treat (Art. 2 a ii), undertake research (Art. 2 d) and for the purposes of education and training (Art. 2 e). It requires that there be a good general environment for the animals (Art. 5) and expects that the best standard in procedures is used (Art. 6). Further stipulations concern the appropriate choice of species for the experiment (Art. 7) and the anaesthetic that is to be used (Art. 8). There are specific licencing (Art. 9) and authorisation rules (Art. 13), and the Convention also addresses the sacrifice (Art. 11) and setting free (Art. 12) of experimental animals. Interestingly, the Convention also addresses the possibility of deviating from the rules where the experimental setup requires this (Art. 10). The Convention distinguishes the rules for breeding establishments (Arts. 14–17) and those for user establishments (Arts. 18–24), outlines training and education requirements (Arts. 25 and 26) and sets reporting standards (Art. 29). In summary, the Convention provides member states with a strong, top-down framework for those issues that ought to be addressed when encapsulating the experimental use of animals in domestic rulesets. Whilst the rules are by and large generalised, they are also to the point in a fashion which makes it difficult to see any later implementation significantly deviate from them.

As the Convention was signed and ratified by the European Union (Council Decision 1999/575/EC), it stands to reason that EU instruments would also not deviate from the spirit of the Convention. Indeed, the current Directive's predecessor (Directive 86/609/EEC *on the approximation of laws, regulations and*

administrative provisions of the member states regarding the protection of animals
used for experimental and other scientific purposes) was virtually identical to the
Convention.

We have seen the balancing act which is performed in the Convention's preamble, and we started off this chapter discussing the generally very difficult situation
legislators find themselves in when trying to find the right normative place for animals in biotechnology. When we move on from Convention level to look at how the
use of animals for experimental purposes is governed in the EU, we immediately
come across an example of this, namely, in para. 10 of the preamble of Directive
2010/63/EU, which I have already alluded to above:

> [...] this Directive represents an important step towards achieving the final goal of full
> replacement of procedures on live animals for scientific and educational purposes as soon
> as it is scientifically possible to do so.

The Directive's preamble is silent on how this is to be achieved, given the pleth-
ora of novel challenges and responses thereto which we encounter every year. At
this point in time, it is difficult to see this assertion as anything more than symbol-
ism or wishful thinking. It does, however, perform one important task: it makes the
pragmatic baseline for the use of animals in science clear. 'We would rather not use
animals but, for the time being, we must'. It is also obvious that the EU's preamble
is a stronger version of the carefully worded Convention preamble. This may be due
to the comparably strong wording of the treaty (see Art. 13, quoted in full above)
and with the balancing exercise prescribed by Art. 191 TFEU:

> [...] Union policy on the environment shall contribute to pursuit of the following objectives:
> preserving, protecting and improving the quality of the environment, **protecting human**
> **health, prudent and rational utilisation of natural resources** [...] (My emphasis.)

The relevant Directive (2010/63/EU) on the protection of animals used for scientific purposes sets the scene for EU-level regulation of the scientific use of animals.
It is based on the principles of the 3R (reduce, replace, refine) (Russel and Burch
1959), which in turn is the reflection of the desired world without animal experimentation hinted at in the Directive's preamble: the idea is that a consequent deploy-
ment of refinement, reduction and replacement will lead to a diminishing need for
animals in this field. The scope of the Directive is slightly wider than one would
ordinarily expect, extending to foetuses of mammalian species in their last trimester
of developments, as well as to cephalopods. The type of activity the Directive covers is basic research, higher education and training (i.e. the first half of the biotechnology spectrum). Much like the Convention, the Directive lays down minimum
standards for housing and care and stipulates a systematic project evaluation. The
Directive expects regular, risk-based inspections and mandates non-technical project summaries as well as retrospective assessments.

Art. 8 of the Directive concerns itself with the use of non-human primates, which
is of particular interest from a legal perspective. The ordinary way of regulating an
activity would be to work on the assumption that any activity which is not actively

proscribed is therefore allowed. In the case of non-human primates, the Directive works on the assumption that there is a general prohibition of using such primates for research and then opens up individual doors for exemptions from this prohibition. These doors are where the work proposed concerns debilitating or life-threatening conditions in humans and where such work could not be achieved using other species. This norm, prima facie aiming at protecting non-human primates, essentially creates a system in which it is permissible to use animals to help humans but prohibited to use animals to help animals: veterinary research is, in this context, precluded.

At EU level, there are a number of additional instruments which make up the supranational regulatory landscape. Decision 2012/707/EU deals with the common stipulated format for the information that has to be submitted on the use of animals for scientific purposes (i.e. there is an established reporting structure for controlling animal use in the EU), which has started collecting data from 1 January 2014. Recommendation 2007/526/EC contains guidelines for the accommodation and care of animals used for experimental and other scientific purposes.

11.4   National Governance

Both the Convention and, where applicable, the Directive are instruments which will have been implemented in domestic legal systems in those states which have signed up to either the first or both of the systems in question. In an ideal world, this should mean that there is at least widespread convergence of norms in relation to animals used for scientific purposes, possibly even harmonisation. At the same time, it is part and parcel of the nature of Directives that they give a great deal of discretion to member states how to implement the aims of the legislation. In areas of ethical volatility, such as this, this often leads to fragmentation across different legal and ethical cultures.

In Germany, the norms of the Directive are combined with the pre-existing 1970s legislative provisions of the Animal Protection Act (*TierSchg*). The *TierSchG* is an act which contains preventative prohibitions and stipulates that, whilst the use of animals for scientific purposes is permitted, this must be done under the mantle of appropriate licences (i.e., those who wish to experiment using animals cannot do so unless they have expressed official permission in each individual case). This, prima facie, collides with the researchers' constitutional freedom of research (*Art. 5(3) GG*). A derogation from a constitutional right or freedom is only possible where an equally weighty right is in play. We have seen at the outset that, from 2002 onwards, the German constitution contained a provision for the protection of animals (*Art. 20a GG*), which has put the provisions of the *TierSchG* on much more solid constitutional ground than it was in the previous 30 years.

The *TierSchG* has a very broad scope. It is concerned with where animals are used to generate original knowledge under conditions which may cause pain, suffering or harm to the animal (or, interestingly, where a germ-line modification may lead to pain, suffering or harm to subsequent generations). Section 1 of the TierSchG contains the blanket prohibition of causing pain, suffering or harm to animals without reasonable grounds (the reasonable grounds being licenced activities or other statutory defences (such as self-defence or necessity)). Acceptable, and therefore licensable, activities include human or veterinary biomedical research, generating knowledge of new drugs, environmental hazards, toxicity of chemicals and foodstuffs, as well as basic research. Unacceptable uses are experiments aiming to test the efficacy of weapons, ammunitions and related items. Generally unacceptable (i.e. exceptions are possible) are tobacco product testing, detergents and cosmetics. The *TierSchG* further provides for a framework for outlining the experimental setup, issues in relation to accountability and reporting, and the question who is in fact entitled to work with animals at all. Art. 8 of the Directive (on non-human primates) was implemented into German domestic law by way of secondary legislation (*§23 TierSchVersV*) rather than in primary law such as the *TierSchG* (with mixed results— the interpretation and continuing application of the *TierSchVersV* is rather debatable).

The law here also requires researchers to make a number of cases in relation to their methodology, the choice of species and the effects on the animal and also in relation to the ethics of the experiment proposed. There is an expectation that a balancing of benefits and burdens is performed and this is part of the basis of the licence which is eventually granted (though this aspect of the licence application is subject to scrutiny (Röcklingsberg et al. 2014)). Licence contraventions are punishable by a fine up to €35,000 or even imprisonment. It is, to say the least, an exotic regulatory approach to ask the potential infringer to (a) pre-emptively argue his defence, (b) act as the expert witness in his own case and (c) attach significant criminal law norms to an area of inherent scientific uncertainty. It can be said that the law in this particular area tries to achieve too many things: transpose supranational law in the spirit of international treaties, make animals available for use but treat them with respect, regulate access to animals on the basis of applicants' own assessments and attach criminal sanctions to situations of immense scientific uncertainty. This is, of course, hugely unsatisfactory and creates a great deal of legal uncertainty for all involved. The path to a better way of regulating here is, however, not immediately obvious.

11.5   Other Areas of Law and Jurisdictions

There are a number of other areas of law and features of other jurisdiction which one should look at briefly. Agricultural law, for example, is the body of norms, nationally and internationally, which regulates the way natural resources are used. Notions of agricultural law will play a role in at least two different settings: where there is the production of an animal for the downstream purpose of agricultural use and where there is agricultural production of animals for a different downstream purpose (such as industrial-scale production of experimental animals or for xenotransplantation). The regulation of agriculture also aims at setting appropriate standards for the output of products of comparable quality and safety. The treaty bases generally cited by the EU instruments in this area are those of consumer protection and environmental protection. Common standards for the safety of products which are shared in a common market are an important mechanism for ensuring the economic viability of the EU as an economic union.

11.5.1 Precautionary Principle, GMO and CRISPR

There has, traditionally, been significant concern about the safety of genetically modified products, and it is this area particularly which has given rise to the precautionary principle (contained in Art. 191 TFEU). The precautionary principle is an unusual type of regulation, as it comes rather close to the codification of a slippery slope argument. It provides that an activity which *may cause harm* to the environment or the public is deemed to be harmful, and it is the obligation of those wishing to carry out that activity to prove otherwise. It is essentially a reversal of the burden of proof in cases where there is, ever so diffuse, fear that an activity may give rise to undesirable consequences.

At European level, genetically modified organisms (GMO) are subject to a whole bundle of instruments of differing intensity. There are three directives, 2001/18/EC (deals with the deliberate release of GMO into the environment), 2015/412/EU (amendment providing for member states to make their own rules on cultivation of GMO) and 2009/41/EC (which genetically modified microorganisms). In addition, there are three regulations: EC/1829/2003 (on genetically modified food and feed), EC/1830/2003 (on tracing and labelling requirements for genetically modified organisms and products) and EC/1946/2003 (on transboundary movements of GMO). The density of regulations, which we recall from the discussion of normative hierarchies is the hardest instrument in the EU's arsenal, shows that this is an important area of governance for the EU. At the same time, the EU's definition of a genetically modified organism is rather restrictive and seems to miss out on capturing novel technologies, such as CRISPR in its ambit. Article 2(2) of Directive 2001/18/EC states:

> [For the purposes of this Directive:] [...] 'genetically modified organism (GMO)' means an organism, with the exception of human beings, in which the genetic material has been altered in a way that does not occur naturally by mating and/or natural recombination.

It is subject to intense debate in the ethical and legal implications research community, whether a CRISPR intervention which removes a segment without introducing anything else to the organism's genome is caught by this definition. In addition, the Directive provides an annex with a list of different methods that amount to genetic modification (which is an unusual way to regulate in a highly innovative and fluid field). This is, to a certain extent, mirrored in a general stakeholder uncertainty about the differences in using traditional breeding, GM or genomics techniques to produce certain animals (Coles et al. 2015). This is not unlike the legal confusion that is created by the context of breeding human-animal chimera for xenotransplantation purposes. Most IVF norms presuppose the use of a human ovum (e.g. the German *ESchG*). Using, for example, a porcine ovum and introducing human material to create a chimera does not, immediately, trigger legislation in relation to IVF or embryo protection. Similar to the context of genetically modified organisms, the legislator seems to have missed a context here. Not so the bioethics context; there is some evidence that in this field in particular, bioethics expertise plays an elevated role and is likely to influence future regulation (Salter and Harvey 2014).

11.5.2 Product Classification

Another interesting legal issue arises when an animal is either a product in and of itself (such as in the case of genetically modified animals destined for research) or where it is the production unit for a subsequent regulated entity (e.g. where a xenotransplant is generated within an animal). Products derived from animal biotechnology often straddle the divide between agricultural products and health products. This issue is additionally compounded where, at the end of the translational spectrum, we consider an industrial-scale (agricultural) production of these entities. There are, additionally, a number of post-marketing vigilance instruments (derived in general from the broad area of consumer protection, see, e.g. FDA 2006); the main aim of which is to ensure that a licenced animal-derived biotechnology product does not harm consumers or the environment. Where the product is the animal itself, there are also general animal welfare rules which provide for the appropriate keeping of the animals. In all cases, the general laws of liability and product liability apply equally to animal biotechnologies as they do to other products: where there is a product and a consumer, there is a duty to ensure that the consumer is not harmed by the product.

11.5.3 Intellectual Property

Finally, and well outside the scope of this chapter, there is the discussion surrounding intellectual property rights in living things. Ever since the US Supreme Court decision in Diamond v. Chakrabarty (*1980, 447 US 303*) in which a General Electric engineer wanted to patent an oil-slick-devouring bacterium, there is an ongoing debate about 'patenting life'. The debate reached EU courts with the case of Brüstle v. Greenpeace (Case C-34/10, Oliver Brüstle v. Greenpeace e. V., 2011 E.C.R. I-9849), which turned on the question whether embryonic stem cells were legally the same as an embryo. There is significant literature available on this topic from the start of the debate (Raines 1990; O'Connor 1992; Morin 1997) all the way to today (Liddell 2012), and it suffices to say that issues in relation to intellectual property hinge greatly on where one is and what one wished to patent there. The disconnect between a generally felt moral attitude to intellectual protecting property rights in biotechnological innovations and the actual moral aspects of the patenting system is of particular relevance in this consideration (Forsberg et al. 2017).

Conclusion

The legal landscape surrounding the use of animals for the purposes of biotechnology is complex. It is, as we have seen, an amalgam of a pragmatic acknowledgement of the need to use animals for certain purposes and the realisation that animals are not merely lifeless things which can be owned and used to our hearts' content.

As the law is merely a reflection of society, the regulatory frameworks which we have explored in this chapter are, to a great extent, also a reflection of this societal challenge. Centuries of domestication and commodification have left their traces in the laws which are designed to protect animals from being used as means to an end in a context where this would be inappropriate. The balancing act performed by the law in this field is difficult: the German constitutional approach is an unusual way to enshrine the importance of interacting with living things in a respectful way, as it forces the rest of the regulatory system to move along with its legislative aim. At the same time, the objectives we have distilled from the preambles of international conventions and supranational instruments have shown that this is by no means a singular legislative occurrence.

It is also clear that good legal governance in this setting would lead to a better quality of research and a greater degree of reproducibility across the field. It is therefore likely that the legal system of governance of animal biotechnologies would greatly benefit from an overhaul. This is, unfortunately, not to be expected and we may anticipate having to work with the rulesets we have inherited for the foreseeable future. At the same time, there is increasing evidence that it may be a better course of action, in a regulatory sense, to acknowledge that biotechnological innovation generates new regulatory targets, which require bespoke approaches (Faulkner and Poort 2017).

Acknowledgements I am grateful for the kind assistance of Yvonne Stöber, Dennis Peters and Nikita Bangalore in preparing this manuscript.

References

Brom FWA, Schroten E (1993) Ethical questions around animal biotechnology. The Dutch approach. Livest Prod Sci 36(1):99–107

Coles D, Frewer LJ, Goddard E (2015) Ethical issues and potential stakeholder priorities associated with the application of genomic technologies applied to animal production systems. J Agric Environ Ethics 28(2):231–253

Diamond J (2002) Evolution, consequences and future of plant and animal domestication. Nature 418(6898):700–707

Faulkner A, Poort L (2017) Stretching and challenging the boundaries of law: varieties of knowledge in biotechnologies regulation. Minerva:1–20

Food and Drugs Administration (FDA). Animal Health and Consumer Protection (2006) https://www.fda.gov/aboutfda/whatwedo/history/productregulation/animalhealthandconsumerprotection/default.htm. Accessed 30 Aug 2017

Forsberg EM et al (2017) Patent ethics: the misalignment of views between the patent system and the wider society. Sci Eng Ethics:1–26

Kellert SR, Westervelt MO (1983) Historical trends in American animal use and perception. Int J Study Anim Probl 4(3):133–146

Liddell K (2012) Immorality and patents: the exclusion of inventions contrary to *ordre public* and morality. In: Lever A (ed) New frontiers in the philosophy of intellectual property. CUP, Cambridge

Mohr BJ et al (2016) The governance of animal care and use for scientific purposes in Africa and the Middle East. ILAR J 57(3):333–346

Morin E (1997) Of mice and men: the ethics of patenting animals. Health Law J 147:5

O'Connor KW (1992) Patenting animals and other living things. South Calif Law Rev 65:597

Raines LJ (1990) Public policy aspects of patenting transgenic animals. Theriogenology 33(1):129–140

Röcklinsberg H, Gamborg C, Gjerris M (2014) A case for integrity: gains from including more than animal welfare in animal ethics committee deliberations. Lab Anim 48(1):61–71

Russell WMS, Burch RL (1959) The principles of humane experimental technique. Universities Federation for Animal Welfare, London

Salter B, Harvey A (2014) Creating problems in the governance of science: bioethics and human/ animal chimeras. Sci Public Policy 41(5):685–696

Twine R (2010) Animals as biotechnology: ethics, sustainability and critical animal studies. Routledge, London

United States Department of Agriculture. Animal Biotechnology (2017) https://nifa.usda.gov/ animal-biotechnology. Accessed 30 Aug 2017

Wong JC (2017) Monkey selfie photographer says he's broke. The Guardian. https://www.the-guardian.com/environment/2017/jul/12/monkey-selfie-macaque-copyright-court-david-slater. Accessed 30 Aug 2017

Ethical Aspects of Animal Biotechnology 12

Dirk Lanzerath

Abstract

An ethical perspective on a field like the application of genetic engineering on animals in farming and food production is much more than mere technology assessments of risks and benefits. Rather ethics must approach the relationship between principles and practical moral challenges as a kind of reflective equilibrium. Against this background, applied ethics is not merely the application of principles to practice but a process in which practical experiences reciprocally influence the content and interpretation of ethical principles. Ethical theory learns from moral practice. Given that the interventions of biotechnology and genetic engineering principally affect the existence of nonhuman animals, ethics not only addresses questions relating to human health or social cooperation but takes a particularly fundamental interest in the ever-changing normative relationship between humans and animals as well as humans and nature. Therefore in this paper the analysis of the ethical challenges posed by biotechnology applied on animals reveals several problem areas that must be considered in developing ethical criteria for the investigation of biogenetic activities. Firstly, the paper will address the basic relationship of humankind and its technologies to nature (Chap. 1) in addition to the development of relevant evaluative criteria in environmental ethics and the ethics of nature (Chap. 2). The latter aid in the normative evaluation of technologies when weighing up which ends and means can be considered justified in relation to the goods a society recognises (Chap. 3). Against the background of biotechnological contributions to food production, the anthropological question as to what role food plays in a current culture and lifestyle—and what kind of change that culture and lifestyle might admit—will be analysed (Chap. 4). Yet the evaluation of biotechnology, to the extent that it is used on animals,

D. Lanzerath
Deutsches Referenzzentrum für Ethik in den Biowissenschaften (DRZE) (German Reference Centre for Ethics in the Life Sciences), University of Bonn, 53225 Bonn, Germany
e-mail: lanzerath@drze.de

© Springer International Publishing AG, part of Springer Nature 2018 251
H. Niemann, C. Wrenzycki (eds.), *Animal Biotechnology 2*,
https://doi.org/10.1007/978-3-319-92348-2_12

covers more than just the well-being of humans and society; transgenic animals also pose a great challenge to animal protection and welfare (Chap. 5). Finally the paper discusses that a society dealing with new technologies must constantly consider what form risk assessment should take and what kind of tolerance for environmental, health, and economic risks that implies (Chap. 6). Processes that advance sustainability, justice, food quality, and animal welfare alike should be placed at the centre of modern agriculture and food production. From an ethical perspective, the gene technological and biotechnological production of animals must be measured according to the extent of its contribution to these complex and highly relevant goals.

12.1   Introduction

The ethical evaluation of the uses of modern biotechnology involves analysis of the relationship between a given technology's means and ends and the values and moral principles of the society in which that technology is or is to be implemented. As a philosophical discipline, then, ethics is fundamentally concerned with the normative demands that confront societies faced with the question of whether to introduce or newly regulate technologies like biotechnology and genetic engineering or even to limit their usage. What does biotechnological activity mean for our *society* and *self-image*? And what are the ramifications of such technologies for our relationship to *animals* but also to *nature* as a whole?

In addressing questions such as these, ethical reflection goes significantly beyond the kind of risk analysis undertaken in the field of technology assessment. An ethical judgement ought primarily to be guided by careful analysis of the moral sanctions and motivations prevalent in a community. A particular challenge in this regard is the fact that - as agents in a global society - we find ourselves confronted with individual traditions emphasizing very different norms and values. This is especially relevant with respect to applied technologies like genetic engineering and biotechnology, which are of transnational scientific, economic, and social significance. In scrupulously weighing up the pros and cons, the possibilities, and the limits associated with the implementation of new technologies, it is the goal of ethical discourse to strike an appropriate balance between *naive faith in progress* on the one hand and *paralysing fear of technology* on the other. An entirely different kind of fear is the 'heuristic' variety proposed by Hans Jonas in his ethics of technology, where he advocates the reasoned application of the so-called precautionary principle in dealing with technological phenomena (Jonas 1984, pp. 26–27).

The precautionary principle is an approach to decision-making that seeks to avoid or reduce as much as possible the infliction of strain or harm on health, society, and the environment by agents with access to only *incomplete information* (see van den Daele 2007; Ammann et al. 1999; Ahteensuu 2008; Mepham 2008; Munthe 2011; Deblonde 2010; Andorno 2004). The *task of ethics* against this background is not to diminish the responsibility of scientists, engineers, farmers, manufacturers, politicians, or consumers for their actions, but rather to *provide an analytical framework* for the formation of individual *ethical judgements* (Harremoës et al. 2002; Pielke 2002; Lanzerath 2014a).

In order to avoid restricting itself to the moral aspects of technology assessment, ethics must approach the relationship between principles and practical moral challenges as a kind of reflective equilibrium (Rawls 1951). In other words, applied ethics is not merely the application of principles to practice, but a process in which practical experiences reciprocally influence the content and interpretation of principles. And given that the interventions of biotechnology and genetic engineering principally affect the existence of nonhuman animals, ethics not only addresses questions relating, for instance, to human health and social cooperation but takes a particularly fundamental interest in the ever-changing normative *relationship* between humans and *animals* as well as humans and *nature*.

Analysis of the ethical challenges posed by biotechnology reveals several problem areas that must be considered in developing *ethical criteria* for the investigation of gene technological activities. Firstly, it is necessary to address the basic relationship of humankind and its technologies to nature (Chap. 1) in addition to the development of relevant evaluative criteria in *environmental ethics* and the *ethics of nature* (Chap. 2). The latter aid in the normative evaluation of technologies when weighing up which *ends* and *means* can be considered justified in relation to the *goods* a society recognizes (Chap. 3). Against the background of biotechnological contributions to food production, the anthropological question as to what role *food* plays in our *culture* and *lifestyle*—and what kind of change that culture and lifestyle might admit—is of fundamental importance (Chap. 4). Yet the evaluation of biotechnology, to the extent that it is used on animals, covers more than just the well-being of humans and society; transgenic animals also pose a great challenge to *animal protection* and *animal welfare* (Chap. 5). Finally, a society dealing with new technologies must constantly consider what form *risk assessment* should take and the tolerance for environmental, health, and economic risks that implies (Chap. 6).

12.2 The Relationship Between Humans, Technology, and Nature

Under ideal circumstances, biotechnological interventions in nature—in the nature around us, which we encounter in the form of animals, plants, and microorganisms but also in our own nature—are planned in accordance with specific intended goals. The fact that humans intervene in nature is not *in itself* new to genetic engineering and biotechnology; humans have been doing so from the moment *Homo sapiens* appeared on earth. What is new about biotechnology is the *range* and *depth of the intervention* in nature. It uses gene technology in particular to transform the nature of organisms in a way that breeding and selection programmes have not traditionally been capable of. Mutation that once occurred spontaneously is now induced intentionally, with the possibility of exchanging genetic material between organisms belonging to different species. Central characteristics of organisms can be fundamentally altered in this way. Moreover, synthetic biology is not merely expanding the scope of biotechnology but allowing engineering technology to be interfaced in or with organisms (Serrano 2007; Schmidt 2012; Baldwin et al. 2012). Depending on the degree of biotechnological construction and modification they have undergone, these biological systems can differ considerably from their original forms in nature. Some of them can neither be unambiguously categorised as naturally developed organisms nor as technologically produced machines, with certain biological constructs that have come into being in this way straddling the border between *living being* and *artefact, nature*, and *technology*.

The various processes of biotechnology are contributing to the creation of a new, artificially produced 'nature' in areas such as the breeding of livestock and laboratory animals or within food production, which is becoming increasingly more *artificial* as synthetic biology develops. Thinking about this 'artificial nature' as part of modern culture entails certain basic ethical considerations. As part of nature, every living organism intervenes in the natural world around it in diverse ways, both as a producer and a consumer. But unlike other natural creatures, humans are not wholly driven by nature or led by instinct in doing so. In order to survive, mankind must intervene in nature via the creation of culture, which in turn becomes its 'second nature' (Mcdowell 1996; Kertscher and Müller 2017). As this Aristotelian expression indicates, culture always involves the interpretation and reshaping of the original, biological human nature, in which every human's ability is rooted. Technology and culture thus always remain connected to their origins in nature, meaning that nature transformed into culture is never entirely artificial. Humans belong to *natural history* as they belong to *cultural history*. This is not only of genealogical but also of normative significance, for even as moral subjects, we remain living organisms evolving *in* nature rather than *at a remove from* nature (McDowell 1996, pp. 112–125). One essential aspect of humans' second nature is to define ethical norms for controlling actions.

Animal breeding, plant breeding, and the utilization of microorganisms are part of mankind's *cultural productivity within nature*. While in a basic sense genetic engineering and biotechnology can be seen as advancing the notion of breeding, the differences between traditional breeding and biogenetic intervention are impossible to ignore. Where traditional breeding must wait for the recombination and mutation it initiates to take effect and has been stymied in its efforts by the more or less clear limit divisions between species place on gene recombination, the essential similarity of all organisms' DNA has allowed gene technology to overcome this limit. Hundreds of years of coincidence-driven work is thus being replaced or supplemented—as far as possible—by targeted, planned intervention initiated by humans in an effort to bring about mutation themselves. This is considerably increasing the speed at which breeding takes place, i.e. the length of time that elapses between the initiation of an intervention and the successful breeding of new organisms is being shortened, even though phenotypes and thus the success of breeding depend on many things other than genes. Thanks to genome editing technologies—such as CRISPR/Cas9—the relevant procedures are becoming simpler, more economical, and possibly even more precise.

Although synthetic biology started to incorporate unnatural amino acids into organisms, admittedly even ordinary gene technology only modifies and recombines what is found in nature; it does not—as the use of the term 'technology' suggests—invent or manufacture from scratch. This may accelerate the results of breeding and production, but it also introduces new types of risk, since small changes in an organism's genetic building blocks can have great effects both on the organism itself and the environment. It remains to be seen whether the new precision tools of genome editing will help in preventing the side effects. Its impact on our ability to predict unintended consequences is highly relevant to the ethical evaluation of all kinds of biotechnology.

12.3 Criteria in Environmental Ethics and the Ethics of Nature

As knowledge of the natural world has increased, humans have gained an increasing amount of power over it along with the ability to shape nature, constantly surpassing and displacing natural barriers to action. Aristotle and Thomas Aquinas understood nature as both predetermined by fate *and* ready for redesign, and power over the natural world is central to the early modernity proposed by Francis Bacon in his *Novum Organum* (1620). Descartes even describes humans in his *Discours de la méthode* (1637) as 'masters and possessors of nature' (on the historical development of the concept of nature, see Hager et al. 1984; Schäfer and Ströker 1993-96; Lustig et al. 2008; Mittelstraß 2003). With the advent of late modernity, nature increasingly becomes a mere object to be processed, shaped, and manipulated in research and practice. At the same time, as a of result of mankind's growing domination of the natural world, 'wild nature' loses something of its threatening, demonic element, and there is a corresponding intensification of the aesthetic appreciation with which humans respond to the contemplation and experience of both wild and cultivated nature. Yet the threatening aspect of nature repeatedly catches up with technological civilisation in negative and partially irreversible side effects of its actions such as climate change, the spread of monocultures, and declining biodiversity (Lanzerath 2014b).

Nature, however, is more than just an object and adversary of human endeavour. For despite the many ways modern man has of controlling the natural world, he cannot free himself from it. Even modern man *is part of nature* and *lives in dependence on it*. His intake of air and nutrients places him in a constant metabolic exchange with nature. Its *intactness* is therefore a *necessary condition* for the survival and cultural production of humankind. The threats to which modern man is exposed are often associated with a failure to adequately control nature. If *nature* is only ever understood as an *object of domination and adversarial other*, and no longer as the foundation of metabolic entwinement, the accompanying *estrangement* from it threatens to result in human self-harm.

Consciousness of our own nature and its integration into the natural world around us not only enables us to perceive the unity of nature but also its great diversity—a diversity that not only pertains to the use that mankind derives from it but is inherent in it as the result of millions of years of evolution. Within this naturally occurring diversity, an intrinsic hierarchy of living organisms has taken shape. Encapsulated in the concept of the *scala naturae*, it has defined the human understanding of nature and our interaction with it from the earliest of times (Siep 1998, p. 28). As criteria for differentiation, we encounter the distinction between inanimate and animate nature and within animate nature distinctions according to the degree of selfhood manifested in a life, i.e. according to the level of *individuality* and *self-organisation* it exhibits. An *organism's level of consciousness* plays an important role here. All that we know of physiology and ethology tells us that conscious perception is a prerequisite for the capacity to feel pain and experience *suffering*. The graduated distinctions we draw between organisms are relevant to assessing what *protective duties* we are bound to observe with respect to the well-being of other living creatures, especially when, as in the case of agricultural or laboratory animals, humans take care for them on farms or in labs.

Environmental ethics, animal ethics, and the ethics of nature, then, discuss possible justifications for human behaviour towards nonhuman nature and the *value* (or even *intrinsic value*) that can and must be attributed to nature in the process *without* taking nature itself as normative. The moral debate in this field focuses primarily on the *scope* of claims to protection. In what, exactly, does the worth of individual organisms lie? Is it *merely instrumental* and thus to be located in the utility of an organism to humankind, or do nature and individual aspects thereof possess their own *intrinsic* value?

The various possible answers to these questions lead to different modes of ethical reasoning, the most common types of which are *anthropocentric, pathocentric, bio-centric*, and *holistic* in approach (Kabasenche et al. 2012; Attfield 2008).[1] These

---

[1]*Anthropocentric* (focused exclusively on humans) approaches to environmental ethics assume that other animate beings only possess value to the extent that they are important or useful to humans. From this perspective, only humans have intrinsic value. Moral duties only exist in relation to other humans; no direct protective obligation exists towards nonhuman organisms. The issue of justice is central to anthropocentric ethical debates, especially in reference to the actual or potential utility of animals, for instance, as well as the question of who profits from it and to what extent. From the *pathocentric* perspective, value is accorded to those creatures capable of feeling pain and expressing this through observable behaviour, e.g. trembling or attempting to escape. At the very least, it acknowledges the claims of higher animals and humans to protection. It also implies an indirect obligation to protect the environment, to the extent that both humans and animals suffer from environmental destruction. *Biocentrism* goes further in arguing for protection than *pathocentrism*, relating as it does to all living things rather than extending human responsibility only to organisms that are of interest to humans or capable of suffering. Justifications of this approach frequently start from the assumption that all living creatures, whether conscious or unconscious, are subjects of a life and have a pursuit to survive that humans may not disregard without reason. *Holism* attributes moral value to the entirety of animate and inanimate nature. It incorporates not only individual creatures but the whole natural world including natural systems (e.g. ecosystems or ecological niches). According to this line of argumentation, human protection ought to extend to nature as a whole—not because or to the extent that it is useful, sensitive to pain, or animate, but simply because it exists.

four approaches to environmental ethics differ in the range of objects to which they attribute intrinsic value and towards which they thus acknowledge the existence of direct protective obligations. What they all have in common is their *anthroporelationality*, i.e. the fact that the value they attribute to nature always manifests itself in protective claims directed towards humankind. Only humans can be addressees of protective obligations; only they are capable of establishing relevant behavioural rules and taking responsibility for their actions.

No approach, moreover, can avoid defining the human relationship to nature. If we do not consider man as a bipartite being in which nature (body) and mind (soul) are understood as distinct modes of being but rather as the unification of both features, then a human being is itself a *natural creature* (Lanzerath 1998, 2014b). The relationship, in equal parts mediated and unmediated, that mankind has to nature— comprising the nature it represents and that which it encounters as an active subject—is a foundation on which *the ethical evaluation of applied biotechnology* can build. For the very status of humans as subjects and the unity in which moral subject and human organism coexist entail the observance of those claims that the nature both inherent in and surrounding humanity makes of it. In the first instance, this is necessary due to obvious self-interest. For if *humanity can only flourish as part of nature*, which in turn can only flourish if respected in its particular claims, then the protection of ambient nature as both habitat and resource of humankind, but also as part of the cultural and socioeconomic environment it deems valuable, cannot be neglected.

The role that man—as the only rational agent on earth—adopts in relation to other living organisms is of crucial importance to reflections on our interaction with animals and nature. While the core thesis of epistemic anthropocentrism, namely, that humans are fundamentally only capable of reflection and judgement from a human perspective, is rarely challenged, it does not necessarily entail a special ethical status for the human life form (Sturma 2013, p. 144). Man is a moral subject not only out of interest in his own well-being but also out of the desire to be able to provide reasons for his own actions, i.e. to be able to account for his behaviour in the face of his rationality.

The desire to do what reason deems good, however, not only implies recognition of the same desire in every other rational being but also demands, in accordance with the principle that like cases should be treated alike, the recognition of similar efforts in nonhuman animals (Regan 2004, p. 128; Honnefelder et al. 1999, pp. 308ff). The pain of animals, in other words, must be taken seriously by way of analogy with that of humans and indeed has found respect in mankind's ethical consciousness in the idea of animal protection (Honnefelder 1994, p. 125; see also Chap. 5). Yet the cognitive capacity of humankind differs from that of other higher animals by virtue of man's ability to assume a perspective on his own experience of pain, allowing him, for example, to experience serious or chronic illness not only as a painful state but as an experience of powerlessness and loss of meaning while also enabling his acceptance of such states (Lanzerath 2000, pp. 204–9; Höffe 1993, p. 222).

Biotechnological and gene technological intervention should not only be assessed according to its effect on humans and the well-being of animals capable of suffering but also to how it alters the dynamic structure and order of nature and natural processes. When natural systems suffer damage, humanity loses its natural origins and thus the foundation of its subjectivity. Every human intervention must refer to the entwinement of nature and culture and the many *benefits* associated with it. This requires an analysis of ends and means in relation to the goods recognised in a given society to provide a proper fundament for a normative assessment of modern biotechnology.

12.4 Weighing Up Ends, Means, and Goods

If by introducing genetically modified products derived from medical, biological, and agronomical research a society intends to exploit their advantages while avoiding the risks associated with them, it is essential that they be subjected to an ethical examination of their utility and suitability. This means considering both the *legitimacy of the ends* of genetic engineering and biotechnology and the *defensibility of the means drawn on*, i.e. of biotechnological and gene technological modification itself and their side effects. Two questions are of central importance here: Are the desired *ends* pursued via gene and biotechnology sufficiently *worthy* to justify the risk of undesirable side effects resulting from their implementation? And are genetic engineering and biotechnology *adequate means* for the solution of the problems they address or might there be better ways of achieving the desired ends?

A glance at the modern research landscape reveals that the satisfaction of human theoretical curiosity has long since ceased to be the only goal of scientific research; the objective now is to increase the possible applications of scientific knowledge in response to mankind's practical needs—a notion already present in the Baconian concept of knowledge. While in the first instance basic research into the implementation of gene technological methods seeks to clarify fundamental mechanisms of heredity, gene expression, and gene regulation, these branches of research are nonetheless characterised by their substantial practical bent. At least some of the biosciences feature a remarkably close connection between theoretical and *practical knowledge*.

Every novel technology—and gene technology is no different—is introduced with a range of ends in mind: positively formulated, genetic engineering seeks to improve quality of life with respect to health and nutrition, to increase agricultural yields, to create jobs, to maximise economic profits, etc. The ethics of each of these goals and the fields of research and application they involve must be formally evaluated. Not least among the concerns to be considered are the required financial resources and the persistent need to explain how limited funds ought most sensibly to be used.

When it comes to rating the objectives of genetically modified food development, *food security*, for example, is of crucial importance. As populations grow, this has emerged as life-or-death matter specifically for developing countries. On a global scale, an increase of at least 33% is anticipated in the next three decades (Telugu et al. 2016). Where genetic engineering and biotechnology have the potential to make key contributions to this type of basic aim, their application is typically uncontroversial. But there exist concerns whether genetic engineering will be able to meet those objectives since the causes of starving are rather complex and more related to social and economic structures (Thompson 2007, pp. 220–224; Pinstrup-Andersen and Schiøler 2000, pp. 2–6, 86–105). Other objectives, such as the development of nonessential designer food, are considerably more difficult to justify. The same goes for so-called genetic art, such as fluorescent transgenic fish (e.g. GloFish) (West 2006), when its purpose is purely aesthetic. Even when it is possible to assess the risks to humans and animals posed by such food or art, there is scepticism as to whether societies' need justify taking them. The use of enzymes (chymosin, amylases, pectinases, insulin, etc.) manufactured for food production[2] or drug development using genetically modified microorganisms or, nowadays, obtained from the mammary gland of vertebrates is generally seen as much less problematic. These methods are viewed particularly positively where production using conventional sources is incomparably more laborious, qualitatively worse, and associated with similar risks.[3]

Nowadays therapeutic substances extracted from the mammary glands of *transgenic mammals* are increasingly being approved for use, *turning mammary glands into living bioreactors*. Eggs laid by transgenic hens are likewise being used to produce vaccines. In some EU countries, the first drug produced from transgenic animals came onto the market in 2008. Antithrombin III, or ATryn®, inhibits blood clotting and is intended to protect people with hereditary antithrombin deficiency from life-threatening thromboses during high-risk operations. The gene that encodes for antithrombin is introduced into the genome of goats, and the tissue-specific promotor directs expression of the protein into the mammary gland. In 2010, the drug Ruconest (conestat alfa) for treatment of hereditary angioedema (HAE) was approved for the European market. HAE is a rare hereditary illness, which under certain circumstances can lead to life-threatening swelling of the skin, the mucous membranes, and the internal organs. Patients have insufficient blood concentrations of the plasma protein C1 esterase inhibitor (C1INH). In this case, the mammary glands of transgenic rabbits are the source of the solution. In establishing therapeutic goals, appeals to the *high-ranking good of health* provide an important ethical

---

[2] The enzyme chymosin (rennin), for instance, facilitates the curdling of milk and is indispensable in the production of most types of cheese. Genetically modified chymosin is legal in almost all European countries, including Germany. Due to a shortage of conventionally obtained rennin from calf stomachs, its use is well established nowadays. The cheese made from it can be bought and sold without restriction and does not require any special labelling.

[3] These kinds of applications of gene technology are considered particularly worthy when their simpler and purer production methods are linked to improved recovery chances for sick people. Similarly, the connotations of gene technology used on humans in the form of somatic gene therapy are generally positive. Indeed, patient—and relative—advocacy groups campaign strongly for this kind of research, including on children, even though the methods involved are still fraught with risk and only rarely generate long-term success.

justification for these methods. In contrast with medical applications of gene technology, however, the legitimacy of agricultural applications—particularly in food production—is frequently called into question, especially in cases where there is no lack of alternatives.

Among the applications of gene technology considered desirable in many cases is the breeding of transgenic laboratory animals as *model organisms* for research into new forms of treatment for disease—and particularly for illnesses that are genetic or partly genetic in nature. Once again it is clear that *ends* associated with particularly *high-ranking goods* such as human health are likely to attract support.

For a long time, transgenic farm animals had hardly any practical significance in *agriculture*. With the exception of salmon, food and other products from transgenic animals are not currently available for sale anywhere in the world. There have of course been numerous research studies, but compared with plants, the introduction of new genes into the genomes of animals has not only proven more laborious but also more prone to error. Now the introduction of *genome editing*, using CRISPR/Cas9, for instance, has opened up new perspectives. Now it seems possible to proceed with animal breeding that classical gene technology had almost given up on. There are improvements in research with farm animals like for meat production (increasing meat, elevating omega-3 fatty acids), for milk production (improving milk composition), for fibre production (improved wool), or for the adaptation to new habitats (like fish farming in colder waters) (Telugu et al. 2016, pp. 8–10; Ormandy et al. 2011). These new methodologies have also led to a sudden rise in the number of laboratory animals being used for medical research, a state of affairs that conflicts with the goals of *animal protection* and the standards of animal ethics (see Chap. 5).

12.5   Food as Part of Culture and Lifestyle

One of the goals of producing transgenic animals is the design of new food by improving animal productivity: disease resistance, environmental sustainability, higher yields, and better quality (Ormandy et al. 2011). Yet, the way people deal with food is a central aspect of the human relationship to nature and culture outlined at the beginning of this paper (see Chap. 1). Every stage of the process, from breeding to preparation and consumption, reveals how nature and culture are intertwined in human behaviour. Food brings together attempts to *master nature and to create culture*. It is thus not only a source of nourishment but is closely connected with the *identity* and normative self-image of mankind. Thus, ethical evaluation of the use of genetically modified food not only addresses environmental and health risks but also touches on basic attitudes of humans as cultural and natural beings towards diet and food.

The knowledge we have about how early communities of *Homo sapiens* lived reveals that even in prehistoric times the preparation of gathered or hunted food was more than just a condition of human sustenance and survival. It was a part of the culture within which mankind, transcending its purely natural needs, created a lifestyle that not only allowed humans to *live* but *live well* (see Plessner 1981, p. 383 on man's 'natural artificiality' ('natürliche Künstlichkeit'). Drawing on the cultural anthropology of Claude Lévi-Strauss and his remarks on *The Raw and the Cooked* (Lévi-Strauss 1983), Ludger Honnefelder describes food as a 'means of living'.[4] He discusses how it can be understood as such in a comprehensive sense that interlaces nature and culture (Honnefelder 2011, p. 49).

Today, as a rule, people no longer personally grow what they consume, nor do they have the kind of familiarity with it that comes from proximity to the producer. Instead our food is produced and delivered to market via routes that are practically opaque to the consumer. Products are only familiar to us in their physical form and as part of an established culture of production and preparation. Many people are hesitant to consume agricultural products due to a lack of trust in the long and confusing production and trade chains between producers and consumers. Milk, meat, and eggs are no longer procured from neighbourhood farmers, but at branches of large retail chains. For today's consumer, it is barely possible to comprehend how and according to what division of labour production (animal breeding and feeding, slaughter, processing, etc.) takes place or whether labels provide realistic information about products, given that trade routes can only be retraced under certain circumstances. Production errors and fraudulent labelling of goods lead to a loss of trust in the entire production and supply chain. In an attempt to maintain public confidence, advertisers thus appeal to the *transparency* of products and their close relationship to nature, which in reality signifies less about their *natural* qualities than their affiliation with a familiar, established, and respected *culture* (Honnefelder 2011, p. 55–65; Siep 1993). The kind of division of labour found in highly complex societies like that of modernity is only viable if there are basic conditions (transparency, control mechanisms, etc.) designed to maintain the trust of individual members of society in the system, regain that trust when mistakes are made, and appeal to the familiar in marketing new products. Accurate and transparent *labelling* that provides information about the production processes, sustainability, animal-friendliness, ingredients, genetic modification, and origins of animal products enables consumers to decide in accordance with their personal ethical convictions whether or not to buy a product. One way to support this would be to approve only independently scrutinised, trustworthy labels conforming to set norms (Bütschi et al. 2009; Weirich 2007; Falkner 2007).

The range of food available to societies is expanding today like never before, not primarily because of dramatic innovation in the form of completely new products, but due to the appropriation of foods that have established themselves in other cultures over long periods of time. The process and modes of globalisation can be identified in food production and consumption. Foreign modes of preparation are likewise frequently adapted to new cultures (see, e.g., the Europeanisation of Asian cuisine). The introduction of novel food in the form of different varieties and methods of preparation in fact has a long history; in its current cosmopolitan context, it is accepted and valued. In contrast, biotechnologically and genetically engineered

---

[4]A German concept for food is 'Lebensmittel', or literally 'means to life'.

novel food in its true sense of the word is generally—if at all—only introduced with extreme caution, and compared with the enrichment effected by the appropriation of established foreign foods, it is rather an anomaly (Honnefelder 2011, p. 51; Thompson 2007, pp. 195–220).

If food plays the kind of cultural role outlined above, it is clear that its significance and the resulting claims to ethical protection extend far beyond much-discussed environmental and health risks. Food is not only part of the biosphere (in the form of agricultural crops and animals) and an indispensable condition for human survival (in the form of nourishment) as well as (in its variety and combinations) a requirement for lasting health; its production and consumption are also part of a sociocultural and economic way of life. And if the factors mentioned—the viability of the environment, the health and survival of humanity, and its position in an intact and sound economic, social, and cultural world—are all goods worthy of protection, then food production gives rise to a diverse range of ethical requirements. The central question to be considered will be whether a positive effect on quality will dominate the various negative side effects.

Ultimately such reflections must also be incorporated into a comprehensive concept of modern agriculture and food production that accounts for issues around global food provision and climate change. From the perspective of *theories of justice*—underpinned by a concept of *sustainability*—each individual case must be assessed on whether the use of biotechnologically produced animals can be justified or whether it would instead be preferable to optimise established methods and adjust dietary habits in quite different ways, such as through the reduction of overall meat consumption (Twine 2015, 127–143; Ormandy et al. 2011). The justification of biotechnologically produced food depends on its compatibility with basic goods like good health, sustainable environment, trustful economy, social justice, and—last not least—its contribution to animal welfare.

12.6 Genetically Engineered Animals: A Challenge to Animal Protection and Welfare

If, in questioning the defensibility of a particular technological means, one takes the pain sensibility of higher animals to be similar to that of humans (see above), it seems to be morally questionable to use gene technology to produce animals whose artificially developed anatomy or physiology causes them unnecessary pain, all for the benefit of humans. However, the development of genetically modified laboratory animals for use in medical research when the only alternative would be to conduct research on humans induces a reasonable argument; finding new medical methods or drugs to save human lives is a very worthy end anchored in fundamental human rights and should not easily be compromised. Nonetheless, here too we must reflect on whether the objective of conducting a given experiment or developing a particular laboratory animal is genuinely high-ranking and whether or not alternative methods are available (Sturma and Lanzerath 2016, 63–104).

This set of problems is not new to the age of genetic engineering. In fact, so-called torture breeding is also entirely possible, albeit similarly unjustifiable, using traditional animal breeding methods (see Chap. 6). Owing to the depth of its interventions, gene technology has the potential to increase such problematic pursuits, but it is traditional agricultural breeding that really requires scrutiny. Often breeding and living conditions do not accommodate the species-appropriate needs and natural scope for self-realisation of the animals involved. This not only contravenes the principles of animal welfare but also harms humans and their cultural systems. The animal welfare standards for biotechnological breeding should not be any different than those for conventional animal husbandry. However, the conditions of how animals are kept in labs and at farms are still questionable in many cases.

The well-being of animals is also an important and sensitive topic of various types of animal experimentation, since many experiments cause suffer for the sake of humans. As early as 1986, in the context of the debate on Directive 86/609/EEC, the European Commission made very clear its goal to considerably reduce *animal experimentation* to avoid unnecessary suffering. Since then substantial funding has been directed towards research into alternatives. In practice, however, exactly the opposite has taken place. While the number of animals involved dropped by approximately 30% compared to 1999, it has risen considerably since then. In Germany, for instance, the number of genetically modified animals used in experiments almost tripled to nearly 950,000 animals per year between 2004 and 2013 (Sturma and Lanzerath 2016, 37–57).

The increasingly extensive production of transgenic laboratory animals is not only responsible for a rising number of animal experiments but also for much pain and suffering in animals. Transgenic animals whose genes can be activated and deactivated are playing a bigger and bigger role in biomedical research, from drug development to xenotransplantation. And alongside medical research, the use of genetically modified farm animals is also becoming more prevalent in agricultural experimentation, with the aim of developing new milk compositions, for example, of increasing meat production or to reduce agricultural pollution.

It is currently to be expected that CRISPR/Cas9 and similar technologies will be used on growing numbers of animals in the future (Nuffield Council on Bioethics 2016). Consequently, however, the number of experiments conducted on animals is increasing. Owing in particular to the *patenting* of genetically modified animals, the production of transgenic animals with pathogenic mutations that cause them suffering has become a *business model* (Twine 2015, pp. 95–113)—one that is completely unconducive to animal welfare. The Nuffield Council states:

> The implications of introducing and deleting specific genes cannot usually be predicted and the effects on welfare can be difficult to detect and measure. One report suggested that ten percent of GM animals experienced harmful effects. Another found that 21 percent experienced minor discomfort, 15 percent experienced severe discomfort and 30 percent had an increased risk of death and disease. Another concern is that most methods of producing GM animals are inefficient, and large numbers of animals are required to produce individual strains. (Nuffield Council on Bioethics 2016, p. 80)

A specific case of engineering technologies applied on animals is the *somatic cloning* technique as somatic cell nuclear transfer that started successfully in mammals with the sheep 'Dolly' 1996. The first cloned cat 'CC' was created 7 years later. Clients have started to require cloning services, in particular the cloning of deceased pets. But often clients have wrong expectations and mix up individuality and identity with 'having the same genome'. Another application is the cloning of extinct species. There are plans to clone the thylacine and the woolly mammoth. From the perspective of conservationism, those technologies are suspicious concerning their ends (Cottrell et al. 2014). There are various causes for the extinction of animals. Very often the habitat has been destroyed, and cloning will not bring it back. Speaking from an environmental ethics perspective, the efforts to protect endangered species need to be improved instead of enhancing the application of cloning technologies to bring them back to life. Also, an animal population needs a certain genetic variability to survive that is missing in cloning procedures. Apart from the observation that many indivudual animals need to be used in the process of experimentation to be successful in one case, there may also be negative side effects for the surviving cloned animals—such as increased embryonic and foetal mortality rates or other health risks. Thus a convincing reason to justify the cloning of pets or wild animals is often missing (Ormandy et al. 2011; Honnefelder et al. 1999; West 2006; Holt et al. 2004).

Animal experimentation and artificial production for human utility are test cases for how successfully modern society manages the relationship between humans and animals, something that poses a great normative challenge to the natural and life sciences, ethics, and law. Although there is general agreement in contemporary discussion that animals should not be regarded as mere things and that they have claims to ethical and legal consideration (Beauchamp and Frey 2013), the constant increase in the practice of animal experimentation seems to contradict this basic assumption.

Humans as moral subjects often oscillate concerning their position on the moral status of animals and the way how they interact with animals: humans either treat animals as automatons who exist only for human benefit or, in contrast, they anthropomorphise them by acting as if they were persons like themselves. In order to avoid such extreme positions and beliefs, the modern society must develop normative methods of evaluation—rather than simply assessing individual animal experiments on a case-by-case basis—that neither morally neglect animals by rendering them wholly instrumental nor assume an excessive moral burden by assigning them exaggerated protective claims.

Of course, despite many successes with alternatives to animal experimentation (cell cultures, computer simulation, etc.), under the conditions of current life sciences, researchers cannot do without it altogether (Sturma and Lanzerath 2016, 63–104). Nevertheless, modern society must subject it, and in particular the current economic conditions created for it by patenting and industrial interests, too much greater scrutiny (Twine 2015, pp. 101–113).

As moral subjects, humans have obligations both to themselves and to other living beings—especially those to which evolutionary biology and culture place them in close proximity. This is especially true if it can be assumed that there are species of animals with the makings of a consciousness not much dissimilar to our own. At the very least, this should play a central role in ethical considerations of our treatment of highly developed animals like primates. Going forward researchers need to work harder to consistently apply the accepted principles of animal ethics—the 3Rs of replacement, reduction, and refinement (Nuffield Council on Bioethics 2016, pp. 185–216; Buck 2007; Brønstad and Berg 2011)—in determining whether or not the specific reasons given for individual animal experiments are ethically convincing. Researchers can be required to explain precisely how they intend to meet the demands of the 3Rs to decrease the numbers of animals used in research.

12.7   Risk Assessment and Tolerance: The Environment, Health, the Economy, and Society

In addition to evaluating the aims of genetic engineering, it is necessary to ask whether gene technology as a means is even capable of adequately solving the problems it addresses and whether there are risks associated with it that could lead to new and possibly even bigger problems in the future. For instance, when pondering the use of gene technology to combat food shortages in developing countries, the specified goals seem less contentious than the question of whether and to what extent gene technology represents an adequate means of securing basic food supplies, adapting agricultural plants and animals to the conditions of climate change, or preventing deficiency diseases. The last of these is a particular source of critical discussion because the reasons for malnutrition lie not in food supplies generally, but in economic relations and the conditions necessary for a better and more equal distribution of food. Furthermore, the introduction of biotechnology and genetic engineering can—above all in developing countries—lead to new forms of economic dependency (Thompson 2007, pp. 220–224).

In determining whether to approve the use of gene technological methods, however, we must not only question their genuine utility and thus the adequacy of gene technology as a means to specific ends but also whether despite worthy goals and discernible benefits its use causes severe and/or possibly irreversible harm. To answer this question, it is necessary to investigate whether the application of gene technological methods carries the *risk of harmful consequences*, however doubtlessly unintended. This means that, taking account of the relationship between humans and nature outlined above, when seeking to avoid interventions in nature that cause lasting harm or seriously endanger both humans and their natural surroundings, we must be able to precisely describe the consequences of the intended action. But in a field as complex as gene technology, the task of extrapolating all such consequences is fraught with difficulty; gene technology thus always involves uncertainty and risk.

In the context of nonmedical applications, a range of risks posed by transgenic organisms are being addressed in numerous research projects around the world: possible toxicological effects of modified metabolic patterns with the potential to harm both humans and other creatures or give rise to new allergies, the irreversible penetration of harmful transgenic organisms into ecological communities, and the unleashed developmental potential of transgenic organisms that increase their evolutionary fitness over that of natural species. These are only a few examples of the frequently discussed risks of gene technology, some of which are comparable with potential risks of conventional breeding (for a detailed discussion of this point, see van den Daele 2007; Ammann et al. 1999; Ahteensuu 2008; Mepham 2008; Munthe 2011; Deblonde 2010; Andorno 2004).

Given the relative novelty of gene technology in comparison with traditional breeding methods, it is entirely natural to suspect that in addition to discernible risks and those that can be extrapolated from them, transgenic organisms also pose other risks that have not yet been determined. This is particularly true of areas where knowledge remains limited and any prognoses are therefore tinged with uncertainty and ambiguity. In contradistinction to known risks, these are referred to as 'theoretical' or often 'speculative' (van den Daele 1996, p. 263). In contrast with 'real' or 'hypothetical risks', they are underpinned by the knowledge that suspicion of risk does not need to be based on familiar processes and that it is possible to postulate hitherto unheard-of processes and events. But to cite them alone as grounds for a ban on relevant technologies would hardly be universally appropriate and would be difficult to defend against counterclaims to basic liberties. For when the comparison of conventional and genetic engineering reveals no differences with respect to risk, the only reason for conjecturing that there are concealed differences is ignorance. Yet such conjecture could with equal justification be applied to the continued use of old technology. Related exclusively to new technologies, it would result in an undiscriminating ban on them all (Honnefelder 2011, p. 60). 'The generation of a new genetically engineered line of animals often involves the sacrifice of some animals and surgical procedures (for example, vasectomy, surgical embryo transfer) on others. These procedures are not unique to genetically engineered animals, but they are typically required for their production' (Ormandy et al. 2011, 547). When drawing analogies with conventional breeding, we must ultimately consider that our knowledge of it is limited too, meaning that we must always account for unforeseeable consequences. The comparably greater 'proximity to nature' of conventional breeding might be the reason to assume that the associated risks are lower, but it cannot be ruled out per se. Especially when harmful conventional breeding, as seen, for example, in the case of Belgian Blues, in which a genetic defect is exploited but not artificially produced, causes animals to suffer. This cattle provides meat producers with higher yields, but at the expense of the animals, which as a rule can no longer even calve naturally (see Chap. 5 on animal welfare).

In contrast, the novelty of genetic engineering in combination with the *greater scope and depth of its interventions* in nature and the lack of experience on which to basic assessment of its safety and appropriateness has been taken as grounds to implement *safety precautions* going beyond the examination of risks also associated with conventional breeding. The justificatory requirements for certain applications of gene technology thus necessitate a special form of licencing test.

On the whole—at least as indicated by the results of various studies (van den Daele 2007; Kjellsson et al. 1997; Ammann et al. 1999)—the risks associated with the use of transgenic organisms lose some of their drama when compared with those of traditional breeding. At the same time, however, the comparison highlights potential sources of harm associated with modern agriculture *in general* more strongly than before. There is thus a crucial need for discussion of issues beyond the debate on gene technology in a narrow sense. Just as the discussion around medical applications of human genetics (gene therapy, genetic diagnosis, genetic tests, etc.) is constantly pointing out abuses and misguidedness within conventional medicine (Lanzerath 2000, pp. 83–85, 146, 247), the debate on the application of gene technology in the agricultural sector draws attention to corresponding problems with conventional breeding practices. After all, increased safety precautions with respect to health and environmental risks, appropriate consideration of the criteria governing species-appropriate animal husbandry (e.g. with regard to feeding and living conditions), careful and comprehensible labelling, and traceable production chains are not only sensible demands in context of new gene technological methods in breeding. Increased risk management is desirable for all kinds of breeding in the agricultural and food production sectors. The risks of gene technology substantially affect the entire context of application and are thus more than just a matter for the natural sciences. An ethical discussion that only focusses on gene technology with reference to its methods, irrespective of the context in which it is applied, bears the inherent risk of failing to consider important ethical problems affecting that context as a whole.

In the international debate of the social and ethical risks posed by technologies like biotechnology and genetic engineering, the precautionary principle (see Sect. 12.1 on the 'heuristics of fear') in particular has established itself as a basic standard for action. It also features in the normative texts of the European Union (van den Daele 2007; Ammann et al. 1999; Ahteensuu 2008; Mepham 2008; Munthe 2011; Deblonde 2010; Andorno 2004). The principle states that the introduction of a new technology or product must be prohibited if risks cannot be completely ruled out (Rio Declaration 1992, principle 15). It is frequently used as a basis for evaluating genetically modified foods. An alternative principle is to prohibit new technologies or products only when the anticipated harm is extremely likely to materialise and is also extremely grave (Rippe 2001, p. 15). Critics object that the precautionary principle prevents economic growth and technological progress that could be of great benefit (Beckerman 2006; Van den Daele 2001, pp. 103ff.). As the ethical debate stands, it is not possible to discern any unified stance for or against the application of the precautionary principle in the evaluation of transgenic animals, though different risk scenarios are of course discussed—especially when it comes to contrasting the risks and possible benefits to the environment, health and nutrition, the economy, and society.

If the precautionary principle is to be taken seriously, risk research into the prevention of dangers associated with genetically modified organisms must become an integral component of genetic research itself. *Attitudes to risk cannot, however, be adopted at will* when the utility of and risks to others are at stake. How a decision is to be made under risky circumstances depends on the general attitude to risk brought to the situation by those making the decision and those affected by it. Risk assessment is thus only a partly technological matter. Integrating it into a given context requires ethical and social evaluation and its incorporation into a public discourse in which decision-making involves diverse actors as a prerequisite for the development of trust in new technologies rather than being left to commissions of experts alone. The aim of this is not merely to promote acceptance but to foster an interactive discourse and shared sense of responsibility.

As genome editing is increasingly introduced to breed transgenic animals for use in agriculture and laboratories, one of the greatest challenges facing our society in the coming years will be how to facilitate the kind of discussion between *experts* and *public* necessary to the formation of social consensus over the exploitation of new possibilities in the modern life sciences and associated reflections on the bioethical principles guiding action in this area. The quality and outcomes of the social discourse on norms and standards governing the work processes and objectives of the life sciences affect important political and economic decisions. A successful discourse of this kind is a prerequisite for political effectiveness. Moreover, a lack of contextual knowledge and failure to reflect on basic ethical questions lead many laypeople to fear biotechnology and genetic engineering. Their fundamentally critical and often emotional stance towards discussions of these methods fails to respect what experts in science and technology see as the proper rules of discussion. The latter expect the standards of rationality and proof commonly observed in the sciences to be adhered to even in discussion with nonspecialists (Twine 2015, pp. 43–47, 63–65). When that proves impossible, they expect their expert authority to go unchallenged. But laypeople with an interest in the subject cannot meet such expectations without giving up any claim to an opinion. Conversely, laypeople often make unrealistic demands on the knowledge and prognostic abilities of experts. Experts cannot accept these demands without casting doubt on their own trustworthiness. But as long as the arguments for wise goals do not dominate potential negative side effects, the public enthusiasm for biotechnology outside the medical sector will remain rather low.

Representatives of the ethical field also frequently find themselves in a contested role: on the one hand, it is demanded that they act as voices of warning and caution, while on the other they are assumed to exist simply to provide alibis for science and industry. However, *ethicists* themselves must focus on their analytical role, part of which involves finding ways to communicate to laypeople that moral questioning is not or need not be a matter of subjective attitude, but the result of argument and reflection. Against this background, bioethical discourse has the task of scrutinising and improving communication between public and experts.

Improved communication between various stakeholders with respect to biotechnology and genetic engineering is also a precondition for significantly increasing the social visibility of scientific work processes, the everyday relevance of science and technology, and their impact and significance in solving future problems. It would be facilitated by the establishment of certain basic conditions such as institutional requirements encouraging experts to seek training in public communication and exploiting the fundamentally critical stance of various members of civil society as an opportunity to develop new forms of discussion between the scientific and public spheres (Wray 2016; Wohlers 2010; Thompson 2007, 286–290). In this context, laypeople should be addressed in acknowledgment of their genuine powers of ethical judgement and specific expertise as representatives of the civil society. Biotechnological experts are called on as being partly responsible for ensuring that social discussion of the results of scientific research and development is fit for purpose.

Conclusions

The process of ethical assessment of biotechnology is subject to constant change. At present, after weighing all the risks and possible alternatives, there are good reasons for advocating the implementation of genetic engineering in many fields pertaining to the high-ranking good of *health*. This specifically pertains to the development of *model organisms* for research into the treatment of disease. More questionable at this juncture, however, is the increasing use of laboratory animals, which thanks to patenting practices has encouraged commercial dealings that are often lacking any justification. The researchers should in particular consider the 3Rs to protect animals and use alternative procedures to avoid unnecessary suffering of the animals. With regard to *food* production and with it the use of biotechnology on farm animals, we must intensely consider the anthropological question of whether in the long run leaving the traditional paths in favour of genetically modified food will change our culture more deeply than we would currently anticipate. Modern societies might not be prepared for this kind of permanent change towards an increased artificial way of farming. Just as an environment containing a single type of tree or single type of cow would be aesthetically unacceptable to us, so too would we find the genetic modification of food aesthetically and ethically unacceptable if it were to result in reduced variety. The global and regional goal of protecting biodiversity refers not only to wild species but also to farm animals. Ultimately, monocultures in the field lead to monocultures on the plate.

Yet at the same time, the comparison of genetically modified with conventionally produced food sheds light on significant ethical problems of modern agriculture as a whole—problems relating to the environment, animal welfare, the economy, social responsibility, and cultural sensitivities. In attempting to provide solutions to these challenges, we immediately encounter the more fundamental question of what criteria we should adhere to in our interaction with nature—both our own and the nature that surrounds us. The question of what kind of nature we want to live in and to what extent we are willing to let it be manipulated by humans must be addressed not only in a discussion between ethics and the sciences but also in public discourse for which suitable forms are yet to be found. The ability to genetically modify food brings with it an awareness of the problems of industrial food production in general, thus making it evident that the increase in opportunities for action afforded to humans by new technologies that intervene in nature will only remain human if we can muster the strength to set appropriate limits for it. The more we see animals as helpful machines (models of disease, objects of experimentation, bioreactors, etc.) (Thompson 2007, p. 129) rather than our fellow creatures—even in the secular sense of the word—the more the application of biotechnological methods to them becomes a test case for our ethical relationship to the animal world. This relationship is an indicator of our humanity. Such a comprehensive perspective is not only of importance to ethics in economics or relevant to environmental and health risks but reflects in a very fundamental way the relationship between humans, nature, and animals.

The mechanisation brought about by biotechnology can be seen as the continuation of a long tradition of breeding but may also be a further step in our estrangement from nature and animals. The differences between it and conventional animal husbandry do, however, seem to be more quantitative than qualitative. Ethical examination of human interaction with genetically modified animals in medicine and agriculture provides an impetus for renewed reflection on utility, estrangement, and artificiality in conventional dealings with animals in these fields. The great challenges of modernity in the areas of climate protection, food supply, public health, and economic balance demand a sensitive approach to the issue of whether certain forms of naturalisation are in fact desirable given our own natural status and in many respects artificiality has gone too far. Synthetic biology in particular will continue to open up new aspects of this issue, with the possible manufacture of hybrid animal machines and further mechanisation of nature. Furthermore, when new technologies are introduced in given economic structures, the power of regional economies and those of global players seems to be unequally distributed.

Processes that advance sustainability, justice, food quality, and animal welfare alike should be placed at the centre of modern agriculture and food production. From an ethical perspective, the gene technological and biotechnological production of animals must be measured according to the extent of its contribution to these complex and highly relevant goals.

References

Ahteensuu M (2008) The precautionary principle and the risks of modern agri-biotechnology. In: Launis V, Räikkä J (eds) Genetic democracy. Philosophical perspectives, vol 37. Springer, Dordrecht, pp 75–92

Ammann K, Jacot Y, Simonsen V, Kjellsson G (eds) (1999) Methods for risk assessment of transgenic plants, vol 3. Birkhäuser, Basel

Andorno R (2004) The precautionary principle. A new legal standard for a technological age. J Int Biotechnol Law 1:11–19

Attfield R (2008) The ethics of the environment. (The International Library of Essays in Public and Professional Ethics series). Ashgate, Aldershot

Baldwin G, Bayer T, Dickinson R, Ellis T, Freemont PS, Kitney RI, Polizzi K, Stan G-B (eds) (2012) Synthetic biology. A Primer, London

Beauchamp TL, Frey RG (eds) (2013) The Oxford handbook of animal ethics. Oxford Univ. Press, Oxford

Beckerman W (2006) Ein Mangel an Vernunft. Nachhaltige Entwicklung und Wirtschaftswachstum. Liberal Verlag, Berlin

Brønstad A, Berg A-GT (2011) The role of organizational culture in compliance with the principles of the 3Rs. Lab Anim 40(1):22–26

Buck V (2007) Who will start the 3Rs ball rolling for animal welfare? [letter]. Nature 446(7138):856

Bütschi D, Gram S, Haugen JM (eds) (2009) Genetically modified plants and foods. Challenges and future issues in Europe. Final report. EPTA, Berlin

Cottrell S, Jensen JL, PeckEmail SL (2014) Resuscitation and resurrection: the ethics of cloning cheetahs, mammoths, and Neanderthals. Life Sci Soc Policy 10:3

Deblonde M (2010) Responsible agro-food biotechnology. Precaution as public reflexivity and ongoing engagement in the service of sustainable development. In: Gottwald F-T, Ingensiep HW, Meinhardt M (eds) Food ethics. Springer, New York, pp 67–85

van den Daele W (ed) (1996) Grüne Gentechnik im Widerstreit: Modell einer partizipativen Technikfolgenabschätzung zum Einsatz transgener herbizidresistenter Pflanzen. Weinheim, VCH

van den Daele W (2001) Zur Reichweite des Vorsorgeprinzips – rechtliche und politische Perspektiven. In: Lege J (Hg.): Gentechnik im nicht-menschlichen Bereich – was kann und was sollte das Recht regeln? Arno Spitz GmbH, Berlin. pp 101–125

van den Daele W (2007) Legal framework and political strategy in dealing with the risks of new technology. The two faces of the precautionary principle. In: Somsen H (ed) The regulatory challenge of biotechnology. Human genetics, food and patents. Cheltenham, Edward Elgar, pp 118–138

Falkner R (ed) (2007) The international politics of genetically modified food. Diplomacy, trade and law. Palgrave Macmillan, Basingstoke

Hager FP, Gregory T, Maierù A, Stabile G, Kaulbach F (1984) Natur. In: Ritter J, Gründer K (eds) Historisches Wörterbuch der Philosophie, vol 6. Schwabe, Basel, pp 421–478

Harremoës P, Gee D, MacGarvin M, Stirling A, Keys J, Wynne B, Vaz SG (eds) (2002) The precautionary principle in the 20th century: late lessons from early warnings. Earthscan, London

Höffe O (1993) Moral als Preis der Moderne. Frankfurt am Main, Suhrkamp

Holt WV, Pickard AR, Prather RS (2004) Wildlife conservation and reproductive cloning. Reproduction 127:317–324

Honnefelder L (1994) Elemente einer philosophischen Anthropologie. In: Honnefelder L, Rager G (eds) Ärztliches Urteilen und Handeln: Zur Grundlegung einer medizinischen Ethik. Insel, Frankfurt am Main

Honnefelder L (2011) Welche Natur sollen wir schützen? Berlin University Press, Berlin

Honnefelder L, Lanzerath D, Hillebrand I (1999) Klonen von Tieren: Kriterien einer ethischen Urteilsbildung. Jahr Wiss Ethik 4

Jonas H (1984) The imperative of responsibility: in search of an ethics for the technological age. Trans. Hans Jonas and David Herr. University of Chicago Press, Chicago and London

Kabasenche WP, O'Rourke M, Slater MH (eds) (2012) The environment. Philosophy, science, and ethics (Topics in contemporary philosophy). MIT Press, Cambridge, MA

Kertscher J, Müller J (eds) (2017) Praxis und 'zweite Natur': Begründungsfiguren normativer Wirklichkeit in der Diskussion. Mentis, Paderborn

Kjellsson G, Simonsen V, Ammann K (eds) (1997) Methods for risk assessment of transgenic plants, vol 2. Birkhäuser, Basel

Lanzerath D (1998) Natürlichkeit der Person und mechanistisches Weltbild. In: Dreyer M, Fleischhauer K (eds) Natur und Person im ethischen Disput. Alber, Freiburg im Breisgau, pp 181–204

Lanzerath D (2000) Krankheit und ärztliches Handeln: Zur Funktion des Krankheitsbegriffs in der medizinischen Ethik. Freiburg im Breisgau, Alber

Lanzerath D (2014a) The use of genetic knowledge: ethical problems. In: Lanzerath D, Rietschel M et al (eds) Incidental findings. Scientific, legal and ethical issues. Ärzte-Verlag, Köln, pp 93–108

Lanzerath D (2014b) Biodiversity as an ethical concept. In: Lanzerath D/Friele M (eds) Concepts and values in biodiversity (Routledge biodiversity politics and management series). Routledge: Abingdon, NY, pp 1–19

Lévi-Strauss C (1983) The raw and the cooked. Vol 1 of mythologiques. Trans. John and Doreen Weightman. University of Chicago Press, Chicago

Lustig BA, Brody BA, McKenny GP (eds) (2008) Altering nature, vol 1. Concepts of "nature" and "the natural" in biotechnology debates (Philosophy and medicine 97). Springer, Dordrecht

McDowell J (1996) Mind and world. Harvard University Press, Cambridge

Mepham B (2008) Risk, precaution and trust. In: Mepham B (ed) Bioethics: an introduction for the biosciences, 2nd edn. Oxford University Press, Oxford, pp 327–349

Mittelstraß J (2003) The concept of nature. Historical and epistemological aspects. In: Ehlers E, Gethmann CF (eds) Environment across cultures. Springer, Berlin, pp 29–35

Munthe C (2011) The price of precaution and the ethics of risk (The international library of ethics, law and technology 6). Springer, Dordrecht

Nuffield Council on Bioethics (2016) Genome editing. An ethical review. Nuffield Council, London

Ormandy EH, Dale J, Griffin G (2011) Genetic engineering of animals: ethical issues, including welfare concerns. Can Vet J 52(5):544–550

Pielke R Jr (2002) Is the precautionary principle a useful guide to action? Review of *the precautionary principle in the 20th century: late lessons from early warnings*, ed. Poul Harremoës, David Gee, Malcolm MacGarvin, Andy Stirling, Jane Keys, Brian Wynne, and Sofia Guedes Vaz. Nature 419:433–434

Pinstrup-Andersen P, Schiøler E (eds) (2000) Seeds of contention: world hunger and the global controversy over GM crops (International Food Policy Research Institute). John Hopkins Univ Press, Baltimore

Plessner H (1981) Die Stufen des Organischen und der Mensch. Vol 4 of Gesammelte Schriften. Frankfurt am Main, Suhrkamp

Rawls J (1951) Outline of a decision procedure for ethics. Philos Rev 60(2):177–197

Regan T (2004) The case for animal rights, 3rd edn. University of California Press, Berkley

Rio Declaration on Environment and Development of the United Nations Conference on Environment and Development (UNCED) (1992) UN Doc. A/CONF.151/26 (vol I); 31 ILM 874 (1992)

Rippe K-P (2001) Vorsorge als umweltethisches Leitprinzip. Bericht der Eidgenössischen Ethikkommission für die Gentechnik im ausserhumanen Bereich. http://www.ph-karlsruhe.de/uploads/media/Vorsorgeprinzip.pdf. Accessed 28 May 2013

Schäfer L, Ströker E (1993–96) Naturauffassungen in Philosophie, Wissenschaft, Technik. Alber, Freiburg im Breisgau

Schmidt M (ed) (2012) Synthetic biology. Industrial and environmental applications. Wiley–Blackwell, Weinheim

Serrano L (2007) Synthetic biology: promises and challenges. Mol Syst Biol 3(158):1–5. https://doi.org/10.1038/msb4100202

Siep L (1993) Ethische Probleme der Gentechnologie. In: Ach JS, Gaidt A (eds) Herausforderung der Bioethik. Stuttgart-Bad Cannstatt, Frommann-Holzboog

Siep L (1998) Bioethik. In: Pieper A, Thurnherr U (eds) Angewandte Ethik: Eine Einführung. Beck, Munich

Sturma D (2013) Naturethik und Biodiversität. In: Sturma D/Honnefelder L (eds) Jahrbuch für Wissenschaft und Ethik, Bd 17. de Gruyter, Berlin, pp 141–155

Sturma D, Lanzerath D (eds) (2016) Tiere in der Forschung. Naturwissenschaftliche, rechtliche und ethische Aspekte. (Ethik in den Biowissenschaften - Sachstandsberichte des DRZE, 17), Freiburg i.Br. München, Alber

Telugu BP et al. (2016) Genome editing to the rescue: sustainably feeding 10 billion global human population. Natl Inst Biosci J 1. https://doi.org/10.2218/natlinstbiosci.1.2016.1743

Thompson PB (2007) Food biotechnology in ethical perspective. (The international library of environmental, agricultural and food ethics 10), 2nd edn. Springer, Dordrecht

Twine R (2015) Animals as biotechnology. Ethics, sustainability and critical animal studies. Earthscan, Abingdon

Weirich P (ed) (2007) Labeling genetically modified food. The philosophical and legal debate. Oxford Univ. Press, Oxford

West C (2006) Economic and ethics in the genetic engineering of animals. Harvard J Law Technol 19:413–442

Wohlers AE (2010) Regulating genetically modified food. Policy trajectories, political culture, and risk perceptions in the U.S., Canada, and EU. Politics Life Sci 29(2):17–39

Wray B (2016) Public engagement in synthetic biology. "experts", "diplomats" and the creativity of "idiots". In: Hagen K (ed) Ambivalences of creating life. Societal and philosophical dimensions of synthetic biology, Cham, Heidelberg, pp 177–197

Public Perception of Animal Biotechnology

13

Alison L. Van Eenennaam and Amy E. Young

Abstract

The commercialization of any product hinges on consumer acceptance. Genetic engineering has faced an uphill battle in this regard since the introduction of genetically engineered (GE) crops in the 1990s. Public perception of GE animals is generally negative, with biomedical applications being more positively perceived than agricultural applications. To date most GE animals have been developed in private or university laboratories for research purposes. Opposition to GE animals is often conflated with opposition to use of animals in research in general, as well as opposition to aspects of intensive animal agriculture. In general, concerns about animal biotechnology are influenced by (1) views around the moral status of animals, the boundary between "natural" and "unnatural," and perceived risks and benefits of GE animals to health and the environment (personal and cultural characteristics); (2) the purpose of the application, the method(s) being used, and the motivation of the research group making the genetic modification (research characteristics); (3) the species being modified (animal characteristics). As such, it is difficult to generalize about public perception of GE animals as a discrete category. The first GE food animal approval, the AquAdvantage salmon, in 2015, followed years of regulatory delay partially resulting from the negative public perception of genetic engineering. There are a number of new animal applications in development, enabled by new methods, which specifically target traits for animal health and well-being. A nuanced consideration of these applications by those that are not intrinsically opposed to the technology may positively impact public perception of GE animals.

A. L. Van Eenennaam (✉) · A. E. Young
Department of Animal Science, University of California, Davis, Davis, CA 95616, USA
e-mail: alvaneenennaam@ucdavis.edu; ayoung@ucdavis.edu

© Springer International Publishing AG, part of Springer Nature 2018
H. Niemann, C. Wrenzycki (eds.), *Animal Biotechnology 2*,
https://doi.org/10.1007/978-3-319-92348-2_13

13.1   Introduction

Public acceptance of emerging technologies is a dynamic, complex process. New technologies are frequently met with suspicion or hostility as they are often hard to understand because they are associated with complicated scientific principles and unfamiliar terms. In some cases, new technologies become coupled or conflated with existing societal controversies and are initially rejected by the public. Gupta et al. (2011) observed, "It is important to note that on one hand a technology may bring about radical changes in society, while on the other hand the fate of that technology rests with the society in which it is being applied." Much of the research into determinants of public acceptance of emerging technologies takes place after the public has already rejected a particular application (Frewer et al. 2011).

Biotechnology has been a popular topic for public perception research in recent years. As with many other emerging technologies, the public perception of biotechnology is complicated and the public generally does not have a uniform view of all types of biotechnology (Lusk et al. 2015). This is perhaps understandable as there are a variety of biotechnology applications and many definitions of biotechnology, often with unclear or overlapping meanings. Even definitions of the so-called modern biotechnologies, including "genetic modification" (GM), genetic engineering (GE), and transgenesis, are often imprecise.

The Convention on Biological Diversity defines biotechnology as "any technological application that uses biological systems, living organisms or derivatives thereof to make or modify products or processes for specific use" (Secretariat of the Convention on Biological Diversity 2005). The broad definition includes technologies that are routinely used to make pharmaceuticals, food additives, enzymes, vaccines, and hormones but are uniquely controversial when used to modify the plants and animals we use to produce food. In general, there is little concern over medical applications of biotechnology as compared to the global furor over food applications (Sanchez 2015); there is more concern about GE animals than GE plants and greater acceptance of applications that provide clear benefits for the consumer (Lusk et al. 2015).

The first GE mouse (Gordon et al. 1980) predated the first GE plant, and the first GE farm animals followed soon thereafter (Hammer et al. 1985); however, the development of GE animals has proceeded much more slowly than for GE crops (Mora et al. 2012). In 2015, twenty-six countries planted biotech crops on a total of 185.1 million hectares, making it the "fastest adopted crop technology in recent times" (James 2016).

GE animal applications are as diverse as the species involved, and each comes with its own specific set of risks, benefits, concerns, and considerations. To date the vast majority of GE animals, primarily mice, rats, rabbits, and pigs, have been developed for research purposes in private or university laboratory settings (Mora et al. 2012). A small number of applications have been successfully commercialized including GE animals as pets (GloFish®) and GE animals that produce pharmaceutical products in their milk or eggs (Table 13.1).

Table 13.1 Examples of genetically engineered (GE) food animals that have been produced for biomedical and agricultural applications. Products approved for market are shown in bold

Species	Biomedical applications		Agricultural applications	
	Products/Targets	Goal/Trait	Targets	Goal/Trait
Cattle	lactoferrin	innate host defense	a-casein, a-lactalbumin,b casein, k-casein, Omega -3 (Fat-1)	milk composition
	lysozyme	antimicrobial	lysostaphin	mastitis resistance
			myostatin	increased muscle yield
			prion protein (PrP)	spongiform encephalopathy resistance
			SP110	bovine tuberculosis resistance
Chicken	a-interferon	hepatitis	avl6 envelope glycoprotein	avian leukemia virus resistance
	Kanuma®/sebelipase alta	**lysosomal acid lipase deficiency**	lacZ	nutrient utilization
			short hairpin RNA	avian influenze resistance
Goat	**Atryn®/Antithrombin**	**anticoagulant**	beta-defensin 3	milk composition
	factor IX	haemophilia	monosaturated fatty acid	mastitis resistance
	lactoferrin	innate host defense	myostatin	increased muscle yield
	lysozyme	antimicrobial	prion protein (PrP)	spongiform encephalopathy resistance
	lysosomal acid b-glucosi dase	Gaucher disease	SCD	improved milk fat
	MSP(1)42	Malaria vaccine		
Pig	a(1,3)galactosyltransferase, albumin,b-Mamose, CD59, DAF, GnTIII, hHO-1, N-glycolylneuraminic acid	organ transplantation	a-lactalbumin, lysozyme	piglet survival
	factor VIII	haemophilia	CD163SRCR5	PRRS resistance
	fibrinogen	tissue sealant	cSKI	muscle development
	haemoglobin	transfusion	FAD2	improved milk fat
	protein C	blood coagulation	FMDshRNA	foot and mouth disease resistance
			growth hormone	growth rate
			Mx1	influenza resistance
			Omega -3 (Fat-1)	meat composition
			phytase	feed uptake/decreased environmental impact
			RELA	African Swine Fever Virus resistance
Rabbit	calcitonin	osteoporosis		
	erythropoietin	anemia		
	factor VII, von Willebrand factor	haemophilia		
	growth hormone	HGH insufficiency		
	Interleukin-2	cancer treatment		
	Ruconeste®/C1-Esterase Inhibitor	**hereditary angioedema**		
	superoxide dismutase	blood purification		
	tissue plasmogen activator	anti-clotting agent		
	VP2, VP6	rotavirus vaccine		
Sheep	a-1-antitrypsin	cystic fibrosis	CsK, IGF-1	wool growth
	factor VIII,factor IX	haemophilia	IF	wool quality
	GGTA1	organ transplantation	Omega -3 (Fat-1)	meat composition
			prion protein (PrP)	spongiform encephalopathy resistance

Despite the fact that arguments for or against GE crops are largely applicable to GE animals, with some modifications (Sandler 2015; Tizard et al. 2016) (Fig. 13.1), at present not a single GE food animal product has been successfully commercialized in the United States. The commercialization of the fast-growing AquAdvantage GE salmon following its groundbreaking, albeit lengthy, regulatory approval for food purposes (US Food and Drug Administration 2015), continues to be thwarted by congressional interference, and it will likely take several years for the product to come to market (Box 13.1). Some groups have called for a ban on the sale of the fish and have agitated grocery stores to pledge not to sell it. Clearly the use of GE animals for food is a controversial topic. The FDA's public comment call for the AquAdvantage regulatory approval application elicited 360,000 comments, with commenters expressing many of the issues and concerns that are covered in this chapter.

Public perception of animal biotechnology is far from straightforward, and the lines between animal biotechnology and other issues related to animal use are often blurred. Opposition to GE animals frequently goes hand in hand with opposition to research involving animals or even use of animals more generally, echoing "fundamental disagreements about what our attitudes and behavior towards animals should be" (Biotechnology and Biological Sciences Research Council 1999). Pets are considered as members of the family by many in modern society, and this, coupled with the

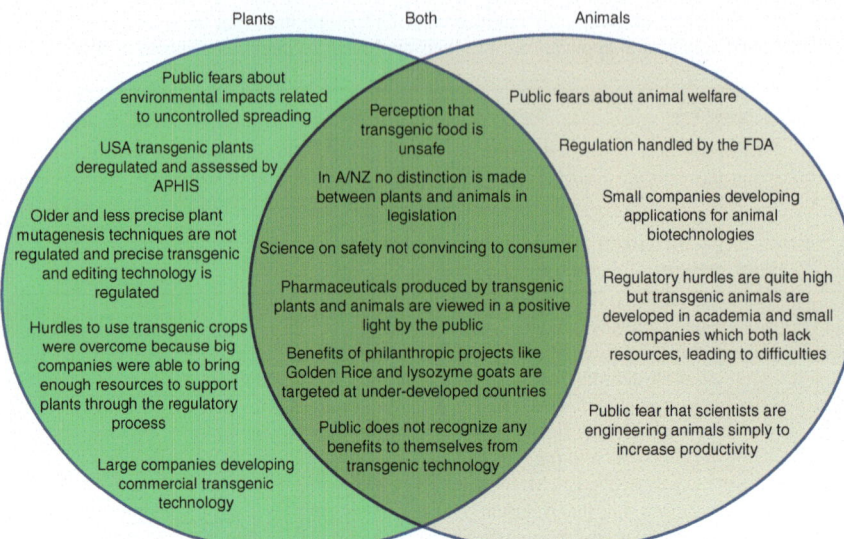

Fig. 13.1 Public perception issues posed by plant and animal biotechnology (Reproduced with permission from Tizard et al. 2016)

Box 13.1 AquAdvantage Salmon

The AquAdvantage fast-growing, genetically engineered Atlantic salmon founder individual was generated almost 30 years ago by introducing a growth hormone gene from Chinook salmon into the genome of an Atlantic salmon. A promoter sequence from an ocean pout was also introduced as a switch that turns, and keeps, the growth hormone gene on, meaning the fish grow rapidly year-round. The fast-growth trait has since been faithfully transmitted to subsequent generations by conventional reproduction. By significantly increasing the growth rate, the fish reach market weight in 18 months, as opposed to the conventional 3 years. This also translates to a reduction in the amount of feed consumed. The fish are contained in inland facilities and are all sterile females, meaning that they are unable to breed with other salmon. AquaBounty, a science-based aquaculture venture, submitted its application for the AquAdvantage salmon to the US Food and Drug Administration (FDA) in 1995 for ruling on safety, performance, and environmental impacts. The application faced extreme challenges from many fronts, including politicians and activists, from the beginning. The company submitted the last FDA-required regulatory study in 2009. Almost 20 years after the initial application, and at a cost of more than $60 million, the FDA approved the salmon for US markets in 2015 (Waltz 2016). Health Canada and the Canadian Food Inspection Agency followed with their approvals in 2016. Sales of the AquAdvantage salmon in Canada began in 2017 (Waltz 2017). However, the product has continued to face a complicated and confusing regulatory landscape in the United States where legal and political actions, as well as questions over labeling, have put its path to market on hold once again.

A genetically engineered, fast-growing AquAdvantage salmon is shown alongside a smaller, conventional Atlantic salmon of the same age. Image courtesy of AquaBounty Technologies, Inc.

increased advocacy of animal rights and welfare groups, makes the status of GE animals of particular concern to the mainstream public (Agriculture and Environment Biotechnology Commission 2002). Oftentimes, public reactions to GE of animals are not specific to genetic engineering per se, but rather are concerned with production methods associated with intensive animal agriculture. Some traits generated through genetic engineering, such as faster growth, have also been achieved through traditional selective breeding for hundreds of years, in the absence of public outcry. One global study reported that 62% of respondents specifically did not approve of biotechnological applications focused on increasing farm animal productivity (Mora et al. 2012).

Opposition to GE animal applications is strong and widespread. Activist organizations make an effort to be visible and take advantage of the public platforms offered by social media. Through these avenues, they have demonstrated the ability to strongly influence the public. Perhaps one of the best examples of the effect of anti-GE rallying of the public to halt a GE animal application is the case of the Enviropig. Scientists in Canada genetically engineered pigs that produced manure with reduced levels of phosphorus. This GE animal was intended to be an environmentally friendly alternative to traditionally bred animals as excessive phosphorus produced by swine facilities is known to contaminate groundwater and lead to algal growth, which in turn has negative effects on fish populations (Forsberg et al. 2013). Despite years of research and positive progress within the regulatory review system in the United States and Canada in the late 2000s, anti-GE activists vigorously condemned the project as a "technofix" and an excuse to farm pigs more intensively. The lack of public acceptance caused the long-time funder of the project to withdraw its support. In the absence of other funding sources, the project was halted, withdrawn from regulatory review, and the animals were euthanized.

The case of the Enviropig highlights the intuitive appeal of opposition to genetic engineering (Blancke et al. 2016). People often reject GE plants and animals based on disgust and absolute opposition to genetic engineering (Scott et al. 2016), irrespective of any potential benefits that might be associated with the application. People who are genuinely concerned about the environment often reject GE applications that have been demonstrated to address environmental problems (Blancke et al. 2016). This outright rejection of genetic engineering is often associated with concern that it is unnatural and "violates species boundaries" or is equivocal to "playing God." It has been argued that these concerns are spurious from both scientific and ethical standpoints as species are not fixed nor unchanging and that when we domesticated animals we effectively changed their genetics in an unnatural way as evoked by the term "artificial," as distinct from "natural," selection (Rollin 2014).

In general public concerns about animal biotechnology are influenced by the purpose of the application, methods used to achieve the genetic modification, the species being modified, moral status of animals, boundary between "natural" and "unnatural," and consequences to human health and the environment (Council for Agricultural Science and Technology 2010). To facilitate discussion, these concerns can be separated into personal and cultural characteristics, animal characteristics, and research characteristics, but it is important to note that these categories are interconnected on many levels (Fig. 13.2).

13.2 Personal and Cultural Characteristics

13.2.1 Perception of Risks and Benefits

People are usually more afraid of uncertain risks and hazards than optimistic about uncertain future benefits. Proponents of animal biotechnology tend to focus on the potential benefits, whereas opponents highlight potential risks (Knight et al. 2007).

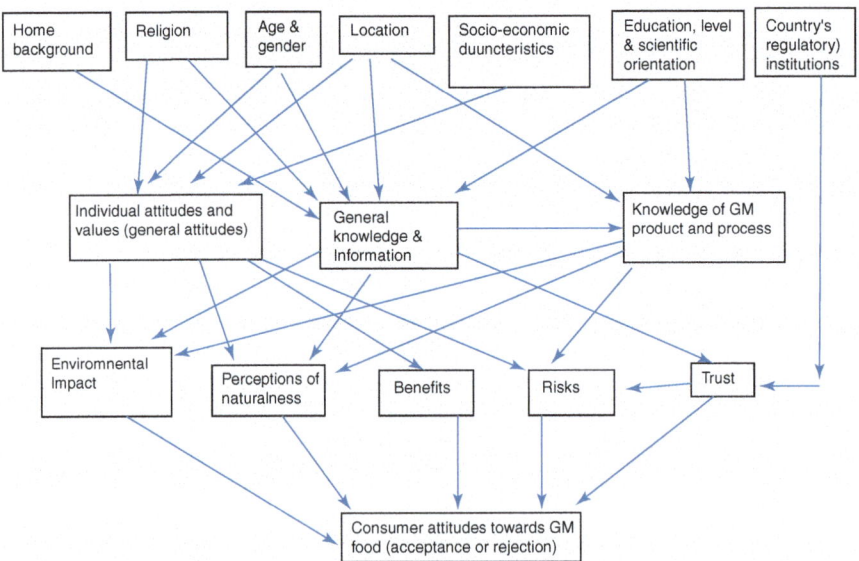

Fig. 13.2 The determinants of consumers' attitudes to GE foods (Reproduced with permission from Hudson et al. 2015)

In evaluating the risk-benefit balance, it is important to acknowledge that some activities inevitably carry more risk than others, but it doesn't necessarily follow that low-risk activities are better than high-risk ones. There is no such thing as 100% safe; there is no way to guarantee that any one process or activity will never present any risk (Biotechnology and Biological Sciences Research Council 1999). For the public, it comes down to a sense of control and choice regarding the outcome. People tend to perceive involuntary risks as more threatening than voluntary risks, even if the likelihood of resulting harm is the same, or even lower. Potential loss evokes more emotional significance than benefits of equivalent strength (Sanchez 2015), and perceived technological hazards are thought of as more threatening than naturally occurring risks (Frewer et al. 2004).

Two issues in particular come into play when talking about the risk-benefit balance as it applies to animal biotechnology: knowledge (familiarity) and trust. It has been widely observed that only a small percentage of the public adequately understands the techniques involved in genetic engineering (Curtis and Moeltner 2007; Gaskell et al. 2000; Steinhart 2006). Genetic engineering represents complex technology that is mired in conflicting information in the public domain, much of which is misinformation (Bode and Vraga 2015). This exacerbates public unease and has led to the perpetuation of the wait-and-see approach of the precautionary principle prevalent in Europe. Lack of knowledge about science, or familiarity with scientific concepts, has been proposed and investigated as underlying causes for the largely negative public perception of animal biotechnology. This so-called deficit model "refers to the idea that acceptance of newly emerging technologies can be achieved through a more scientifically informed public" (De Witt et al. 2015). Studies have shown that this model is flawed in that, whereas scientific knowledge is positively correlated with support for science in general, this does not necessarily translate to support for specific technological applications (Allum et al. 2008; De Witt et al. 2015; Moerbeek and Casimir 2005). When applied to genetic engineering, knowledge does appear to make people more likely to differentiate between medical and agricultural applications, for example, but not necessarily to make a distinction between different methods of achieving genetic modifications (Mielby et al. 2012). In a US study, an increase in approval for GE plants was associated with higher formal education (Puduri et al. 2005), but other studies could not establish links between GE acceptance and education level (Ganiere et al. 2006; Priest 2000). A 2016 YouGov and Huffington Post survey of 1000 American adults showed that respondents with college degrees were more likely to consider GE foods to be "generally safe" (49%) than those that had completed some college (36%) or those that had completed high school or less (22%) (YouGov 2016).

The contradictory results that have stemmed from a number of studies could in part be due to how knowledge is measured. Study respondents generally report high awareness of biotechnology, usually due to media sources such as television, newspapers, and the Internet, but it has been argued that awareness is not the same as knowledge (Sheehy et al. 1998). Other methods of measuring knowledge include self-reporting and questions that directly assess objective knowledge. When participants were asked to make a list of GE food products, the majority believed that they had eaten GE foods, but many of the items on their lists were not GE. Some also incorrectly stated that they could differentiate between GE and non-GE foods based on appearance, such as size or uniformness, and taste (Knight 2009).

In the absence of knowledge, trust functions as a substitute (Leahy and Mazur 1980). Social trust, which focuses on reliance upon institutions and experts, is often used as a heuristic to simplify complicated management decisions involving science and risk management for individuals lacking detailed knowledge about biotechnology who feel they have no control over issues such as GE food (Critchley 2008; Frewer et al. 2004; Siegrist 2000). For information regarding food biotechnology, US consumers report the highest levels of trust in health organizations (50%), government agencies (45%), health professionals (45%), and farmers (40%) (International Food Information Council 2014). When considering variation in attitudes about GE foods between Europeans and North Americans, differences appear to be more related to trust patterns and not necessarily to knowledge in science (Priest et al. 2003). Trust in regulatory institutions, the perceived motives of institutions and scientists, and information about the risks and benefits of particular applications are known to be of great importance in the public acceptance of GE foods (Frewer et al. 2004).

Some researchers have concluded that institutional failures to address concerns voiced by the public have had negative impacts on public trust as well as on the commercialization of GE foods. A 2014 US study by the Pew Research Center found a

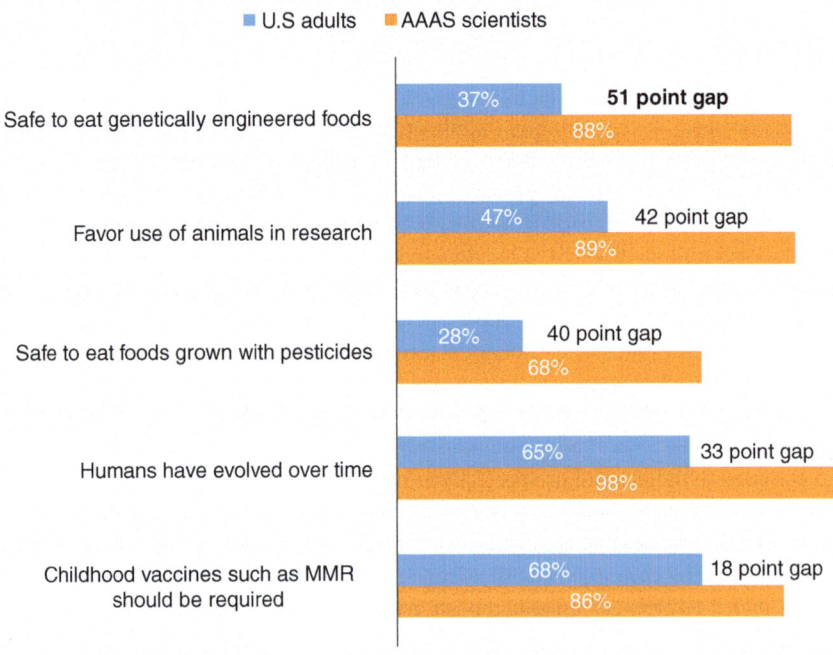

Opinion Differences Between Public and Scientists

■ U.S adults   ■ AAAS scientists

Safe to eat genetically engineered foods — 37% / 88% — **51 point gap**

Favor use of animals in research — 47% / 89% — 42 point gap

Safe to eat foods grown with pesticides — 28% / 68% — 40 point gap

Humans have evolved over time — 65% / 98% — 33 point gap

Childhood vaccines such as MMR should be required — 68% / 86% — 18 point gap

Fig. 13.3 Opinion differences between public and scientists that are members of the American Association for the Advancement of Science (AAAS) on issues related to biomedical sciences. The largest gap (51 points) between the two groups is on the issue of the safety of consuming GE foods (Pew Research Center 2015)

gap of 51 percentage points, the largest difference in the study, between US adults and American Association for the Advancement of Science (AAAS) scientists on the question of whether it is safe to eat genetically modified foods (Fig. 13.3). Additionally, 67% of adults said that they do not think that scientists clearly understand the health effects of GE crops (Pew Research Center 2015). A 2016 YouGov survey found that 33% of 1000 American adults polled think it is generally safe to eat GE foods, 39% think it's generally unsafe, and 27% reported that they are not sure if it's safe. The same poll found that 12% of respondents think scientists have a very clear understanding of the health effects of GE foods, with 40% reporting a somewhat clear understanding, 31% not a very clear understanding, 9% not a clear understanding at all, and the remainder reporting that they are not sure (YouGov 2016).

When it comes to benefits, medical applications of biotechnology are typically perceived as useful, relatively risk-free, and therefore widely accepted. However, biotechnology in food production is highly controversial (Mielby et al. 2012); agricultural applications are generally regarded as less useful and more risky (Gaskell et al. 2003). This is often the result of uncertainty as far as who will benefit directly from the application: the consumer, the farmer, or the corporation (Gaskell et al. 2004; Pin and Gutteling 2009). Perceived societal benefits, such as medical applications, are more important to the public than perceived economic benefits, such as agricultural applications (Gaskell et al. 2003). In a 2014 report, unfavorable impressions of animal biotechnology were found to be due to lack of information (55%) and a poor understanding of the potential benefits (42%) (International Food Information Council 2014).

13.2.2 Food Safety and Food Security

Risk-benefit considerations are intertwined with societal experience with both food safety and food security. The World Health Organization (WHO) defines food security as existing "when all people at all times have access to sufficient, safe, nutritious food to maintain a healthy and active life" (World Health Organization 2016). The WHO and Food and Agriculture Organization of the United Nations (FAO) define food safety through the Codex Alimentarius as "assurance that food will not cause harm to the consumer when it is prepared and/or eaten according to its intended use" (Food and Agriculture Organization of the United Nations 2009).

Public views on food safety and security largely depend on the country of residence, the country's industrialization status, and historical experience with issues of food safety and security. Generally, studies with a focus on biotechnology and food safety are from developed countries, where concerns are focused on issues such as bacterial food poisoning, whereas developing countries, in which food access is often limited and widespread hunger is prevalent, are more concerned with the need for food security (Rodriguez and Abbott 2007). In the United States, overall consumer confidence in the food supply has remained consistently high since 2008 (International Food Information Council 2014). However, food safety scares such as the bovine spongiform encephalitis (BSE) crisis in Britain in the mid-1980s had a persisting negative effect on public perception of the food supply in Europe, making residents skeptical of biotechnology as it applies to food (Finucane 2002). In other parts of the world, for example, for resource-poor farmers in Africa, improvements in production, enhancements to nutritional value, resistance to pest-driven diseases, and better integration into the global economy by use of biotechnology may be received positively as one means to combat chronic food security issues (Abah et al. 2010; Food and Agriculture Organization of the United Nations 2013; Ruane and Sonnino 2011; Smyth et al. 2015).

In general, products from GE animals that have been studied have equivalent protein, nucleic acid and lipid composition to their non-GE counterparts, meaning that there should not be any unique issues in terms of food safety. Similarly, there is no evidence that foods from GE animals examined to date are more allergenic than food from non-GE animals. In fact, genetic engineering may actually be able to remove allergens from foods such as eggs (McColl et al. 2013). Of course, safety evaluations need to be carried out on a case-by-case basis dependent upon the characteristics of the novel protein being expressed, if any, in the GE animal.

A meta-analysis of studies on GE crop adoption in select countries to date demonstrates income gains and positive impacts on food security (Qaim and Kouser 2013), and it is envisioned that GE animals modified for traits like disease resistance could enable similar outcomes. Worldwide population growth, especially in developing countries, along with the need to significantly increase animal protein production in response to the rising number of middle class consumers, will likely necessitate the inclusion of many different technologies in agricultural production systems, including the use of biotechnology in animal genetic improvement programs. Public perceptions may shift as the demand for sustainably produced animal protein becomes increasingly pressing.

13.2.3 Sociodemographic and Socioeconomic (Individual Differences)

13.2.3.1 Religious Values

Although religious traditions, beliefs, and practices can draw ethical implications about the role of animals in general, including animal husbandry and breeding and the inclusion of specific animals in the human diet, there are few clear rules among the world's largest religions regarding animal biotechnology. Traditional religious sources, many of which are hundreds of years old, do not directly address this recent technological development (Council for Agricultural Science and Technology 2010). Therefore, benefits and concerns about animal biotechnology can often be argued either way from a religious perspective.

The major Western religions—Christianity, Judaism, and Islam—generally permit animal biotechnology. They specify that animals are God's creatures, but as long as there is a sufficient human benefit and animal welfare is respected, animals are at the service of humans. However, some religious leaders have opposed animal biotechnology as a violation of God's role as Creator or its potential threat to biodiversity or "the integrity and ecological balance of creation" (Council for Agricultural Science and Technology 2010). The World Council of Churches opposes GE as it "messes with our common inheritance and health," whereas the Catholic Church supports genetic engineering of food as a solution to poverty and malnutrition (Melodlesi 2011).

Some religions specify dietary considerations that could have implications with respect to animal biotechnology. Certain animal species may be revered, such as cows in Hinduism, which require followers to adhere to strict vegetarian diets. In some religions, followers are forbidden from consuming foods containing genetic material from certain animals (Ormandy and Schuppli 2014). However, the implications of the transfer of small amounts of genetic material on the identity of an animal are largely open to interpretation within these religious contexts.

In Eastern religions—Buddhism, Hinduism, and Confucianism—animals are considered to have a moral status almost equal to humans. There are also beliefs in cross-species reincarnation and a balance in nature of human, plant, animal, and environmental interactions (Epstein 1998). Similarly to the Western religions, animal suffering is balanced equally against human benefit (Crawford 2003). There is no general consensus within these religions regarding the religious permissibility of animal biotechnology (Council for Agricultural Science and Technology 2010).

Animal biotechnology, along with many other forms of technology, is often criticized as "playing God." Some people hold religious views that modern biotechnology, as a whole, is blasphemous (Biotechnology and Biological Sciences Research Council 1999). These views stem from the belief that God has created a perfect natural order and these technologies disrupt that order. Some have concerns that humans are taking over powers that are thought to belong only to God (Pew Research Center 2015). Agnostics and atheists may share this concern that biotechnology is "unnatural" and therefore somehow wrong. With regard to the "playing God" argument, bioethicist Bernard Rollin questions, "Where is the moral problem in playing God? In fact, what does 'playing God' even mean? When we change the course of rivers, domesticate animals, build flying machines, create clothing that can withstand Arctic cold, or, in general, exhibit human inventiveness, are we not 'playing God', but in a way that no one finds morally objectionable?" (Rollin 2014).

13.2.3.2 Natural Versus Unnatural

The "playing God" argument is akin to concerns that animal biotechnology is unnatural, also known as the appeal to nature fallacy. This is cited as an intrinsic concern meaning that the very use of genetic engineering is wrong, rather than evaluating the consequences of its application, which are referred to as extrinsic concerns. This viewpoint is controversial as not everything natural is good and the fallacy that natural is always better has allowed the organic and non-GE markets to exploit consumer perceptions and preferences (Sanchez 2015).

Although individual perceptions of these terms may vary, as well as the context in which they are used, the concept of natural generally evokes positive associations; synonyms for natural include normal, usual, common, logical, ordinary, reasonable, and wild, and the term is commonly contrasted with artificial or man-made. However, based on this usage, practically every aspect of modern lifestyles is artificial or unnatural. Natural is often equated with "good" and unnatural with "bad." But just because something is natural does not mean that it is necessarily beneficial or even benign.

Plants that produce natural toxins, bacteria, and viruses that naturally cause diseases, and earthquakes and tornadoes are collectively termed natural disasters.

Concerns about animal biotechnology as unnatural revolve around the concept of species integrity or the perceived breaching of natural species boundaries. The problem with this is that the concept of species is not very clear-cut, and even biologists are not sure about the definition of natural species boundaries since the crossing of some species occurs "naturally," and the genetic makeup of species can change over time.

Even if natural species boundaries can be identified in the future, the fact that they exist does not tell us what, if anything, should be done about them from an ethical standpoint. This suggests that it's the involvement of animals, not the crossing of species boundaries, that people find objectionable (Biotechnology and Biological Sciences Research Council 1999), and for these critics, GE animals will be considered unacceptable despite any potential benefits (Knight 2009; Shaw 2002).

13.2.3.3 Gender and Age

Gender has been shown to be the strongest correlate of opposition to animal research, which is closely tied to opinions of animal biotechnology (Pifer 1996). Globally, females are less likely to support animal-based research (Driscoll 1995; Navaro et al. 2001; Ormandy and Schuppli 2014; Swami et al. 2008). A number of theories have been proposed to explain this effect, including that females are more likely to attribute mental states to animals and males do not believe as strongly in the mental abilities of animals (Herzog and Galvin 1997; Knight et al. 2004). It has also been suggested that women may have more direct contact with animals through household tasks since they are typically the primary family caretakers (Kendall et al. 2006). Additionally, mothers in particular tend to report having read or heard "a lot" about food biotechnology, will pay more for products perceived as sustainable, and want more information on food labels, especially with regard to ingredients, making them an active and vocal community in these discussions (International Food Information Council 2014).

The effect of age is less clear than the effect of gender. Studies report conflicting results on impact of age on perception of animal research. Several studies found that older participants show greater levels of support (Driscoll 1992; Hagelin et al. 2003; Ormandy and Schuppli 2014), but a 2010 study found that younger participants were more supportive of animal-based research (Schuppli and Weary 2010). A 2014 US report found millennials to be more favorable toward food biotechnology than other age groups (International Food Information Council 2014).

13.2.3.4 Country/Region

In addition to effects of gender and age, public perception of biotechnology is also subject to country and regional influences. Given the strong consumer opposition to GE crops and foods that developed in Europe in the 1990s, along with unique political, religious, cultural, historical, and social differences between countries of the world, regional differences are not surprising. For example, in the United States, GE applications for common diseases of farm animals, such as mastitis, are largely perceived as efforts to improve animal welfare. The same applications in Europe are considered excuses to worsen housing conditions in animal agriculture, resulting in negative impacts on animal welfare (Vazquez-Salat and Houdebine 2013).

Overall, greater negative press coverage of genetic engineering in Europe is no doubt partly fueled by public health concerns such as the BSE crisis, as well as the historical consumer backlash against GE foods. Perhaps as a result of these incidences, Europeans exhibit low levels of trust in regulators and institutions that are responsible for protecting consumers and environmental interests with respect to food production (Houghton et al. 2008). Conversely, it has been proposed that the greater availability of GE foods and crops in the United States results in greater positive consumer experience with GE foods (Ceccoli and Hixon 2011). The European Union has a low GE crop adoption rate, with the main GE crop, *Bt* corn, primarily grown in Spain (James 2016), although it should be noted that the EU is a major importer of GE feed for their livestock populations.

Most studies have found that North American and Asian consumers have more positive attitudes to GE food production compared to Europeans, but North Americans also perceived more benefits associated with genetic engineering overall than either of the other two populations. One study found that ethical and moral concerns in North America and Asia were greater as compared to Europe, and North American, South American, and Asian study participants perceived fewer overall risks than Europeans (Frewer et al. 2013). In 2013 an Australian study found that respondents were overall significantly less willing to eat GE food than they were to eat other foods and were the least willing to eat meat and other products from GE animals (Ipsos Social Research Institute 2013).

Along with the United States, Canada and some Latin American countries, primarily Brazil and Argentina, have widely adopted biotech crops. Brazil, which is second only to the United States in hectares of biotech crops grown (James 2016), has been described as "the engine of biotech crop growth globally." A British company, Oxitec, which has genetically engineered mosquitoes to pass on a lethal gene to offspring, effectively reducing the population and the chances for the spread of diseases like dengue fever, is based in Brazil. Field trials of these mosquitoes have been carried out in Panama, but proposed field trials in the Florida Keys have met with public resistance (Adalja et al. 2016), demonstrating how regional public perception issues can impact technology adoption.

Views on science and technology in Africa have improved in recent years as a result of various efforts of the African Union. However, translation of these efforts to the public has been lacking, and as a result, residents are largely unaware of biotechnology. In a 2005 study in South Africa, 80% of a total 7000 respondents did not know the meaning of the word biotechnology (Network of African Science Academies 2015). Regulations in Africa take an extreme precautionary approach and are restrictive to development and adoption of biotechnology despite calls from political leaders to boost productivity and competitiveness through the use of modern biotechnology tools. In 2016, just two African countries, South Africa and Sudan, planted GE crops, although additional countries granted environmental release approvals or evaluated field trials (James 2016). In 2002, Zambia, Zimbabwe, Mozambique, and Malawi refused donated emergency food, citing uncertainty about human and environmental safety that could be caused by GE ingredients. More recently, Egypt and Kenya banned GE crops after the publication of the highly criticized Serálini study (Arjó et al. 2013) that purportedly revealed tumor growth in rats that consumed GE feed (Network of African Science Academies 2015). Fear of cross-pollination with local varieties, distrust of the intentions of the United States and private companies in promoting GE food, uncertainty about the ultimate usefulness of the technology in various microclimates, and ethical concerns about "animal genes" have also been reported (Cooke and Downie 2010).

Conversely, Asia, especially China, was an early leader in GE technology. In 2008, China committed $3.8 billion to a 10-year R&D program on GE crops and animals, and initial studies showed Chinese citizens were not as concerned about GE crops and animals as other world regions (Vazquez-Salat 2013). Despite the fact that China was predicted to be the first country to commercialize GE rice, public opposition resulted in a hesitancy to issue the appropriate permits for release of GE crops (Jayaraman and Jia 2012). In 2015, China announced its intention to support more research and development of GE products, as well as a commitment to supporting public education around these products. The overall goal is to regain the country's position as a global leader in GE technology (Li et al. 2015), and the country is poised to become a global leader in the GE animal field (Vazquez-Salat and Houdebine 2013).

Despite growing GE crops in many instances, other Asian countries have variously faced bans on GE products (*Bt* eggplant in India), as well as field trial plots of GE crops being vandalized by groups such as Greenpeace (Philippines) (Hallerman and Grabau 2016). Interestingly, the country of Bangladesh has become a model for the adoption of GE crops, having had great success with *Bt* brinjal/eggplant and with late blight-resistant potato, golden rice, and *Bt* cotton currently in the pipeline (James 2016).

13.2.3.5 Life Experiences

Many of these country and regional differences are linked to industrialization status, as well as the distribution of the population between rural and urban communities. Links between a nation's level of industrialization and urbanization and attitudes toward animal research have been well established. Animal use differs greatly between rural and urban areas, and people with rural backgrounds tend to exhibit greater acceptance of animal use and greater support for animal experimentation than their urban counterparts (Jasper and Nelkin 1992; Kalof et al. 1999; Pifer et al. 1994). In one study, pet owners rated animal-based research as less acceptable than non-pet owners (Driscoll 1992), highlighting the fact that previous or existing experience with animals has an influence on perception of animal use (Knight and Barnett 2008; Wells and Hepper 1997).

In addition to experience with animals, other personal experiences and circumstances can also shape an individual's perception of biotechnology. For example, "parents of children suffering from cystic fibrosis may well have different perceptions about the use of genetically modified animals in the search for a cure, than will those with no personal experience of the condition" (Biotechnology and Biological Sciences Research Council 1999). Similarly, perceived environmental risks can outweigh perceived benefits in an industrialized country largely unaffected by vector-borne diseases, for example, whereas those that live in areas impacted by diseases such as malaria may have more favorable views of the release of GE mosquitoes to combat the disease.

Vegetarianism and veganism are specific food views involving animal welfare, the environment, and health. From an animal welfare perspective, these viewpoints have specific objections to intensive animal agriculture as a whole, namely as it involves concentrated animal feeding operations (CAFOs). Vegans and vegetarians have been reported to be less accepting of the use of animals in research and less accepting of GE in general than their counterparts (Furnham et al. 2003; Hallman et al. 2003; Kendall et al. 2006; Ormandy et al. 2012). Gender has been shown to be associated with vegetarianism, with more women represented than men (Gabriel et al. 2012; Hagelin et al. 1999), consistent with reports of gender effect in attitudes about animal research. A UK study found that vegans express animal rights positions more strongly than vegetarians, especially those that had been vegan for more than 1 year. This difference was explained by noting that some people choose a vegetarian diet for reasons such as health more than for ethical reasons about animals. The same study also found considerations of animal welfare to be prevalent in meat eaters as well but with "a wider range of ethical frameworks" than the other groups (Lund et al. 2016). Although overall many vegetarians and vegans are opposed to GE animals, and genetic engineering in general, a minority is supportive of potential benefits.

13.2.3.6 Consumer Perception Research

Consumer attitudes toward biotechnology are affected by all of the aforementioned factors and are also subject to change over time. More than one study has suggested that consumers reject the technology overall based on their general sociopolitical attitudes and values in a "top-down" process, rather than in a "bottom-up" process based on an evaluation of the characteristics of the specific product (Bredahl 2001; Novoselova et al. 2007; Scholderer and Frewer 2003). However, it has also been suggested that experiencing a "real" GE product that has strong, observable benefits could change consumer attitudes (Grunert et al. 2004).

In evaluating consumer attitudes toward biotechnology, it is useful to acknowledge that how consumers are polled or interviewed can greatly influence their answers (Hallman et al. 2003; Hess et al. 2013; Stephan 2015). The precise wording of questions, methods used to select respondents, and background information provided can greatly affect the outcomes. Social science research is tied to the socioeconomic context in which it is conducted. Researchers and participants alike are sensitive to conditions surrounding surveys and polls. It has been reported that public discourse and strong negative opinions expressed by some European policymakers have led more European researchers to ask survey questions that tend to be critical of biotechnology (Hess et al. 2013; Pin and Gutteling 2009). A large 2015 US survey highlighted the influence of lack of knowledge about biology on the perception of GE foods, reporting that 82% of respondents supported "mandatory labels on foods produced with genetic engineering." However, an almost equal number, 80%, supported "mandatory labels on foods containing DNA" suggesting a lack of understanding regarding the ubiquitous presence and safety of consuming the DNA that is present in much of our food (Lusk and Murray 2015). Lastly, survey results may be unintentionally biased. Low response rates are common, meaning that respondents are often those with strong knowledge, interest, and/or opinions on the subject (Townsend and Campbell 2004).

13.2.3.7   Role of Media and Social Media

In our modern world, we cannot ignore the influential role of media and social media in shaping public perceptions of many topics, including biotechnology. In one study, focus group respondents reported that they had heard of the term "biotechnology" from multiple media sources (Knight 2009). A report from the FAO stated that "mass media represents the main sources of information for consumers on all nutrition and food safety topics, including biotechnology" (Hoban 2004).

In light of the close relationship between public attitudes of animal biotechnology and animal research in general, the ability of animal rights organizations to be successful in raising public awareness of issues related to animal research undoubtedly has some bearing on public perceptions of animal biotechnology. Additionally, social media has made it very easy for false or misleading information to spread online (Vis 2014). The ability of food activists, in particular, to mobilize the public through the use of social media to pressure companies, whether the alleged wrongdoing is factual or fabricated, is becoming almost commonplace. By the time the facts are released, the damage to the company's reputation is done, and they often have no recourse but to acquiesce to the demands (Veil et al. 2015). It is well known that people tend to focus on information that confirms prior-held beliefs, and the Internet provides a wealth of credible sources of information based on evidence alongside low-quality data, anecdotes, and personal opinions presented as facts. Although misperceptions are notoriously difficult to correct, a study exploring the use of Facebook to combat misinformation revealed positive results for the topic of GE organisms and illness, although it was less successful for the issue of vaccines and autism (Bode and Vraga 2015).

Public attitudes do fluctuate over time in relation to the volume of media coverage (Frewer et al. 2002). Studies have reported that support for GE plants and animals is lower in the face of greater media coverage of GE food, but that support for genetic engineering increases when media coverage is lower (Flipse and Osseweijer 2013; Marques et al. 2015). Flipse and Osseweijer observed that "media articles generally cover the controversy rather than the debate." Media attention is usually greatest in response to an initial event or "scandal." It gradually decreases from there, and by the time scientific validations appear, the media attention has decreased to a minimum. Unfortunately, it can take years in some cases for the scientific community to respond to an issue, usually due to the fact that it takes time to generate data and come to a consensus, whereas the media responds in mere days (Flipse and Osseweijer 2013). The attention spans of both the public and the media are limited to short periods of time which peak and then decrease. Since these "scandals" tend to dominate media reports and subsequent public discourse, they can lead to "GM phobia" (Jayaraman and Jia 2012).

13.3 Animal Characteristics

In addition to the considerations outlined above that occur at a personal level, when it comes to animal biotechnology, the public is also greatly concerned with the animals themselves. Issues of animal welfare, differences between species and their perceived sentience, as well as concerns about hybrids can all influence public attitudes.

13.3.1 Species and Sentience

It is much easier to rally followers and funds for animals that are perceived as "cute" or attractive (Hagelin et al. 2003; Herzog and Galvin 1997; Knight and Barnett 2008). For example, it is easier to generate public approval of conservation efforts directed toward more attractive animals like sea otters and pandas than toward insects, reptiles, or fish. Studies have reported clear distinctions between popular and unpopular animals, with large mammals (especially primates and companion animals) being the most popular and biting invertebrates (mosquitoes) the least popular. Driscoll reported that "non-mammalian species are clearly devalued when compared to mammals on dimensions of smartness, responsiveness, and lovableness." An almost perfect correlation was reported between ratings of perceived usefulness and importance of various species. Additional criteria for the public's evaluation of animals are based on how the species has been regarded historically, the utility of the species for human use, and emotional reactions to the species (Driscoll 1995).

Activities involving companion animals or large, attractive mammals are less acceptable to the public than the same activities when they involve rodents or non-mammalian species. It has been well established that such public attitudes about various species correlate to attitudes about research animal use (Driscoll 1992; Herzog and Galvin 1997; Ormandy et al. 2012). The most commonly used species for research are mice, rats, and zebrafish (Ormandy 2009).

People tend to rate animals classified as pets or nonhuman primates as having higher mental abilities compared to other species such as fish and mice. It follows, then, that people are more supportive of using smaller-brained animals, such as mice and rats, in research and are less supportive of using animals with "higher" mental abilities, often classified as those that can use tools, solve problems, and be self-aware, or sentient (Driscoll 1992; Herzog and Galvin 1997; Knight and Barnett 2008). Sentience refers to an organism's ability to experience pain, suffering, happiness, and pleasure. It is sometimes referred to as "belief in animal mind," or BAM, and has been shown to be "a relatively consistent predictor of attitudes toward the human use of animals" (Herzog and Galvin 1997; Knight et al. 2004; Schuppli 2011).

13.3.2 Phylogenetic Distance and Hybrids

As mentioned previously, the concept of species integrity, or breaking species boundaries, is problematic since species change on a regular basis and the definition of a species continues to be debated by biologists (Samadi and Barberousse 2015). Rollin observed that "species are in fact spatio-temporal slices of a changing and dynamic process, at best snapshots of what is in constant flux" (Rollin 2014).

However, these issues still bother the public. In particular, the concept of hybrids seems to be especially concerning, conjuring images of chimeras, creatures from Greek mythology that were part lion, part goat, and part snake. Indeed, whereas the public appears to have few reservations about common hybrids such as grapefruit, tangelos, mules, and beefalo, unfamiliar hybrids are "imagined more negatively" (Kronberger et al. 2013), and hybrids deemed to be of incompatible kinds are often perceived as having "no essence of their own." They are also reported to frequently evoke "the yuck factor" and feelings of disgust (Wagner et al. 2010).

Hybrids have been "used by authors to evoke images that raise suspicion, apprehension and unease about science, biotechnology, government and human nature." For example, HG Wells used hybrids in his 1896 novel *The Island of Doctor Moreau* to question scientific limitations and the meaning of being human. More recently, Margaret Atwood featured hybrid creatures in her dystopian work *Oryx and Crake*, a doomsday story in which biotechnology is the norm but ends up destroying the world (Sanderson 2013). Although the methods and practicalities of how hybrid creatures are created in such stories is usually unclear, the creatures effectively capture public concern over how far science *could* go and whether, as a society, we should even start down the path if there is even the remotest possibility that's where it could lead.

Although these fantastical creatures inhabit fictional worlds, the images that they conjure have real-world effects on animal biotechnology. It is common to find pictures depicting parts of one animal pasted to another in anti-GE propaganda, often accompanied by a Frankensteinian title. For example, BioSteel, a high-strength fiber made from spider silk protein extracted from transgenic goats (Lazaris et al. 2002), can be found depicted online with a picture of a goat head on a spider body. Such images are effective in preying upon the public's fears of hybrids and perpetuate misinformation about what is real versus imaginary.

With the public's distrust of hybrids in mind, researchers have begun to explore the possibility that cisgenics, or organisms that contain genetic material from sexually compatible donor species, would be more acceptable than transgenics, organisms that contain genetic material from species that are sexually incompatible. Overall, the public is more positive about cisgenics than transgenics for a number of reasons, including that they are expected to cause less harm for the environment (Kronberger et al. 2013). The potential acceptance of cisgenics may be tested in the near future with the recent production of dairy cattle from one breed that carry an allelic form of a gene from a beef breed that prevents the growth of horns, a condition known as polled. Polled animals are desirable because dairy cattle that have horns usually have their horns removed by mechanical means since they are a danger to handlers as well as other cattle. This process of dehorning has become a lightning rod for animal activist groups, and farmers themselves express their distaste for the process. Since the snippet of DNA that confers the polled trait is from another breed of cattle, so it is within the same species, as well as the fact that the application of genetic editing in this case addresses a welfare issue (Carlson et al. 2016), this example changes the discussion from the usual concerns expressed about transgenics. This distinction between transgenics and cisgenics will have increasing relevance with respect to regulation of animal biotechnologies and the pursuit of future applications (Marchant and Stevens 2016).

13.3.3 Ethics and Animal Welfare

Some of the most frequent objections to animal biotechnology revolve around issues of animal ethics and welfare due to perceived pain and suffering resulting from genetic engineering and imbalances between the human beneficiaries of genetic engineering and the consequences for the GE animal. It is difficult to generalize about the welfare of GE animals since different species and different applications will present unique considerations. For those opposed to animal agriculture overall, genetic modifications to improve food animal productivity can be viewed as exacerbations of current problems with intensive agriculture that could lead to increased stress and performance-related issues for the animals. For example, it has been suggested that using genetic engineering to produce medicines in milk could increase pressure to extend the length of lactation for individual animals or increase the frequency of milking. Similarly, changes that modify growth rate in embryos could create additional stress for the mothers. Genetic engineering aimed at disease prevention could be argued as addressing diseases that are endemic to intensive farming methods which could be prevented by altering or eliminating certain farming practices or the use of animals for food production entirely.

Bioethicist Rollin uses the term "telos" to describe the "nature" of an animal, which he refers to as "the pig-ness of a pig, the dog-ness of a dog." He argues that it is ethically acceptable for humans to change the telos of an animal as long as the animal is respected and, in the case of genetic engineering, that the transgenic animal should be no worse off than its parents, a principle he terms "conservation of welfare" (Pew Initiative on Food and Biotechnology 2007). Specific GE applications to reduce disease incidence, reduce or eliminate undesirable practices such as dehorning, prevent or treat genetic disorders, and save endangered species from extinction would arguably alter the telos of the subject for the better.

Bioethicists have discussed ethical quandaries related to this premise extensively. One of the most often mentioned examples is "the blind hen problem," in which blind hens would be more docile, less stressed, and less prone to pecking each other so they would not need to be debeaked. Similarly, turkeys could be engineered without brooding impulses and hens to have no desire to nest or to want to nest in a cage. On the most extreme end, "microencephalic pigs and chickens that have brain function that is sufficient for maintaining growth but not for supporting mental states or psychological experiences" would theoretically never suffer because they would lack the capacity. The public clearly has qualms about these types of hypothetical situations. Critics argue that these examples would simply cover up the root problem and that the ethical responsibility is to eliminate the conditions creating the problem (Pew Initiative on Food and Biotechnology 2007; Sandler 2015). In addition to the intended effects motivating the specific modification, there are always considerations of unintended effects that contribute to the public's view of the unpredictable nature of genetic engineering. With all of these things in mind, Kaiser observed, "welfare aspects of biotechnology in animal production require a close case-by-case and step-by-step evaluation in order to avoid negative impacts of the technology" (Kaiser 2005).

13.4 Research Characteristics

13.4.1 Method and Level of Invasiveness or Harm

Issues of animal welfare are often tied to the research methods employed and the actual or perceived levels of invasiveness or harm of those methods. Increased levels of invasiveness result in decreased acceptance of animal-based research (Hagelin et al. 2003). One study found respondents to be more supportive of genetic engineering than N-ethyl-N-nitrosourea (ENU) mutagenesis, a method commonly used to induce mutations in zebrafish, because they believed that genetic engineering is more accurate, more efficient, and less painful than ENU mutagenesis (Ormandy et al. 2012).

Along with these concerns, people's underlying beliefs about the availability of alternative methods and the number of animals used in particular experiments also shape attitudes about animal research (Knight et al. 2003). Institutional Animal Care and Use Committees, which oversee animal research at many institutions, focus on "the three Rs," replacement, reduction, and refinement (Russell and Burch 1959). Animal use is more likely to be supported when participants perceive there to be no alternative to using animals (Knight et al. 2003). These issues reflect concern about the "waste of animal lives" and are in line with historical concerns about high mortality rates and birth defects in transgenic and cloned animals (Gjerris 2012).

13.4.2 Application

Numerous studies have established the important role of the specific GE application in public and consumer acceptance or rejection. Much of the research on public attitudes has focused on consumer views toward the use of GE animals for food production, but biomedical applications for GE animals appear to have greater support, even in traditionally anti-GE regions such as the EU (Schuppli and Weary 2010). Interestingly, the first GE animal-produced pharmaceutical protein, ATryn, was approved in the EU before it was approved in the United States. To date, three biomedical applications of GE animals have been approved for the US market (Table 13.1), whereas only a single application, the AquAdvantage salmon, has been approved for food production. Despite the FDA's approval, this product is still not available in the US market (Box 13.1).

In general people are much more opposed to genetic engineering of food animals, partly due to issues of species and sentience, as discussed, as well as application (medical vs food), than genetic engineering of laboratory animals. In one large, international study of consumer attitudes toward biotechnology, with some 35,000 respondents representing 35 countries, almost 75% of global consumers were opposed to genetic engineering of animals to increase productivity (Hoban 2004). In an FAO global study, 62% of respondents worldwide opposed applications of biotechnology to increase farm animal productivity (Mora et al. 2012). In another study, 65% of consumers disagreed with creating transgenic fish in order to improve efficiency of production (Logar and Pollock 2005).

Concerns have been voiced over the "reduction of complex, natural beings to single-purpose, utilitarian objects" (Pew Initiative on Food and Biotechnology 2007). This idea that we are turning animals into machines for industrial use echoes public concerns over intensive agriculture as a whole, not just GE applications.

These issues lead to questions such as: Because we can, does that mean we should? How far can animal biotechnology go? Are some genetic alterations intrinsically wrong? The problem is that we could ask the same questions of selective breeding as well. These issues take on new dimensions with respect to animal models of disease, such as the Harvard OncoMouse, a cancer-susceptible GE mouse. Is genetically engineering an animal to develop cancer different than inducing cancer in a laboratory animal? New gene editing techniques make it increasingly likely that animals will be specifically designed to become sick so as to be models of human diseases. The perceived and realized benefits to human society will be instrumental in determining public acceptance of these applications.

13.4.3 Unknown Consequences and Unintended Effects

Another frequently expressed concern is about unexpected and potentially harmful effects that might result from GE animals. Perceived harms include disease transmission from animals to humans (zoonoses), environmental and ecological effects from accidental or deliberate release, and concerns that GE organisms might have evolutionary advantages and outcompete natural species or have side effects on nontarget species. Regulatory bodies are in place to undertake assessments of these issues during the development phase of specific applications. As mentioned previously, there's no way to prove that something is 100% safe. The intrinsic uncertainties associated with applications of GE animals consequently leads to an invocation of some form of the precautionary principle, which can morph into generalized prohibition (Carroll and Charo 2015). Of course, not adopting technology comes with risks also. The Biotechnology and Biological Sciences Research Council (BBSRC) of the United Kingdom noted, "It is possible that 'playing safe' by abandoning research and development in all forms of animal biotechnology might deny us a technique or product which could prevent an environmental disaster in fifty years' time, or could prove invaluable in the treatment of serious diseases" (Biotechnology and Biological Sciences Research Council 1999).

The use of GE animals for xenotransplantation, which is the transplantation of organs, tissues, or cells from one species into another, represents a unique application of animal biotechnology. There are specific concerns and considerations related to xenotransplantation itself, such as disease transmission between animals and humans, regardless of whether the donor animal is GE or not. Xenotransplantation tends to be met with public skepticism in general. It has been suggested that this response is due to the fact that xenotransplantation involves higher-order organisms, i.e., animals that are more closely related to humans, such as pigs. Since xenotransplantation is perceived as risky on its own, it is difficult to ascertain whether public resistance to using transgenic animals for xenotransplantation is due to concerns about xenotransplantation, transgenic animals, or the combination of the two.

13.4.4 Patenting

Disagreements over intellectual property as it applies to the field of biology are of concern to the general public. US patent law prohibits "patents encompassing 'natural laws, phenomena, or products' or 'abstract ideas'" (Sherkow and Greely 2015). Consumers are generally supportive of the principle of patenting but become increasingly uneasy when it comes to patents involving higher life forms (Einsiedel 2005). Objections include a perceived breakdown of the differences between living and nonliving matter, as well as issues over viewing animal life as having purely mechanical functions (Pew Initiative on Food and Biotechnology 2007). The public often expresses disapproval over the concept of "owning life." This argument weakens somewhat, however, if we consider that humans "own" pets, which are living beings. In the case of GE crops, public opinion has been soured by what is perceived as a monopoly on seed patenting and ownership by commercial companies such as Monsanto.

Several landmark cases, beginning in the 1980s and lasting until recently, have challenged the interpretation of patent laws as they apply to biology, genetics in particular. These have included debates over patents on genes, nonhuman animals, and genetically engineered organisms. In 1984, Harvard University famously filed a patent for the "Harvard OncoMouse." Despite challenges, the patent was upheld in the United States but faced rejection or modification in other countries, thereby emphasizing disparities in international patent approval and enforcement. In Canada and Europe, this mouse sparked the "No Patents on Life" movement. In the United States, several bills were introduced into Congress to ban patents on animals, but none passed (Sherkow and Greely 2015).

Recently, arguably one of the most influential cases involving biological patents, *Association for Molecular Pathology v. Myriad Genetics, Inc.*, which challenged the latter's patents on the *BRCA1* and *BRCA2* genes for human breast cancer susceptibility, went through a number of appeals at the Federal level, including the Supreme Court. In mid-2013, the Supreme Court ruling held that Myriad was able to keep their patent claims on cDNA because "it is not naturally occurring," but almost none of its original seven patents were upheld because "a naturally occurring DNA segment is a product of nature and not patent eligible merely because it has been isolated." The outcome in this case has essentially led to the end of gene patents in the United States, except for those of novel DNA sequences not found in nature, but they could continue in various forms in other countries (Sherkow and Greely 2015).

Intellectual property rights, such as patents, play a role in incentivizing companies to invest in research and development of products as they provide a means to profit from the innovations (Caswell et al. 2003). The reality is that recent history has proven that successfully getting GE plants, and their associated products, into the marketplace comes with hefty regulatory price tags, hence the appeal of patents for such products to businesses. Patents are a way for companies to recoup upfront costs due to research and development and clearance of regulatory hurdles. GE animals face even more challenges than GE plants due to lengthy reproductive cycles for livestock, complex biology that comes with greater technical challenges, and high costs associated with animal losses, as well as more diverse applications that have regional, niche end user communities (Vazquez-Salat and Houdebine 2013). For many applications, these complications result in higher upfront costs, even before seeking regulatory approval.

Conclusions

At the current time, overall public support for GE animals is generally low, with biomedical applications being more positively perceived than agricultural applications. Individual views on animal biotechnology are extremely complex and are dependent on personal factors as well as the species being modified, the purpose of the modification, along with the individual's weighting of the potential risks and benefits of each application. Determining what level of risk is acceptable and ethically justifiable is difficult, but it is even more complicated and controversial when applied to animal biotechnology since the costs and benefits must be weighed for both humans and animals, two different groups with different, sometimes opposing, interests. To be effective, this must be done on a case-by-case basis, making it almost impossible to generalize about GE animals as a category or animal biotechnology as a whole. With the 2015 regulatory approval of the fast-growing AquAdvantage GE salmon for food purposes, GE animals (and by extension a lot of this public perception research that has largely been based on hypothetical examples) may get their first real test in the marketplace. The salmon is available and has been selling Canada (Waltz 2017), but it likely won't appear in US retail stores until at least 2019 (Gallegos 2017), and at that time, its success or failure will truly be up to the public. New breeding methods like gene editing are likely to result in novel animal applications such as disease-resistant animals and modifications that are explicitly focused on animal welfare traits. Additionally, the developers of such products are increasingly academic researchers and small companies. Such examples are likely to significantly expand the discussions around animal biotechnology and possibly the public perception depending upon whether the public is generally willing to deliberate the potential of such modifications to address problems ranging from human health to animal health and well-being.

Acknowledgments The authors acknowledge research funding support from the National Institute of Food and Agriculture and the Biotechnology Risk Assessment Grant (BRAG) program, US Department of Agriculture, under award numbers 2011-68004-30367, 2013-68004-20364, 2015-67015-23316, 2015-33522-24106, and 2017-33522-27097-0.

References

Abah J, Ishaq MN, Wada AC (2010) The role of biotechnology in ensuring food security and sustainable agriculture. Afr J Biotechnol 9(52):8896–8900

Adalja A, Sell T, McGinty M, Boddie C (2016) Genetically Modified (GM) mosquito use to reduce mosquito-transmitted disease in the us: a community opinion survey. PLOS current outbreaks, 2016 May 25. Edition 1. https://doi.org/10.1371/currents.outbreaks.1c39ec05a743d41ee3939 1ed0f2ed8d3

Agriculture and Environment Biotechnology Commission (2002) Animals and biotechnology. http://webarchive.nationalarchives.gov.uk/20100419143351/http://www.aebc.gov.uk/aebc/pdf/animals_and_biotechnology_report.pdf. Accessed 25 Jan 2018

Allum N, Sturgis P, Tabourazi D, Brunton-Smith I (2008) Science knowledge and attitudes across cultures: a meta-analysis. Public Underst Sci 17:35–54

Arjó G, Portero M, Piñol C, Viñas J, Matias-Guiu X, Capell T, Bartholomaeus A, Parrott W, Christou P (2013) Plurality of opinion, scientific discourse and pseudoscience: and in depth analysis of the Séralinin et al. study claiming that Roundup™ Ready corn or the Herbicide Roundup™ cause cancer in rats. Transgenic Res 22(2):255–267

Biotechnology and Biological Sciences Research Council (1999) Ethics, morality and animal biotechnology. http://www.bbsrc.ac.uk/documents/animal-biotechnology-pdf/. Accessed 25 Jan 2018

Blancke S, Van Breusegem F, De Jaeger G, Braeckman J, Van Montagu M (2016) Fatal attraction: the intuitive appeal of GMO opposition. Trends Plant Sci 20(7):414–418

Bode L, Vraga EK (2015) In related news, that was wrong: the correction of misinformation through related stories functionality in social media. J Commun 65:619–638

Bredahl L (2001) Determinants of consumer attitudes and purchase intentions with regard to genetically modified food—results of a cross-national survey. J Consum Policy 24(1):23–61

Carlson DF, Lancto CA, Zang B, Kim ES, Walton M, Oldeschulte D, Seabury C, Sonstegard TS, Fahrenkrug SC (2016) Production of hornless dairy cattle from genome-edited cell lines. Nat Biotech 34:479

Carroll D, Charo RA (2015) The societal opportunities and challenges of genome editing. Genome Biol 16:242

Caswell M, Fuglie K, Klotz C (2003) Agricultural biotechnology: an economic perspective. Novinka Books, New York

Ceccoli S, Hixon W (2011) Explaining attitudes toward genetically modified foods in the European Union. Int Polit Sci Rev 33(3):301–319

Cooke JG, Downie R (2010) African perspectives on genetically modified crops: assessing the debate in Zambia, Kenya, and South Africa. http://csis.org/files/publication/100701_Cooke_AfricaGMOs_WEB.pdf. Accessed 25 Jan 2018

Council for Agricultural Science and Technology (2010) Ethical implications of animal biotechnology: considerations for animal welfare decision making. http://www.cast-science.org/publications/?ethical_implications_of_animal_biotechnology_considerations_for_animal_welfare_decision_making&show=product&productID=2952. Accessed 25 Jan 2018

Crawford SC (2003) Hindu bioethics for the twenty-first century. State University of New York Press, Albany, NY

Critchley CR (2008) Public opinion and trust in scientists: the role of the research context, and the perceived motivation of stem cell researchers. Public Underst Sci 17(3):309–327

Curtis KR, Moeltner K (2007) The effect of consumer risk perceptions on the propensity to purchase genetically modified foods in Romania. Agribusiness 23(2):563–278

De Witt A, Osseweijer P, Pierce R (2015) Understanding public perceptions of biotechnology through the "Integrative Worldview Framework". Public Underst Sci. https://doi.org/10.1177/0963662515592364 0963662515592364, E-pub ahead of print July 3, 2015

Driscoll JW (1992) Attitudes towards animal use. Anthrozoös 5:32–39

Driscoll J (1995) Attitudes toward animals: species ratings. Soc Anim 3(2):139–150

Einsiedel EF (2005) Public perceptions of transgenic animals. Rev Sci Tech 24(1):149–157

Epstein R (1998) Buddhism and biotechnology. http://online.sfsu.edu/repstein/GEessays/Buddhism%20and%20Biotechnology.htm. Accessed 25 Jan 2018

Finucane ML (2002) Mad cows, mad corn and mad communities: the role of socio-cultural factors in the perceived risk of genetically-modified food. Proc Nutr Soc 61(1):31–37

Flipse SM, Osseweijer P (2013) Media attention to GM food cases: an innovation perspective. Public Underst Sci 22(2):185–202

Food and Agriculture Organization of the United Nations (2009) Codex Alimentarius. Rome, Italy

Food and Agriculture Organization of the United Nations (2013) Biotechnologies at work for smallholders: case studies from developing countries in crops, livestock and fish

Forsberg CW, Meidinger RG, Liu M, Cottrill M, Golovan S, Phillips JP (2013) Integration, stability and expression of the E. coli phytase transgene in the Cassie line of Yorkshire Enviropig. Transgenic Res 22(2):379–389

Frewer L, Miles S, Marsh R (2002) The media and genetically modified foods: evidence in support of social amplification of risk. Risk Anal 22(4):701–711

Frewer L, Lassen J, Kettlitz B, Scholderer J, Beekman V, Berdal KG (2004) Societal aspects of genetically modified foods. Food Chem Toxicol 42(7):1181–1193

Frewer L, Bergmann K, Brennan M, Lion R, Meertens R, Rowe G et al (2011) Consumer response to novel agri-food technologies: implications for predicting consumer acceptance of emerging food technologies. Food Sci Technol 22:442–456

Frewer L, van der Lans I, Fischer A, Reinders M, Menozzi D, Zhang X et al (2013) Public perceptions of agri-food applications of genetic modification: A systematic review and meta-analysis. Trends Food Sci Tech 30:142–152

Furnham A, McManus C, Scott D (2003) Personality, empathy and attitudes to animal welfare. Anthrozoös 16(2):135–146

Gabriel KI, Rutledge BH, Barkley CL (2012) Attitudes on animal research predict acceptance of genetic modification technologies by university undergraduates. Soc Anim 20:381–400

Gallegos J (2017) GMO salmon caught in U.S. regulatory net, but Canadians have eaten 5 tons. The Washington Post. https://www.washingtonpost.com/news/speaking-of-science/wp/2017/08/04/gmo-salmon-caught-in-u-s-regulatory-net-but-canadians-have-eaten-5-tons/?utm_term=.0d6ec3f269fc. Accessed 25 Jan 2018

Ganiere P, Chern WS, Hahn D (2006) A continuum of consumer attitudes toward genetically modified foods in the United States. J Agr Resour Econ 31(1):129–149

Gaskell G, Allum N, Bauer M, Durant J, Allansdottir A, Bonfadelli H et al (2000) Biotechnology and the European public. Nat Biotechnol 18(9):935–938

Gaskell G, Allum NC, Stares SR (2003) Europeans and biotechnology in 2002: Eurobarometer 58.0. European Commission, Brussels

Gaskell G, Allum N, Wagner W, Kronberger N, Torgersen H, Hampel J, Bardes J (2004) GM foods and the misperception of risk perception. Risk Anal 24(1):185–194

Gjerris M (2012) Animal biotechnology: the ethical landscape. In: Brunk CG, Hartley S (eds) Designer animals: mapping the issues in animal biotechnology. University of Toronto Press, Toronto

Gordon JW, Scangos GA, Plotkin DJ, Barbosa JA, Ruddle FH (1980) Genetic transformation of mouse embryos by microinjection of purified DNA. Proc Natl Acad Sci U S A 77(12):7380–7384

Grunert KG, Bech-Larsen T, Lahteenmaki L, Ueland O, Astrom A (2004) Attitudes towards the use of GMOs in food production and their impact on buying intention: the role of positive sensory experience. Agribusiness 20(1):95–107

Gupta N, Fischer A, Frewer L (2011) Socio-psychological determinants of public acceptance of technologies: a review. Public Underst Sci 21(7):782–795

Hagelin J, Hau J, Carlsson HE (1999) Undergraduate university students' views of the use of animals in biomedical research. Acad Med 74(10):1135–1137

Hagelin J, Carlsson HE, Hau J (2003) An overview of surveys on how people view animal experimentation: some factors that may influence the outcome. Public Underst Sci 12:67–81

Hallerman E, Grabau E (2016) Crop biotechnology: a pivotal moment for global acceptance. Food Energy Secur 5(1):3–17

Hallman WK, Hebden WC, Aquino HL, Cuite CL, Lang JT (2003) Public perceptions of genetically modified foods: a national study of American knowledge and opinion (RR-1003-004). Food Policy Institute, Cook College, Rutgers—The State University of New Jersey, New Brunswick, NJ

Hammer RE, Pursel VG, Rexroad CE Jr, Wall RJ, Bolt DJ, Ebert KM, Brinster RL (1985) Production of transgenic rabbits, sheep and pigs by microinjection. Nature 315(6021):680–683

Herzog HA, Galvin S (1997) Common sense and the mental lives of animals: an empirical approach. In: Mitchell RW (ed) Anthropormorphism, anecdotes and animals. State University of New York Press, Albany, NY, pp 237–253

Hess S, Lagerkvist CJ, Redekop W, Pakseresht A (2013) Consumers' evaluation of biotechnology in food products: new evidence from a meta-survey. Paper presented at the Agricultural & Applied Economics Association's 2013 AAEA & CAES joint annual meeting, Washington, DC. http://ageconsearch.umn.edu/bitstream/151148/2/Consumers%20Evaluation%20of%20Biotechnology%20in%20Food%20Products%202013%20final.pdf. Accessed 25 Jan 2018

Hoban TJ (2004) Public attitudes towards agricultural biotechnology (ESA Working Paper No. 04-09). http://ageconsearch.umn.edu/bitstream/23810/1/wp040009.pdf. Accessed 25 Jan 2018

Houghton JR, Rowe G, Frewer LJ, Van Kleef E, Chryssochoidis G, Kehagia O et al (2008) The quality of food risk management in Europe: perspectives and priorities. Food Policy 33:13–26

Hudson J, Caplanova A, Novak M (2015) Public attitudes to GM foods. The balancing of risks and gains. Appetite 92:303–313

International Food Information Council (2014) Consumer perceptions of food technology survey. http://www.foodinsight.org/surveys/2014-food-technology-survey. Accessed 25 Jan 2018

Ipsos Social Research Institute (2013) Community attitudes towards emerging technology issues—biotechnology (ISRI Project 12-025766-01). http://www.industry.gov.au/industry/IndustrySectors/nanotechnology/Publications/Documents/Emergingtechstudybio.docx. Accessed 25 Jan 2018

James C (2016) Global status of Commercialized Biotech/GM Crops 2016. ISAAA brief no. 52. ISAAA, Ithaca, NY

Jasper J, Nelkin D (1992) The animal rights crusade. The Free Press, New York, NY

Jayaraman K, Jia H (2012) GM phobia spreads in South Asia. Nat Biotechnol 30(11):1017–1019

Kaiser M (2005) Assessing ethics and animal welfare in animal biotechnology for farm production. Rev Sci Tech 24(1):75–87

Kalof L, Dietz T, Stern PC, Guagnano GA (1999) Social psychosocial and structural influences on vegetarian beliefs. Rural Sociol 64:500–511

Kendall HA, Lobao LM, Sharp JS (2006) Public concern with animal well-being: place, social structural location, and individual experience. Rural Sociol 71(3):399–428

Knight A (2009) Perceptions, knowledge and ethical concerns with GM foods and the GM process. Public Underst Sci 18(2):177–188

Knight S, Barnett L (2008) Justifying attitudes towards animal use: a qualitative study of people's views and beliefs. Anthrozoös 21:31–42

Knight S, Nunkoosing K, Vrig A, Cherryman J (2003) Using grounded theory to examine people's attitudes towards how animals are used. Soc Anim 11:179–198

Knight S, Vrij A, Cherryman J, Nunkoosing K (2004) Attitudes towards animal use and belief in animal mind. Anthrozoös 17(1):43–62

Knight JG, Mather DW, Holdsworth DK, Ermen DF (2007) Acceptance of GM food—an experiment in six countries. Nat Biotechnol 25(5):507–508

Kronberger N, Wagner W, Nagata M (2013) How natural is "more natural"? The role of method, type of transfer, and familiarity for public perceptions of cisgenic and transgenic modification. Sci Commun:1–25

Lazaris A, Arcidiaconon S, Huang Y, Zhou J-F, Duguay F, Chretien N, Karatzas CN (2002) Spider silk fibers spun from soluble recombinant silk produced in mammalian cells. Science 295:472–476

Leahy PJ, Mazur A (1980) The rise and fall of public opposition in specific social movements. Social Stud Sci 10(3):259–284

Li R, Wang Q, McHughen A (2015) Chinese government reaffirms backing for GM products. Nat Biotechnol 33(10):1029

Logar N, Pollock LK (2005) Transgenic fish: is a new policy framework necessary for a new technology? Environ Sci Policy 8(1):17–27

Lund TB, McKeegan DEF, Cribbin C, Sandoe P (2016) Animal ethics profiling of vegetarians, vegans and meat-eaters. Anthrozoös 29(1):89–106

Lusk J, Murray S (2015) Food demand survey. FooDS 2(9):1–5

Lusk J, McFadden B, Rickard B (2015) Which biotech foods are most acceptable to the public? Biotechnol J 10:13–16

Marchant GE, Stevens YA (2016) A new window of opportunity to reject process-based biotechnology regulation. GM Crops & Food 64(4):233–242

Marques M, Critchley C, Walshe J (2015) Attitudes to genetically modified food over time: how trust in organizations and the media cycle predict support. Public Underst Sci 24(5):601–618

McColl KA, Clarke B, Doran TJ (2013) Role of genetically engineered animals in future food production. Aust Vet J 91(3):113–117

Melodlesi A (2011) Vatican panel backs GMOs. Nat Biotechnol 29(1):11

Mielby H, Sandøe P, Lassen J (2012) The role of scientific knowledge in shaping public attitudes to GM technologies. Public Underst Sci 22(2):155–168

Moerbeek H, Casimir G (2005) Gender differences in consumers' acceptance of genetically modified foods. Int J Consum Stud 29(4):308–318

Mora C, Menozzi D, Kleter G, Aramyan L, Valeeva N, Zimmerman K, Reddy G (2012) Factors affecting the adoption of genetically modified animals in the food and pharmaceutical chains. Bio-based Appl Econ 1(3):313–329

Navaro J, Maldonado E, Pedraza C, Cavas M (2001) Attitudes among animal research among psychology students in Spain. Psychol Rep 89:227–236

Network of African Science Academies (2015) Harnessing modern agricultural biotechnology for Africa's economic development: recommendations to policymakers. http://www.interacademies.net/File.aspx?id=28031. Accessed 25 Jan 2018

Novoselova T, Meuwissen M, Huirne R (2007) Adoption of GM technology in livestock production chains: an integrating framework. Food Sci Technol 18:175–188

Ormandy E (2009) Worldwide trends in the use of animals in research: the contribution of genetically-modified animal models. ATLA-Altern Lab Anim 37:63–65

Ormandy E, Schuppli C (2014) Public attitudes toward animal research: a review. Animals 4:391–408

Ormandy E, Schuppli C, Weary D (2012) Factors affecting people's acceptance of the use of zebrafish and mice in research. ATLA-Altern Lab Anim 40(6):321–333

Pew Initiative on Food and Biotechnology (2007) Options for future discussions on genetically modified and cloned animals. Paper presented at the pew initiative on food and biotechnology workshop, Washington, DC

Pew Research Center (2015) Public and scientists' views on science and society. http://www. pewinternet.org/2015/01/29/public-and-scientists-views-on-science-and-society/. Accessed 25 Jan 2018

Pifer LK (1996) Exploring the gender gap in young adults' attitudes about animal research. Soc Anim 4(1):37–52

Pifer L, Shimizu K, Pifer R (1994) Public attitudes toward public research: some international comparisons. Soc Anim 2:95–113

Pin R, Gutteling J (2009) The development of public perception research in the genomics field: an empirical analysis of the literature in the field. Sci Commun 31(1):57–83

Priest SH (2000) US public opinion divided over biotechnology? Nat Biotechnol 18(9):939–942

Priest SH, Bonfadelli H, Rusanen M (2003) The "trust gap" hypothesis: predicting support for biotechnology across national cultures as a function of trust in actors. Risk Anal 23(4):751–766

Puduri V, Govindasamy R, Lang JT, Onuango B (2005) I will not eat it with a fox; I will not eat it in a box: what determines acceptance of GM food for American consumers? Choices 20:257–261

Qaim M, Kouser S (2013) Genetically modified crops and food security. PLoS One 8(6):e64879

Rodriguez L, Abbott E (2007) Communication, public understanding and attitudes toward biotechnology in developing nations: a meta-analysis. Paper presented at the 11th international conference on agricultural biotechnologies: new frontiers and products—economics, policies and science, Ravello, Italy

Rollin BE (2014) The perfect storm—genetic engineering, science, and ethics. Sci & Educ 23:509–517

Ruane J, Sonnino A (2011) Agricultural biotechnologies in developing countries and their possible contribution to food security. J Biotechnol 156(4):356–363

Russell WMS, Burch RL (1959) The principles of humane experimental technique. Methuen, London, UK

Samadi S, Barberousse A (2015) Species. In: Heams PHT, Lecointre G, Silberstein M (eds) Handbook of evolutionary thinking in the sciences. Springer Science, New York, NY

Sanchez D (2015) Genetically modified crops: how attitudes to new technology influence adoption. Australian Council of Learned Academies. http://www.acola.org.au/PDF/ SAF05/4Genetically%20modified%20crops.pdf. Accessed 25 Jan 2018

Sanderson J (2013) Pigoons, Rakunks and Crakers: Margaret Atwood's Oryx and Crake and genetically engineered animals in a (Latourian) hybrid world. Law and Humanities 7(2):218–239

Sandler RL (2015) Food ethics. Routledge, New York, NY

Scholderer J, Frewer LJ (2003) The biotechnology communication paradox: experimental evidence and the need for a new strategy. J Consum Policy 26(2):125–157

Schuppli CA (2011) Decisions about the use of animals in research: ethical reflection by animal ethics committee members. Anthrozoös 24(4):409–425

Schuppli C, Weary D (2010) Attitudes towards the use of genetically modified animals in research. Public Underst Sci 19(6):686–697

Scott SE, Inbar Y, Rozin P (2016) Evidence for absolute moral opposition to genetically modified food in the United States. Perspect Psychol Sc 11(3):315–324

Secretariat of the Convention on Biological Diversity (2005) Handbook of the convention on biological diversity including its cartagena protocol on biosafety, 3rd edn. Friesen, Montreal

Shaw A (2002) "It just goes against the grain." Public understandings of genetically modified (GM) food in the UK. Public Underst Sci 11(3):273–291

Sheehy H, Legault M, Ireland D (1998) Consumer and biotechnology: a synopsis of survey and focus group research. J Consum Policy 21:359–386

Sherkow JS, Greely HT (2015) The history of patenting genetic material. Annu Rev Genet 49:161–182

Siegrist M (2000) The influence of trust and perceptions of risks and benefits on the acceptance of gene technology. Risk Anal 20(2):195–203

Smyth SJ, Kerr WA, Phillips PWB (2015) Global economic, environmental and health benefits from GM crop adoption. Glob Food Secur-Agr 7:24–29

Steinhart H (2006) Novel foods and novel processing techniques as threats and challenges to a hypersensitive world. In: Gilissen LJEJ, Wichers HJ, Savelkoul HFJ, Bogers RJ (eds) Allergy matters: new approaches to allergy prevention and management, vol 10, Springer, Dordrecht, pp 63–75

Stephan HR (2015) Cultural politics and the transatlantic divide over GMOs: cultures of nature. Palgrave Macmillan, London, UK

Swami V, Furnham A, Christopher AN (2008) Free the animals? Investigating attitudes toward animal testing in Britain and the United States. Scand J Psychol 49(3):269–276

Tizard M, Hallerman E, Fahrenkrug S, Newell-McGloughlin M, Gibson J, de Loos F, Wagner S, Laible G, Han JY, D'Occhio M, Kelly L, Lowenthal J, Gobius K, Silva P, Cooper C, Doran T (2016) Strategies to enable the adoption of animal biotechnology to sustainably improve global food safety and security. Transgenic Res 25(5):575–595

Townsend E, Campbell S (2004) Psychological determinants of willingness to tast and purchase genetically modified food. Risk Anal 24(5):1385–1393

U.S. Food and Drug Administration (2015) FDA has determined that the aquadvantage salmon is as safe to eat as non-ge salmon. https://www.fda.gov/ForConsumers/ConsumerUpdates/ucm472487.htm Accessed 25 Jan 2018

Vazquez-Salat N (2013) Are good ideas enough? The impact of socio-economic and regulatory factors on GMO commercialisation. Biol Res 46(4):317–322

Vazquez-Salat N, Houdebine L (2013) Will GM animals follow the GM plant fate? Transgenic Res 22(1):5–13

Veil SR, Reno J, Freihaut R, Oldham J (2015) Online activists vs. Kraft foods: a case of social media hijacking. Public Relat Rev 41:103–108

Vis F (2014) To tackle the spread of misinformation online we must first understand it. http://www.theguardian.com/commentisfree/2014/apr/24/tackle-spread-misinformation-online. Accessed 25 Jan 2018

Wagner W, Kronberger N, Nagata M, Sen R, Holtz P, Palacios F (2010) Essentialist theory of 'hybrids': from animal kinds to ethnic categories and race. Asian J Soc Psychol 13(4):232–246

Waltz E (2016) GM salmon declared fit for dinner plates. Nat Biotechnol 34(1):7–9

Waltz E (2017) First genetically engineered salmon sold in Canada. Nature 548:148

Wells DL, Hepper PG (1997) Pet ownership and adults' views on animal use. Soc Anim 5:45–63

World Health Organization (2016) Glossary: food security. http://www.fao.org/docrep/005/y4671e/y4671e06.htm. Accessed 25 Jan 2018

YouGov (2016) Survey. https://d25d2506sfb94s.cloudfront.net/cumulus_uploads/document/qcjry-hyo22/tabs_HP_Science_20160410.pdf. Accessed 25 Jan 2018

The manufacturer's authorised representative in the EU is Springer
Nature Customer Service Centre GmbH, Europaplatz 3, 69115 Heidelberg,
Germany. If you have any concerns regarding our products, please
contact ProductSafety@springernature.com

Printed and bound by CPI Group (UK) Ltd, Croydon, CR0 4YY

29/04/2026
02099512-0001